OECD Food and Agricultural Reviews

Agricultural Policies in Viet Nam 2015

Please cite this publication as:
OECD (2015), *Agricultural Policies in Viet Nam 2015*, OECD Publishing, Paris.
http://dx.doi.org/10.1787/9789264235151-en

ISBN 978-92-64-23514-4 (print)
ISBN 978-92-64-23515-1 (PDF)

Series: OECD Food and Agricultural Reviews
ISSN 2411-426X (print)
ISSN 2411-4278 (online)

Photo credits: Cover ©OECD/Andrzej Kwieciński.

Corrigenda to OECD publications may be found on line at: *www.oecd.org/publishing/corrigenda*.

Foreword

This Review of Agricultural Policies: Viet Nam is *one of a series of reviews of national agricultural policies undertaken by the OECD's Committee for Agriculture (CoAg). It was initiated in response to a request from Mr. Bui Ba Bong, then Viet Nam's Vice Minister of the Ministry of Agriculture and Rural Development (MARD), and has been prepared in close co-operation with the Ministry.*

The Review *examines the policy context and the main trends in Viet Nam's agriculture. It classifies and measures the support provided to agriculture using the same method the OECD employs to monitor agricultural policies in OECD countries and a growing number of non-member economies, such as Brazil, China, Colombia, Indonesia, Kazakhstan, Russia, South Africa and Ukraine. On request from the Vietnamese authorities, the* Review *includes a special chapter on the policy environment for investment in agriculture, drawing from the OECD Policy Framework for Investment in Agriculture (PFIA). The* Review *is a precursor towards regular OECD engagement with Viet Nam on agricultural policy issues through the annual monitoring and evaluation of agricultural policy developments.*

The study was carried out by the Development Division of the OECD Trade and Agriculture Directorate (TAD) in co-operation with the Investment Division of the OECD Directorate for Financial and Enterprise Affairs (DAF). Andrzej Kwieciński co-ordinated the report and was one of the authors together with Darryl Jones and Coralie David. Chapter 1 benefited from a background report delivered by Richard Barichello (University of British Columbia, Vancouver, Canada) as well as from contributions by Claire Delpeuch and Gaëlle Gourin (both from TAD). Dao The Anh (Centre for Agrarian Systems Research and Development, CASRAD, Viet Nam), Tran Cong Thang and Dinh Bao Linh (both from the Institute of Policy and Strategy for Agriculture and Rural Development, IPSARD, Viet Nam) provided valuable background information for Chapter 2. Chapter 3 drew from answers to the PFIA questionnaire submitted by Ta Kim Cuc (Center for Informatics and Statistics, MARD) and benefited from inputs from Bishara Mansur (DAF). The database for Producer Support Estimates was developed by Florence Bossard and Andrzej Kwieciński in close co-operation with Hieu Phan Sy (Center for Informatics and Statistics, MARD). Statistical support was provided by Florence Bossard. Anita Lari provided administrative and secretarial assistance. Anita Lari and Michèle Patterson provided publication support. Ken Ash, Carmel Cahill, Jared Greenville, Shingo Kimura, Iza Lejarraga, Silvia Sorescu, Frank Van Tongeren, Trudy Witbreuk (all from the OECD Secretariat), Chris Jackson and Steven Jaffee (both from the World Bank office in Hanoi), Marlo Rankin (FAO), MARD's delegation participating in the peer review, but also Hieu Phan Sy (MARD), Dao The Anh (CASRAD), Nguyen Trung Kien (IPSARD) and many other colleagues in the OECD Secretariat and member country delegations furnished valuable comments on earlier drafts of the report.

The Review *benefited from support provided by MARD. Pham Thi Hong Hanh and Dinh Pham Hien, both from International Cooperation Department of MARD, were the main contacts and liaison persons on all aspects of the study. The study also benefited from the input of staff from MARD and its related entities, from other Ministries and from participants at preparatory meetings and consultations in Hanoi, including researchers from academia.*

The study was made possible through voluntary contributions from Australia, Japan and the United States. It was reviewed at an in-country Roundtable meeting with Vietnamese officials and experts in Hanoi in March 2015. Subsequently, the Vietnamese delegation led by Tran Kim Long, Director General of the International Cooperation Department of MARD, participated in the peer review of Vietnamese agricultural policies by the OECD's Committee for Agriculture at its 164th session in May 2015. Steve Neff (ERS-USDA, USA), Matthew Worrell (DFAT, Australia) and Kunimitsu Masui (Permanent Delegation of Japan to the OECD) kindly agreed to lead the discussion during this peer review. Vietnamese officials and experts have been involved from the initial discussions of the study outline through to the peer review and final revisions, but the final report remains the sole responsibility of the OECD.

Table of contents

Abbreviations ... 11
Executive summary ... 15
Assessment and policy recommendations ... 19
Assessment ... 20
Policy recommendations ... 31
References ... 37
Chapter 1. The agricultural policy context in Viet Nam ... 39
1.1. Introduction ... 40
1.2. General aspects ... 40
1.3. Agricultural situation ... 47
1.4. Factors of production and productivity ... 51
1.5. Farm incomes, poverty and food consumption ... 58
1.6. Agro-food trade flows ... 63
1.7. Agro-environmental situation ... 70
1.8. Agricultural land tenure system ... 74
1.9. Competition and structural change beyond the farm gate ... 77
1.10. Summary ... 83
Notes ... 85
Annex 1.A1. Viet Nam: Production and trade performance for major agricultural commodities ... 88
Annex 1.A2. Viet Nam: Projected production, consumption and trade for major commodities by 2023 ... 103
References ... 106
Chapter 2. Trends and evaluation of Viet Nam's agricultural policy ... 111
2.1. Introduction ... 112
2.2. Agricultural policy framework ... 112
2.3. Domestic policies ... 124
2.4. Trade policies affecting agro-food trade flows and agricultural commodity prices ... 145
2.5. Evaluation of support to agriculture ... 162
2.6. Summary ... 175
Notes ... 178
Annex 2.A1. Policy tables ... 183
References ... 187

Chapter 3. **Viet Nam's policy environment for investment in agriculture** 191
3.1. Introduction 192
3.2. Trends in investment in agriculture 193
3.3. Investment policy 197
3.4. Investment promotion and facilitation 204
3.5. Land tenure policy 209
3.6. Financial sector development 215
3.7. Infrastructure development 221
3.8. Trade policy 224
3.9. Human resources, research and innovation 225
3.10. Responsible business conduct 227
3.11. Summary 231
Notes 233
References 236

Tables

1. Contextual indicators, 1995, 2013 21
1.1. Changes in the composition of the value of agricultural production, 1991-2012, % 50
1.2. Average annual growth rate in Agricultural Total Factor Productivity, % 58
1.3. Agro-food sector's integration with international markets, 2000-13 67
1.4. Forest characteristics and dynamics 71
1.5. Water availability and utilisation 72
1.6. Emissions of CO_2 equivalent from agricultural activities, gigagrams per year 72
2.1. Major state-owned enterprises involved in agriculture 123
2.2. Agricultural co-operatives by sector and region, 2013 124
2.3. Agricultural-related products subject to price stabilisation 126
2.4. Temporary storage procurement policy timing and volume, 2009-13 128
2.5. Outcomes from the pilot agricultural insurance programme as at June 2014 136
2.6. Agricultural land use tax 137
2.7. Government extension system by region, 2013 139
2.8. Tariff-rate quota commitments for eggs, sugar and tobacco 149
2.9. Preferential tariff-rate quotas for Cambodia and Lao PDR, 2008-13 150
2.10. Agricultural related products subject to quantitative import restrictions and phasing out schedule 152
2.11. CEPT time frame for ASEAN member states 160
2.12. Viet Nam's tariff reduction commitments under ASEAN+ agreements 161
2.13. Estimates of support to agriculture in Viet Nam, VND million 166
2.14. Estimates of support to agriculture in Viet Nam, USD million 167
2.A1.1. Hierarchy, content, and numbering and coding of legal documents 183
2.A1.2. Main tasks of units under the Ministry of Agriculture and Rural Development 184
2.A1.3. Selected final bound, MFN applied and preferential tariffs for MPS commodities, 2013 185
2.A1.4. Imported goods subject to line management licensing by MARD 187
3.1. Agricultural capital stock 194

3.2. Number of enterprises and capital stock by economic activity and ownership . . . 195
3.3. Corporate Income Tax incentives, 2013 . . . 206
3.4. Access to and use of financial services, 2012 . . . 218
3.5. Access to electricity, 2014 . . . 223
3.6. Telecommunications subscribers/users per 100 inhabitants, 2013 . . . 224

Figures

1. Main macroeconomic indicators, 1990-2013 . . . 21
2. Agro-food trade, 2000-13 . . . 21
1.1. Viet Nam: Selected macroeconomic indicators, 1990-2013 . . . 45
1.2. Agricultural Growth Enabling Index and its sub-component blocks, early 2010s . . . 46
1.3. The share of agriculture in GDP, employment, total exports and imports, 2000-13 . . . 47
1.4. Evolution of agriculture's share in GDP and in employment in selected Asian countries, 1990-2012 . . . 48
1.5. Growth in gross agricultural output in selected Asian countries, 1990-2013 . . . 49
1.6. Growth in agricultural output in Viet Nam, 1990-2013 . . . 49
1.7. Growth in crop production, 1990-2013 . . . 50
1.8. Growth in livestock production, 1990-2013 . . . 51
1.9. Use of chemical fertiliser in selected countries, averages 1990-92 and 2010-12 . . . 52
1.10. Agricultural land, 1990-2012 . . . 53
1.11. Composition of harvested area, 1990-2013 . . . 54
1.12. Composition of rural employment by sector, 2001-11 . . . 55
1.13. The share of agriculture in rural employment, selected regions, 2001-11 . . . 55
1.14. Growth in land productivity in selected Asian countries, 1990-2010 . . . 56
1.15. Growth in labour productivity in selected Asian countries, 1990-2010 . . . 57
1.16. Agriculture value added per worker in selected Asian countries, USD 2005, 1990-2010 . . . 57
1.17. Growth in total factor productivity in selected Asian countries, 1990-2010 . . . 58
1.18. Monthly income per capita by residence, in 2005 prices, 2002-12 . . . 59
1.19. Monthly income per capita for rural residents by source, 2002-12 . . . 59
1.20. Urban versus rural poverty rates, 2004-13 . . . 60
1.21. Poverty rates by region, 2004-13 . . . 60
1.22. Daily per capita energy intake, 1995-2007 . . . 61
1.23. Share of expenditures on food consumption in total expenditures, 2002-12 . . . 62
1.24. Monthly consumption expenditures per capita on food by residence, in 2005 prices, 2002-12 . . . 62
1.25. Monthly per capita expenditures on food by income quintile, in 2005 prices, 2002-12 . . . 63
1.26. Share of Viet Nam in world's exports of selected commodities, 2000-13 . . . 64
1.27. Viet Nam's fisheries production, 1990-2012 . . . 64
1.28. Fisheries production and utilisation in Viet Nam, 1990-2011 . . . 65
1.29. Viet Nam's agro-food trade, 2000-13 . . . 66
1.30. Composition of agro-food exports, averages 2000-02 and 2011-13 . . . 67

1.31. Main export markets for Viet Nam's agro-food products, 2011-13 average.... 68
1.32. Composition of agro-food imports, averages 2000-02 and 2011-13 69
1.33. Share of imports in Viet Nam's domestic use of selected commodities, 2000-11........ 69
1.34. Main suppliers of agro-food products to Viet Nam, 2011-13 average......... 70
1.35. Distribution of farms by land size, 2001 and 2011........ 76
1.36. Rice export prices, 2012-14........ 83
1.A1.1. Rice: Producer prices and trade flows, 2000-13........ 89
1.A1.2. Coffee: Producer prices and trade flows, 2000-13........ 90
1.A1.3. Rubber: Producer prices and trade flows, 2000-13........ 91
1.A1.4. Cashew nuts: Producer prices and trade flows, 2000-13........ 93
1.A1.5. Cassava: Producer prices and trade flows, 2000-13........ 94
1.A1.6. Black pepper: Producer prices and trade flows, 2000-13........ 95
1.A1.7. Tea: Producer prices and trade flows, 2000-13........ 96
1.A1.8. Pigmeat: Producer prices and trade flows, 2000-13........ 97
1.A1.9. Eggs: Producer prices and trade flows, 2000-13........ 98
1.A1.10. Maize: Producer prices and trade flows, 2000-13........ 99
1.A1.11. Sugar cane: Producer prices and trade flows, 2000-13........ 100
1.A1.12. Poultry meat: Producer prices and trade flows, 2000-13........ 101
1.A1.13. Cattle: Producer prices and trade flows, 2000-13........ 102
1.A2.1. Price trends in real terms for agricultural commodities to 2023........ 104
1.A2.2. Production: Per cent change 2023 compared to 2011-13 average........ 104
1.A2.3. Projected net trade for selected products........ 105
1.A2.4. Meat and fish consumption in Viet Nam, kg/capita, 2001-23........ 106
2.1. Level flow chart of the Ministry of Agriculture and Rural Development, 2013... 120
2.2. Comparison of different types of rice prices in Viet Nam, 2000-13........ 128
2.3. Expenditure on supporting irrigation operations and maintenance, 2000-13........ 130
2.4. Revenue from agricultural land use tax, 2000-12........ 138
2.5. Expenditure on irrigation capital development, 2000-13........ 143
2.6. Average MFN applied tariffs for agricultural and non-agricultural commodities, 2003-13........ 147
2.7. Frequency distribution of agricultural final bound and MFN applied tariff lines and imports by tariff rates, 2013........ 148
2.8. Average bound, MFN applied, ATIGA and ACFTA tariffs by product groups, 2013........ 148
2.9. Level and composition of Producer Support Estimate in Viet Nam, 2000-13.... 168
2.10. Producer Support Estimate in Viet Nam and selected countries, 2011-13 average........ 168
2.11. Level and composition of Market Price Support in Viet Nam, 2000-13........ 169
2.12. Level and composition of budgetary transfers in Viet Nam, 2000-13........ 170
2.13. Producer SCTs by commodity in Viet Nam, averages 2000-02 and 2011-13... 171
2.14. Consumer Support Estimate in Viet Nam and selected countries, 2011-13 average........ 173
2.15. Level and composition of General Services Support Estimate in Viet Nam, 2000-13........ 173
2.16. Level and composition of Total Support Estimate in Viet Nam, 2000-13........ 174

2.17. Total Support Estimate in Viet Nam and selected countries, 2011-13 average 175
3.1. The most problematic factors for doing business, 2014 193
3.2. FDI inflows and stocks in selected ASEAN countries, 1986-2013 196
3.3. Foreign direct investment inflows by sector, 2000-13 196
3.4. Sources of credit in rural areas, 2006-10 216
3.5. Trade facilitation performance: OECD indicators, 2014 225

Abbreviations

ADB	Asian Development Bank
AFTA	ASEAN Free Trade Area
AGEI	Agricultural Growth Enabling Index
ARP	Agricultural Restructuring Plan
ASEAN	Association of Southeast Asian Nations
CEPT	Common Effective Preferential Tariff
CIC	Credit Information Centre
CIEM	Central Institute for Economic Management
CIS	Centre for Information and Statistics
CIT	Corporate Income Tax
CPV	Communist Party of Viet Nam
CSE	Consumer Support Estimate
DARD	Department of Agriculture and Rural Development (provincial level)
DOSTE	Department of Science, Technology and Environment
ECTAD	Emergency Centre for Transboundary Animal Diseases
EIA	Environmental Impact Assessment
EPF	Export Promotion Fund
FAO	Food and Agriculture Organization of the United Nations
FDI	Foreign Direct Investment
FIA	Foreign Investment Agency
FIE	Foreign Invested Enterprise
FMD	Foot-and-Mouth Disease
GDP	Gross Domestic Product
GHG	Greenhouse Gas
GMO	Genetically Modified Organism
GSO	General Statistics Office
GSSE	General Services Support Estimate
HCMC	Ho Chi Minh City
HS	Harmonised System
ICOR	Incremental Capital-Output Ratio
ICT	Information and Communication Technology
IDMC	Irrigation and Drainage Management Companies
IFAD	International Fund for Agricultural Development
IFC	International Finance Corporation
IFPRI	International Food Policy Research Institute
ILO	International Labour Organisation
IMF	International Monetary Fund
IPRI	Intellectual Property Rights Index

IPSARD	Institute of Policy and Strategy for Agriculture and Rural Development
ISF	Irrigation Service Fee
ITU	International Telecommunication Union
JICA	Japan International Co-operation Agency
LURC	Land Use Right Certificate
MARD	Ministry of Agriculture and Rural Development
MFN	Most Favoured Nation
MOF	Ministry of Finance
MOH	Ministry of Health
MOIT	Ministry of Industry and Trade
MOJ	Ministry of Justice
MONRE	Ministry of Natural Resources and Environment
MOST	Ministry of Science and Technology
MOT	Ministry of Transport
MPI	Ministry of Planning and Investment
MPS	Market Price Support
MRD	Mekong River Delta
NA	National Assembly
NAEC	National Agriculture Extension Centre
NEA	National Environmental Agency
NTB	Non-Tariff Barrier
ODA	Official Development Assistance
OECD	Organisation for Economic Cooperation and Development
PCI	Provincial Competitiveness Index
PPC	Provincial People's Committee
PPP	Public-Private Partnership
PPP	Purchasing Power Parity
PRRS	Porcine Reproductive and Respiratory Syndrome
PSE	Producer Support Estimate
PSF	Price Stabilisation Fund
QD	Quyet Dinh, meaning 'Decision'
R&D	Research and Development
REDD	Reduced Emissions from Deforestation and Forest Degradation
RRD	Red River Delta
SBV	State Bank of Viet Nam
SCT	Single Commodity Transfers
SEDP	Socio-Economic Development Plan
SEDS	Socio-Economic Development Strategy
SME	Small and Medium Enterprise
SOE	State-Owned Enterprise
SPS	Sanitary and Phytosanitary
TFP	Total Factor Productivity
TI	Transparency International
TRQ	Tariff-Rate Quota
TSE	Total Support Estimate
TTg	Thu Tuong, meaning 'Prime Minister'
UN	United Nations

US	United States
USAID	United States Agency for International Development
USD	United States Dollar
VAT	Value Added Tax
VBARD	Vietnam Bank for Agriculture and Rural Development
VBSP	Vietnam Bank for Social Policies
VDB	Vietnam Development Bank
VEAM	Vietnam Engine and Machinery Corporation
VFA	Vietnam Food Association
VFU	Vietnam Farmers Union
VND	Vietnamese Dong
WB	World Bank
WDI	World Development Indicators
WEF	World Economic Forum
WFP	World Food Programme
WTO	World Trade Organisation
WUG	Water User Groups
WWF	World Wildlife Fund

Executive summary

Doi Moi, or "Renovation", reforms launched in the mid-1980s marked the beginning of the transition of the Vietnamese economy away from central planning towards greater market orientation. Since then, a long series of policy changes have continued to move the economy in this direction by opening markets, establishing private land use rights, reducing the role of state-owned enterprises and encouraging private investment.

The results to date have been impressive. Strong economic growth has lifted real incomes in both urban and rural areas, reducing poverty and combatting undernourishment. Poverty levels fell in Viet Nam further than in any country in the world except for China. Viet Nam's progress in combatting undernourishment was similarly remarkable. The proportion of undernourished in the total population fell from 46% in 1990-92 to 13% in 2012-14. This represents a decrease of 72%, which is one of the highest rates for all countries.

These reforms also created conditions for a strong agricultural supply response to growing domestic demand and improved international market opportunities. Agricultural production more than tripled in volume terms between 1990 and 2013, with agro-food exports soaring. Viet Nam is now the world's largest exporter of cashews and black pepper, the second largest exporter of coffee and cassava, and the third largest exporter of rice and fisheries.

New challenges are emerging, however. Production growth rates are slowing for a number of commodities. A good part of the past revenue growth was due to higher commodity prices in the 2000s. Prices of many commodities have declined over the last two-three years and are projected to fall in real terms over the next decade. Land available for further expansion is also limited and there is increasing evidence of negative environmental impacts which are constraining production growth. Moreover, agricultural labour costs will increase if non-agricultural job creation carries on along its recent path. While rising labour costs will open opportunities to adopt new technologies and encourage larger farms, they may also reduce the sector's overall competiveness, particularly if newer labour-saving techniques are not readily accessible or adaptable to the dominant small-scale farming.

Private investment in agriculture is increasing, but several constraints still deter investors. Land fragmentation limits scale economies and various restrictions on land use rights raise costs. Large investors can have difficulty accessing long-term financing while small-scale producers continue to rely to a large extent on informal credit. Basic rural infrastructure has significantly improved over the past decade. But this investment has not kept pace with economic growth, resulting in serious infrastructure bottlenecks. Finally, the weak role played by farmer organisations obliges investors to interact with numerous small-scale producers. This increases transaction costs and compounds the uncertainty created by weak contract enforcement.

Viet Nam's agricultural policy seeks to achieve high quality output and competitiveness, raise rural incomes and maintain food self-sufficiency. The Ministry of Agriculture and Rural Development has the primary role in developing and implementing policies to achieve these objectives, but a number of other central government ministries and agencies also have significant roles.

Agricultural producers are supported by a range of input subsidies on irrigation, seeds and credit, amongst others. The budgetary cost of these measures has grown rapidly since the mid-2000s. Several initiatives have been introduced to deal with disease outbreaks and natural disasters. A direct payment per hectare began in 2012, but this is tied to maintaining land in rice production. Irrigation accounts for a relatively large proportion of government spending on agriculture, while other agricultural infrastructure and agricultural research and development remain underfunded. Government funded extension services remain top down driven.

Viet Nam has taken steps to reform its border protection and improve trade openness. Tariffs on agro-food imports have fallen. The average MFN applied tariff has dropped from 24% in 2000 to 16% in 2013, and is significantly lower for imports from ASEAN countries and China. However, tariffs remain relatively high for some commodities, including sugar cane, meat and some fruits and vegetables. Import monopolies, licensing requirements and export restrictions on agricultural products were removed in various stages over the 1990s and early 2000s. However, import requirements imposed for sanitary and phytosanitary reasons are becoming more stringent. They are often implemented in a non-transparent manner and add to the cost of importing. On the export side, concerns exist over the current system for controlling rice exports which reduces competition in the market and potentially works against incentives to grow higher quality rice.

The level of support to farmers as measured by the share of the policy-driven transfers from consumers and taxpayers in gross farm revenues (percentage Producer Support Estimate, %PSE) averaged 7% in 2011-13. Since 2000, the level of support has often varied strongly from one year to the next. This is the result of the government's efforts to stabilise domestic prices and to balance the interests of producers and consumers in the context of price volatility on international markets.

The total value of transfers arising from support to agriculture was equivalent to 2.2% of GDP in 2011-13, one of the highest across all countries covered by this measurement. This shows that for a developing country with a large agricultural sector and low GDP, even if the level of agricultural support as measured by the %PSE is low, the cost of support to the economy can be relatively high. This also highlights the potential burden of the current policy mix on the public budget and the need to ensure that the money is spent effectively.

Key policy recommendations

I. Improve the enabling environment for agriculture

- *ease the re-allocation of factors of production across sectors*
- *ease constraints on investment*
- *improve agricultural institutions and governance systems.*

II. Improve agricultural policy performance

- *pursue food security through a broader range of measures*
- *enhance farm restructuring*
- *improve the efficiency of resource use to minimise negative impacts on the environment*
- *reinforce agricultural innovation systems*
- *further integrate into international agro-food markets.*

Assessment and policy recommendations

This *Review*, undertaken in close co-operation with the Vietnamese Ministry of Agriculture and Rural Development (MARD), assesses the performance of Vietnamese agriculture over the last two decades, evaluates Vietnamese agricultural policy reforms and provides recommendations to address key challenges in the future. The evaluation is based on the OECD Committee for Agriculture's approach that agriculture policy should be evidence-based and carefully designed and implemented to support productivity, competitiveness and sustainability, while avoiding unnecessary distortions to production decisions and to trade. Conducted in partnership with the OECD Investment Committee, the *Review* comprises a special chapter highlighting key challenges to be addressed to improve the investment climate in agriculture, drawing from the OECD Policy Framework for Investment in Agriculture.

Assessment

With a territory of 0.33 million km^2 Viet Nam is a mid-size country in terms of area, roughly on par with Finland, Malaysia and Norway. Its population of 90 million makes it the 13th most populous country in the world. Around two-thirds of the population live in rural areas. Its population density is high, at 271 persons/km^2, which is just above the level of the United Kingdom and slightly below that of the Philippines.

Viet Nam is rich in water, but poor in land resources

While Viet Nam is on average relatively rich in water resources, agricultural land is scarce. With just 0.12 ha of agricultural land per capita, one-sixth of the world average, it is similar in proportion to Belgium and the Netherlands, just above the Philippines and India but less than China or Indonesia. Largely due to deforestation, total agricultural land increased by 61% in 1990-2012. Most of this expansion took place in the 1990s, with the arable land area remaining relatively stable since then. This might indicate that almost all accessible arable land is currently in cultivation and further production growth will need to be achieved through higher yields, which are already high compared to Viet Nam's Asian peers. There are continued pressures to convert agricultural land into higher-value non-farm uses (both urban and industrial). This has created a strong incentive to increase land intensity given the availability of relatively cheap labour, high soil fertility in some regions and relatively good climate conditions.

Strong GDP growth

A variety of reforms, known widely as *Doi Moi* or "Renovation", were launched in the mid-1980s which shifted the Vietnamese economy away from a central planning framework towards greater market orientation. Since that time a long series of policy changes have continued to move the economy, including the agricultural sector, in the direction of open markets for trade and investment, private decision-making, private land use rights, and a greater role for private firms.

Box 1. Viet Nam: Contextual information

Table 1. Contextual indicators, 1995, 2013

	1995	2013[1]
Economic context		
GDP (billion USD)	21	171
Population (million)	72.0	89.7
Total area (thousand km^2)	331	331
Population density (inhabitants/km^2)	217	271
GDP per capita, PPP (USD)	1 490	5 294
Trade as % of GDP[2, 3]	89.5	154.1
Agriculture in the economy		
Agriculture in GDP (%)	27.2	18.4
Agriculture share in employment (%)[4]	70.0	47.4
Agro-food exports (% of total exports)[3]	27.1	17.0
Agro-food imports (% of total imports)[3]	6.3	9.8
Characteristics of the agricultural sector		
Agro-food trade balance (million USD)[3]	2 937	9 459
Crop in total agricultural production (%)	80	73
Livestock in total agricultural production (%)	20	27
Agricultural area (AA) (thousand ha)	7 079	10 842
Share of arable land in AA (%)	76	59
Share of irrigated land in AA (%)	44.5	42.4
Share of agriculture in water consumption (%)[5]	n.a.	95
Nitrogen Balance, Kg/ha	n.a.	n.a.

1. Or latest available year.
2. Ratio of the sum of exports and imports to GDP.
3. 2000 instead of 1995.
4. 1996 instead of 1995.
5. 2005 instead of 2013.

Source: WB WDI (2015); UN (2015), *UN Comtrade Database*; FAOSTAT (2015).

StatLink http://dx.doi.org/10.1787/888933223173

Figure 1. Main macroeconomic indicators, 1990-2013

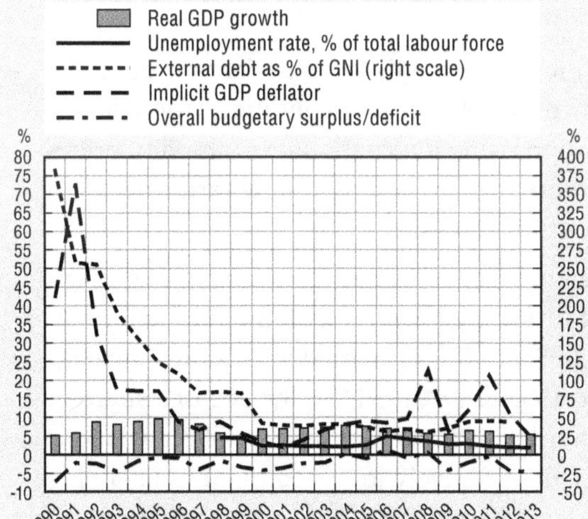

Note: Overall budget surplus/deficit in 1990-95 excluding grants.
Source: ADB (2005 and 2014).

StatLink http://dx.doi.org/10.1787/888933223188

Figure 2. Agro-food trade, 2000-13

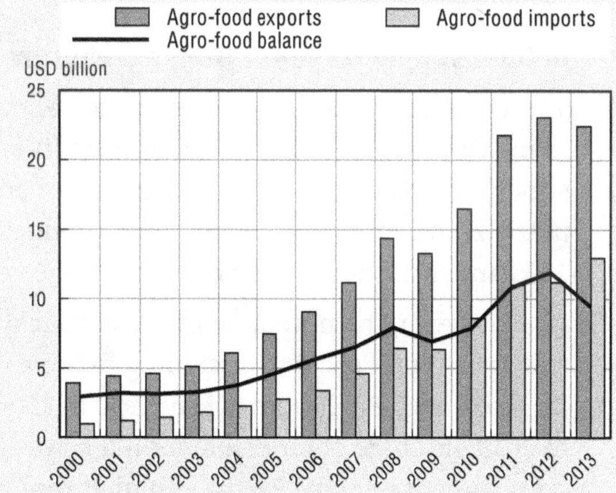

Note: Agro-food trade includes fisheries as well as natural rubber.
Source: UN (2015), *UN Comtrade Database.*

StatLink http://dx.doi.org/10.1787/888933223198

Reforms generated rapid economic expansion. Gross Domestic Product (GDP) growth averaged 7.4% in the 1990s and 6.8% in the 2000s, contributing to a three-fold real increase in GDP per capita over these two decades. Growth has slowed to 5.7% in 2010-14, but still compares favourably with most emerging economies.

Viet Nam's GDP is currently (2013) measured at USD 171.4 billion, which translates into USD 5 294 per capita at purchasing power parity (WB WDI, 2014). It joined the World Bank's category of lower middle-income countries in 2009 – an impressive accomplishment for a country that in the mid-1980s was one of the poorest in the world.

Poverty rates have fallen

Not only did the reforms generate rapid overall growth but the growth was particularly inclusive. Poverty was alleviated in Viet Nam as much as in any country in the world except for China. Real incomes, adjusted for inflation, are steadily rising for both urban and rural residents. While in absolute terms the gap between the two is growing, the relative gap measured as the ratio of urban to rural incomes is closing. However, even by 2012, the income of urban residents was still on average double that of rural residents.

According to the national poverty line definition, rural poverty rates are much higher than those in urban areas. The gap tends to decline, but remains large. This decline in rural poverty rates from 21% in 2004 to 13% in 2013 reflects Viet Nam's success in increasing agricultural productivity for many farm commodities and in diversifying sources of rural incomes.

Undernourishment rates have declined

Viet Nam has made astonishing progress in combatting undernourishment. The proportion of undernourished in the total population fell from 46% in 1990-92 to 13% in 2012-14. This represents a decrease of 72%, which is one of the highest rates for all countries, just after Thailand, and larger than in China. Nevertheless, 11.9 million Vietnamese suffered from undernourishment in 2012-14 (FAO-IFAD-WFP, 2014). Most food insecure people live in rural areas.

Robust supply response from agriculture

Economy-wide and sector-based reforms, including the de-collectivisation of farms mandated in 1988 and the land use rights issuance in 1993, created conditions for a strong supply response to growing domestic demand and to rising international commodity prices in the 2000s. As a result, agricultural production more than tripled in volume terms between 1990 and 2013, outperforming all its major competitors in Asia.

However, the non-agricultural economy has grown substantially faster, pushing down the agricultural sector's (including fisheries and forestry) shares of GDP and employment. Its share of GDP fell from 39% in 1990 to 19% by 2005 and has remained at this high level up to 2013. Its share of employment fell from 70% in 1996 to 47% in 2013. The sector's share of employment remains 2.5 times higher than its share of GDP indicating relatively low labour productivity. This is one of the main reasons for the relatively low income of households dependent on farming.

The agro-food sector is well integrated with international markets. The ratio of total agro-food export value to agricultural GDP was 70-80% in the early 2010s, much higher than in China or Indonesia and equal to the ratio of total Viet Nam's exports to total GDP.

The ratio of agro-food imports to agricultural GDP is only half of that for exports, but it has tripled since 1990.

The annual rate of growth in agricultural production slowed from an impressive average of 5.7% in 1990-2002 to 4.2% in 2002-13. While Viet Nam has maintained growth rates higher than most other countries in the region, the gradual slowdown in more recent years is noticeable. Most likely, the rates would have declined still further in the last period had there not been the agricultural price boom that elevated many world prices by a factor of two. This might be taken as a warning signal that the earlier sources of the sector's boom based on institutional reforms and expanding use of cheap resources have begun to be exhausted.

Rice remains by far the most important commodity, accounting for about 35% of the total value of agricultural production in recent years. However, there has been an important change in the composition of production away from staple foods to other commodities, in particular perennial crops such as coffee and rubber, and livestock production, especially pigmeat. This reflects the strong export orientation of perennial crops and the changing preferences of consumers to higher value products.

Total factor productivity growth slowed in the 2000s

Historically, Vietnamese agriculture has been labour-intensive. The total number of persons employed in agriculture increased up to 2009 and since then stabilised at around 24.4 million (including forestry and fisheries) (GSO, 2014). Agriculture is not yet at the stage of shedding labour in absolute terms, but it might be at the turning point and, according to some projections, farm employment might fall by 9% in the current decade (ILO, 2011).

Viet Nam's agricultural Total Factor Productivity (TFP) growth was strong and sustained over 1990-2010 (averaging 2.65% per year). It was significantly stronger than in the 1980s, clearly reflecting the positive impact of reforms undertaken in the late 1980s and early 1990s. However, while it was stronger than in Indonesia, India and the Philippines and equal to that in Thailand, it lagged behind China and more recently also Malaysia, reflecting a slowdown in the 2000s compared with the highest rates registered in the 1990s (Fuglie and Rada, 2013).

Agro-food exports soared

Prior to 1990, Viet Nam was not a significant player in world agricultural commodity markets, with trade tightly controlled by the government. By 2011-13, Viet Nam had become the world's largest exporter of cashews and black pepper, the second largest exporter of coffee and cassava, the third largest exporter of rice and fisheries, and the fifth largest exporter of natural rubber. Annual exports for these commodities were well above, or very near to, USD 1 billion in the early 2010s. Such trade performance across a relatively wide range of commodities for a country the size of Viet Nam starting from virtually no export market penetration and experience, and within two decades, is unmatched.

Led by expanding exports of the above mentioned commodities, along with aquaculture and fisheries, the total value of Viet Nam's total agro-food exports increased six-fold between 2000 and 2012. Exports are around double the value of agro-food imports, contributing to a positive balance of agro-food trade of about USD 10 billion in 2011-13.

Export prices and value added in exports remain low

Viet Nam's agro-food exports are commonly derived from low-value commodity sales. This "commodity" approach to exports is long on quantity growth, but short on quality and value added. It is recognised in Viet Nam that moving up the value scale in food markets allows exporters, and usually farmers, to capture higher prices without having to increase production or find more inputs such as scarce land. For this reason, it is one of the main pillars of the Government's Agricultural Restructuring Plan (ARP), which includes the improved quality of basic farm commodities, and food processing into innovative products.

The role of SOEs remains strong

While the influence of state-owned enterprises (SOEs) across the Vietnamese economy has declined, their share of GDP remains high at one-third in 2011 and they are an important source of government revenue and export earnings (OECD, 2013). Moreover, many SOEs have been only partly privatised through a so called "equitisation" process, through which they are converted into public limited companies or corporations by selling a part of their equity to the public or a special investor, while the state still holds the majority of shares. In addition to the fact that the newly-created equity shares may be held by the state, the firm may continue to hold advantages from ex-SOE status, such as continued market power and easier access to credit. Within the agro-food sector, SOEs are involved in agricultural input supply firms, processing and storage firms, and marketing including exporting firms.

The food safety regulatory regime needs to be effectively implemented

Viet Nam undertook to comply with the requirements of the Sanitary and Phytosanitary (SPS) Agreement upon its accession to the WTO in 2007. The National Strategy on Food Safety for 2011-20 sets a general objective of implementing master plans on food safety from production to consumption by 2015, and controlling food safety over the entire food supply chain by 2020.

Despite these efforts to set in place a legal framework and structure for quarantine and food safety that conforms to international obligations, further work is required to effectively implement the regulatory regime. The capacity of testing agencies is limited, leading to inconsistent enforcement that adds to uncertainty for producers (Arita and Dyck, 2014). The large number of legal documents relating to food safety (about 400 documents issued by the central government and ministries and about 1 000 documents issued by local governments), result in overlap and lack a clear focus. Co-ordination between agencies, risk analysis and identification systems are poor, both at the central government level and between central and local government.

As a consequence, Viet Nam often experiences difficulties accessing export markets for some commodities. Similarly, exporters of agricultural products to Viet Nam face a complex administrative system, often experience inconsistent requirements and sometimes must comply with standards that are more restrictive than international norms.

Farm structures are dominated by smallholders

Large farms, predominately operated by SOEs, use about 10% of agricultural land focusing on the production of perennial crops. Around 9.6 million households farm on the remaining land; each using about 0.8 ha on average, typically further subdivided into four non-contiguous plots. While the process of farm consolidation in livestock production has started, consolidation of crop production is at its very early stages. Very few farms have grown to the "large scale" category of 2 ha or above.

Restrictions on land use persist

All land is owned by the state and administered by it on behalf of the people. The Land Law of 1993 gave farmers a wide range of rights, including the right to rent, buy, sell, and bequeath land and to use land as collateral with financial institutions for mortgages. Holders of these rights are entitled to Land Use Rights Certificates. By 2012, rights to 85% of agricultural land had been certified.

Revisions of the Land Law made in 1998 and 2003 introduced restrictions on land use stating that changes in land use by the farmer were only allowed within the existing physical planning framework adopted by central and local governments. They mostly confine farmers to growing rice on paddy land at the expense of other crops (or fisheries) that could be grown more profitably on the same land. Farmers can apply at the district level for a change in their designated land use, but in practice changes or removals of these restrictions are rarely allowed.

The 2003 Land Law revisions allowed the state to appropriate land, including farmland, for economic development purposes. While it was introduced to help encourage industrial and urban development, it resulted in a sharp increase in highly contentious land disputes. Farmers are not only involuntarily losing their base for farming, but also they receive very low compensations.

The Land Law passed in 2013 made a number of modest improvements. However, the essential points of controversy in land disputes remain largely unaddressed. In particular, the Law has not given farmers the right to the market price for land expropriated for non-agricultural uses and has not removed restrictions on land use rights.

Environmental pressures risk reducing long-term productivity growth

Rapid economic growth, combined with rising population and expanding agricultural production, is exerting massive pressures on the environment. The deforestation that accompanied the rapid expansion of agricultural land during the 1990s has only been partly remedied by reforestation efforts undertaken over the last 15 years. While the overall forested area has increased, undisturbed primary forests continue to disappear.

Agriculture exerts significant and growing pressure over the country's available water resources with the sector accounting for 95% of freshwater use. Further, due to the excessive use of fertilisers, pesticides and other chemicals, the sector has contributed to a progressive degradation of water and land quality.

Viet Nam is listed among the ten countries potentially the most affected by climate change. Climate change scenarios developed by the Vietnamese government predict increases in average temperature, rainfall and rising sea levels. The potential impacts on agriculture are serious, as floods and droughts are predicted to happen more frequently. In particular, large cultivation areas in the Mekong and Red River deltas are likely to be even more affected by salt water intrusion due to sea level rise (ISPONRE, 2009).

Private investment in agriculture has recently increased

As a result of efforts to improve the business climate, private domestic investment in agriculture has increased since the *Doi Moi* renovation process, accounting for 56% of agricultural investment in 2008 – with the rest coming from SOEs (34%) and foreign investors (10%). A revised Investment Law has been adopted by the National Assembly on 26 November 2014 and entered into force on 1 July 2015. It clarifies the definition of foreign investment, simplifies licensing procedures and reduces the number of sectors where investment is prohibited or conditional.

The government is promoting Public-Private Partnerships (PPPs). In addition to the creation of six PPP task forces formed by MARD around key commodities, a PPP Decree has been approved in February 2015 and covers not only infrastructure but also, more specifically, agricultural infrastructure and rural development services associated with agro-processing and the consumption of agricultural products. To ease investment, the government supports access to credit by providing producers with loans without collateral, subsidised credit for agricultural inputs and assets, and credit guarantees through state-owned banks.

Several constraints continue to deter agricultural investors

Although numerous investment incentives are offered to small and large investors, laws, numerous decrees and provincial regulations lead to a complex web of investment incentives that creates uncertainty for investors who are granted such incentives on an *ad-hoc* basis. The absence of a strong independent Investment Promotion Agency accentuates this complexity. Indeed, promotion activities are performed by a mix of agencies, including the Foreign Investment Agency in the Ministry of Planning and Investment (MPI), Viet-Trade in the Ministry of Trade and Industry, the International Co-operation Department in MARD, and the promotion departments of provinces. Several constraints undermine private investment in the sector (Box 2).

Box 2. **Constraints to private investment in agriculture**

- ***Restrictions on land use rights:*** While the Land Law of 2013 strengthens the development of a land market, it keeps several restrictions on the duration of land use rights, land areas per household, the choice of crops and land transfers and exchanges. Such regulations intend to guarantee equal access to land among the rural population, but they limit land consolidation and hinder long-term investment.
- ***Insecure land use rights:*** Land use planning is not based on a participatory process which opens a possibility for forced conversions that have not been agreed by local communities. Agricultural land acquired in this way is priced at low levels and then rented out at much higher prices for other uses. This process is prone to corruption and involves numerous administrative payments.
- ***Limited access to credit:*** Financial markets in rural areas remain very concentrated. The rural finance market consists of several players in which Vietnam Bank for Agriculture and Rural Development (VBARD) and Vietnam Bank for Social Policies (VBSP) have leading positions, representing 66% of the sources of rural credit in 2010. Although Co-operative Banks – formerly known as Peoples' Credit Funds – and other private financial institutions have been established, so far they have not achieved significant importance in rural finance. Such concentration may explain the limited access to formal banking services in rural areas and high interest rates. The lack of sufficient collateral also limits access to credit by small-scale farmers. As a result, half of rural households were still unable to access banking services in 2010 and the informal sector remains an important source of rural credit.
- ***Inadequate infrastructure in rural areas:*** Viet Nam has made impressive progress in infrastructure development, with now over 90% of the rural population having access to electricity and over 98.5% having access to roads. However, recent rapid economic growth

Box 2. **Constraints to private investment in agriculture** *(cont.)*

has resulted in serious infrastructure bottlenecks. New infrastructure is generally located in urban areas to connect major cities, airports, sea ports, and industrial parks, while rural infrastructure is often in poor conditions and not properly maintained. The implementation of infrastructure projects by local governments and SOEs delays implementation and leads to competition between localities which hinders a holistic development of infrastructure and results in fragmented, suboptimal infrastructure projects with low utilisation rates.

- ***Lack of skilled workers and limited funding of R&D:*** Enterprises underline the mismatch between the supply and demand of skills. Labour productivity remains low, amounting to 23.3% of Malaysia's and 37% of Thailand's in 2010. Extension services face several challenges, including limited human resources, the dominance of a top-down approach, a lack of services tailored to different types of farms, a weak participation of the private sector, and poor monitoring system. Most agricultural research is carried out by state research agencies with limited funding and not able to meet the practical requirements of farmers and private enterprises.

Agriculture is a priority sector for the government

Agriculture is one of the key strategic sectors identified by the government of Viet Nam. An important feature of the policy framework is the establishment of five-year Socio-Economic Development Plans (SEDP). While MARD has the main responsibility for policy development and implementation, a large number of other central government ministries and agencies are involved. Furthermore, since fiscal decentralisation in 2002, local government has been given a greater role in planning and implementing agricultural policy. These factors create co-ordination challenges in agricultural policy development at the central and regional level.

Five distinct phases of policy development

Agricultural policy development since reunification in 1976 can be divided into five stages:

- 1976-86: The role of agriculture during the first decade following reunification was to support the development of heavy industry as part of a centrally planned system. Agricultural production was organised around co-operatives and state farms, with state-owned enterprises providing inputs and controlling output markets. Despite minor reforms to incentivise production, agricultural output failed to meet state targets leading to food shortages and contributing to a major economic crisis.
- 1986-93: As part of the broad *Doi Moi* renovation process to stabilise the economy and develop the private sector, the role of agriculture was elevated to one of primary importance. The focus of agricultural management moved from co-operatives to farm households. Farmland was redistributed to farm households who were given the ability to make their own production decisions provided they met certain production quotas. Broader reforms opened up the market to both greater domestic and international competition. Agricultural production rose sharply, becoming a key driver of overall economic growth.
- 1993-2000: The focus in this period was one of encouraging agricultural expansion. Institutional reforms were introduced to replace the gap left by the collapse of the

co-operative system, for example with the establishment of a national extension service and credit facilities for farmers. Production quota obligations were removed and further regulatory barriers to trade were gradually lifted. A large number of bilateral and regional trade agreements and partnerships were entered into to expand market opportunities. The improved policy environment was supported by a rapid increase in budgetary expenditure. At the same time, a Price Stabilisation Fund was created to stabilise the prices of essential commodities including urea, paddy and rice, coffee and sugarcane.

- 2000-08: This period marked the beginning of the move from expanding production towards greater emphasis on improving yields, quality and value. The goal being to create a modern and industrialised agricultural sector. Previous reforms were locked in and further actions were required as a result of further international integration at the bilateral, regional and multilateral level. The final few quantitative restrictions over agricultural imports and exports were removed.
- 2008-present: Two major resolutions are currently guiding agricultural policy development: Resolution No. 26/2008/NQ-TD and Resolution No. 63/2009/NQ-CP. The first emphasises development based on the market economy with socialist orientation; the second seeks to ensure national food security by guaranteeing adequate food supplies, particularly for rice. There is potential for conflict in achieving both at the same time. These two resolutions have been implemented through a number of documents, including the ARP to restructure the agricultural sector towards improving value-added and sustainable development.

Agricultural policy objectives

Agricultural policy objectives are set out in a number of documents and plans. These often set specific targets and various actions for their achievement. In general, these objectives focus on achieving agricultural production growth through improving productivity, quality and competitiveness; developing infrastructure; improving the living standards of the rural population; strengthening the international integration of the sector; and using and protecting natural resources and the environment in a sustainable and efficient manner.

Agricultural policy instruments

These policy objectives are pursued through the use of output and input subsidies, and payments for the provision of services to agriculture generally (Box 3). Very little use is made of less distorting forms of support such as payments based on land or farm revenue that are not linked to production.

Box 3. **Overview of agricultural policy instruments applied in Viet Nam**

Domestic policy instruments

- *Price support measures:* Farm gate rice prices are supported by a subsidy to rice purchasing enterprises for the temporary storage of rice during harvest and establishment of target prices which vary between regions and crop season with the objective of providing farmers with a profit of 30%.
- *Irrigation service fee exemption:* Prior to 2009, farmers paid a contribution to the cost of managing, maintaining and protecting irrigation works in the upper-level systems. An exemption was provided for most farmers in 2009, leading to a substantial increase in government support to irrigation and drainage management companies.

Box 3. **Overview of agricultural policy instruments applied in Viet Nam** *(cont.)*

- *Seed and livestock breeding subsidies:* Many programmes provide plant genetic and animal breeding material to farmers at subsidised rates. At the national level, these are often provided as part of the package farmers recover from natural disasters or disease outbreaks.
- *Credit schemes:* Since 2009, a number of policy packages have been introduced to provide farmers with cheaper credit to purchase machinery, facilities and materials.
- *Payment based on area:* In 2012, a direct per ha payment was introduced for rice farmers as part of a broad package of measures to protect and support the development of paddy land.
- *Insurance:* A pilot insurance programme was introduced in 2011, providing subsidised premiums to rice, livestock and aquaculture producers in 21 provinces.
- *Income support:* Since 2003, most farming households and organisations have been exempt from paying agricultural land use tax or have had the amount they pay reduced.
- *Extension services:* Central government funding for extension has been allocated through an open bidding process since 2001. It is essentially a top down, supply driven extension system.

General services provided to the agricultural sector as a whole

- *Irrigation:* Funding of irrigation capital works is the largest area of government expenditure supporting agriculture.
- *Research and development:* Despite increasing over the 2000s, expenditure on research is relatively small in comparison to other countries. An attempt to achieve greater co-ordination in research occurred in 2005 with the reorganisation of the various research agencies under the oversight of the Viet Nam Academy of Agricultural Sciences.

Trade policy instruments

- *Tariffs:* The simple average MFN applied agricultural tariff decreased from around 25% in the mid-2000s to 16% in 2013. A MFN applied tariff of 40% applies to a range of commodities including meat or poultry, turkey and duck, tea (green and black), grapefruit, milled rice, refined sugar, and many types of prepared or preserved fruits and vegetables. However, the average agricultural tariff is just 3.4% and 5.4% on imports from ASEAN members and China respectively.
- *Import licensing:* For the purpose of enforcing minimum quality or performance standards, MARD regulates the importation of veterinary medicines, pesticides, plant and animal strains, animal feeds, fertilisers and genetic sources of plants, animals and micro-organisms used for scientific purposes.
- *SPS and food safety:* Since joining the WTO in 2007, Viet Nam has made some progress towards implementing the requirements of the Sanitary and Phytosanitary Agreement. However, the regulatory regime still suffers from limited capacity, poor co-ordination and a large number of overlapping documents.
- *Export taxes:* These are limited to a narrow range of agricultural related products: raw hides, rubber and cashew nuts, although for cashew nuts the tax is zero-rated. Between July and November 2008, a progressive export tax regime was introduced on rice exports with the intent of limiting price increase on the domestic market.
- *Export licensing:* The government maintains a large degree of control over rice exports. Exporters must meet specific milling and storage requirements, and certain administrative functions are given to the Viet Nam Food Association (VFA). The VFA is highly influenced by two large SOEs: Vinafood I and Vinafood II. SOEs play a dominant role in the export of some other commodities such as coffee, rubber and tea.

Box 3. **Overview of agricultural policy instruments applied in Viet Nam** *(cont.)*

- *Regional trade agreements:* Viet Nam is a member of the Association of Southeast Asian Nations (ASEAN), Asia-Pacific Economic Cooperation (APEC), World Trade Organisation (WTO), supports trade liberalisation between ASEAN members and their major trading partners in the region, including China, Japan, India, Korea, Australia and New Zealand and takes part in the Trans-Pacific Partnership (TPP) negotiations.

The level of support to agriculture is relatively low

Developments in agricultural policy can be assessed by changes in the level of support measured by the %PSE (Producer Support Estimate as a share of farmers' gross receipts) and the %TSE (Total Support Estimate as a share of GDP). Over the period 2000-13, the level of support was quite variable without revealing any distinct long-term trend. Nevertheless, the %PSE remained positive over most of this period, indicating that producers generally received moderate support. The level of producer support as measured by the %PSE averaged 7% in 2011-13; less than half the level of support provided to producers in China and Indonesia, and considerably below the OECD average of 18%. Nevertheless, the %TSE at 2.2% for 2011-13 is one of the highest and well above the OECD average at 0.8%. This shows that for a relatively poor country with a low GDP and large agricultural sector, even if agricultural support as measured by the PSE is low, the burden on the economy can be relatively high.

Price support and input subsidies dominate

Market Price Support (MPS) is the dominant form of support to producers. Given the importance of rice within the agricultural sector, the MPS value for rice drives the overall PSE. The dominance of MPS in Viet Nam's PSE explains the annual variations in producer support that are observed because they depend on movement in world and domestic prices, exchange rates and production levels. Furthermore, these swings are relatively greater in Viet Nam and often produce negative values because of the government's efforts to balance the interests between producers and consumers. On the one hand, the government wishes to increase prices received by producers to encourage production and improve farmer incomes. On the other, it wants to keep prices paid by final consumers at an affordable level to help alleviate poverty and avoid social tension.

Budgetary transfers have remained relatively constant at about 20% of producer support on average over the period 2000-13. Expenditure associated with subsidising the irrigation fee exemption remains the dominant payment. A hectare payment with the objective of keeping about 4 million ha in paddy production has been provided since 2012.

General services for the agricultural sector have remained relatively constant as a share of total support transfers, suggesting there has been little re-orientation of policies towards those that can benefit both producers and consumers. The most important GSSE category, representing around 85% of GSSE expenditure, is development and maintenance of infrastructure, which is dominated by expenditure on irrigation systems. Expenditure on some general services such as inspection and control and marketing and promotion receive relatively limited support.

Import-competing commodities are supported

Producers of import-competing commodities such as beef and veal, poultry, eggs and sugar cane are highly supported, receiving prices for their outputs above international prices. This is mainly the result of border protection measures. In contrast, producers of export-competing commodities such as natural rubber, coffee, cashew nuts and tea are implicitly taxed in that producer are paid prices for their outputs that are lower than international prices. However, it would be incorrect to interpret implicit taxation of crop products exclusively as a policy outcome. For example, poor infrastructure can impede market adjustment and exacerbate any policy impact on prices, therefore contributing to the negative results.

Policy recommendations

Over the next ten years, both domestic and international conditions will be more challenging for Viet Nam's agricultural sector than they were in the 1990s and 2000s. Prices of many commodities exported by Viet Nam declined over the last two-three years from the peaks seen in 2007-08 and are projected to fall further in real terms over the medium term, though remaining at or above the pre-peak levels (OECD-FAO, 2014). Most of the easy sources for lifting production, e.g. expanding land area, employing more cheap labour and using higher rates of fertilisers, have been fully exploited and negative environmental impacts are increasingly seen. These will become major challenges for Viet Nam, but will also open opportunities to adopt new technologies, to give incentives for larger farms and to focus attention on quality and higher value added products.

The set of policy reforms suggested below are derived from analysis undertaken in the *Review* and are designed as key building blocks to support increased agricultural productivity, competitiveness and sustainability. These recommendations are not exhaustive and should be interpreted as a starting point for government consideration, refinement and elaboration. In particular, choices will need to be made across this wide range of recommendations as to which policy actions should and can be implemented quickly, and which might be acted upon more gradually.

I. Improve the enabling environment for agriculture

1. Ease the re-allocation of factors of production across sectors

- **Ease constraints on infrastructure development.** According to the MPI, public funding is likely to only cover around 40% of the costs of necessary infrastructure development over the next ten years. Private investment in infrastructure will also be needed and can be attracted by, amongst other things, ensuring a level-playing field between SOEs and private enterprises. The effectiveness of available infrastructure funding would be improved by enhancing co-ordination between national and sub-national governments, avoiding duplication between provincial governments and promoting an integrated approach to infrastructure projects.
- **Enhance labour mobility across sectors and across regions.** The importance of labour moving from agriculture to the non-agricultural sector in maintaining economic growth and in reducing poverty cannot be overstated. Migration from rural to urban areas raises incomes of migrants, contributes to higher incomes of migrants' families through remittances, raises the wage rates of agricultural labour remaining in the countryside as its supply shrinks, enhances information flows and training, and improves land and water availability for those who remain dependent on farming. Even though Viet Nam

has reduced enforcement of the registration system for rural residents that denied migrants access to a variety of public services in locations outside the locality where they were born and registered, it is important that vestiges of these rules do not get applied, and that every effort be made to allow migrants full rights and no restrictions. Stronger integration of farm and non-farm labour markets is required.

- ***Further reform state-owned enterprises.*** While this process is ongoing, reforming agro-business SOEs should be given even more attention. They often possess considerable monopsony or monopoly power in particular sectors, even if there is formally no restriction on new entrants. The use of industry associations such as VFA to implement policy needs to be fully reviewed, as there is a strong possibility for vested interests to limit competition. Efforts to open up various components of the food chain, including importing and exporting, to private firms are unlikely to be successful if the incumbent SOEs have sufficient market power to deter entry. This may delay adjustments to market signals, including those calling for higher-added value products to be supplied to domestic and international consumers. Thus, there is a need to reduce the SOE's role through privatisation, removing explicit and implicit support and guarantees provided to them, and easing entry of truly private domestic and foreign firms to all segments of the food chain to enhance competition and to bring a more innovative and modern processing and marketing environment.
- ***Remove impediments for moving up the value chain.*** In Viet Nam, a number of policies act as impediments to the development of value-added agro-food products for sale on domestic and export markets. For example, the land use restrictions limit the possibility of moving from low value rice to higher value fruit and vegetable production. Other opportunities include enforcing the food safety regulatory regime, which would improve consumers' confidence in products, both processed and unprocessed, originating from Viet Nam. Government's Agricultural Restructuring Plan rightly states as one of the "core principles" that the role of government will shift to being a facilitator, providing an enabling market-based environment for the private sector at farm and agribusiness levels. It is important to allow businesses to identify export opportunities and private firms are generally more aware of the micro data that are important to determine if the benefits exceed the costs for moving up the value chain.

2. Ease constraints on investment

- ***Review investment promotion measures.*** Cost-benefit analyses should be undertaken to evaluate the opportunity cost and the impact of existing investment incentives. Such incentives are currently granted on a case-by-case basis. Investors should be aware of which incentives they would be granted prior to investing which requires clarifying the current design and implementation of such incentives.
- ***Improve access to finance.*** Facilitating access to credit by producers requires the development of a much stronger and more competitive financial market, for instance by supporting the development of Co-operative Banks. Efforts to establish credit reporting systems, credit and assets registry systems (both for movable and fixed assets) and to develop financial services such as equipment leasing and warehouse receipts, should be sustained, while public subsidies should be reduced.
- ***Strengthen the legal framework for PPPs.*** PPPs can enhance the co-operation between public and private actors, thereby increasing returns from public funds through cost and

risk sharing and securing contributions that are more adapted to both public and private demand. The main conditions for forming a successful PPP include: common objective, mutual benefits, complementarity of human and financial resources, clear institutional arrangements, good governance, transparency and public leadership. The new legal framework for PPPs in agriculture, to be refined in two circulars that MARD is developing, should thus clearly state the respective roles and responsibilities of the public and private sectors.

3. *Improve agricultural institutions and governance systems*

- ***Strengthen institutional co-ordination between MARD and other relevant ministries implementing programmes supporting agriculture.*** There are a large number of cases, e.g. in providing financial support to agriculture or in food safety regulatory regime, in which co-ordination between various agencies both at the central government level and between central and local governments is weak. Responsibilities and functions of different agencies as well as of different levels of administration should be clarified to improve the effectiveness of public programmes in meeting stated objectives.
- ***Strengthen transparency and accountability of publicly-funded programmes.*** Coherent data on budgetary support to agriculture combining support from all sources, including various ministries, central and provincial governments, and overseas development assistance are missing. While data on budgetary expenditures on key programmes under the responsibility of MARD are publicly available, data on expenditures to support agriculture from other sources remain sporadic and not necessarily defined in a way allowing comparisons over time and matching them with other funds targeting the same objective. Moreover, while data on budgeted amounts are occasionally released, data on amounts actually spent are missing. It would be advisable to charge MARD with an oversight of overall public expenditures supporting agriculture, including those under the responsibility of other ministries and provincial governments. Transparency would improve: the assessment of the support provided to agriculture and rural areas, the monitoring of sub-national government performance by MARD, the co-ordination of funding to achieve stated objectives, and the reporting process of relevant data to international organisations such as WTO, FAO and OECD.
- ***Base policy decisions on adequate and accurate information and build monitoring and review mechanisms into the policy process.*** Reliable and timely statistics are necessary to assess the results of reforms undertaken so far, formulate policy responses and design policies for the future. While user-orientation of agricultural statistics has been improving, there are still areas which need further attention. The accuracy of data on agricultural commodity prices at the farm gate and wholesale levels, overall farmland versus forest area, farm structures in terms of actual land use pattern (not just legal use rights) is far from adequate. A more comprehensive and coherent system of monitoring, analysing and reporting of Viet Nam's agricultural policies will help analyse, assess and improve policy performance.

II. Improve agricultural policy performance

1. *Pursue food security through a broader range of measures*

- ***Enhance production and income diversification.*** Better infrastructure and unrestricted labour mobility across regions and sectors would be key factors to promote access to off-farm work for farm families, thus providing them with higher incomes and improving their access to food. Diversification from rice production into high-value crops would allow farmers to earn higher incomes from a given amount of land, thus improving their access to food. It would also release resources to increase supply of higher value products for domestic and international markets. Currently, a wide range of agricultural policy measures focus on rice, locking more resources into this activity than otherwise would be the case. In particular, the commitment to provide farmers with a 30% profit on rice production is an unsustainable objective for a major exporter.

- ***Allow market-driven diversification of diet.*** With growing incomes, rice consumption in Viet Nam has started to fall and this process should not be slowed by any interference in relative price ratios across food products.
- ***Assess the effectiveness of current insurance schemes and of alternatives to them.*** Insurance schemes are at the experimental stage in Viet Nam. Such schemes are designed to provide a tool for farmers to deal with income variations caused by pre-defined types of natural disasters and epidemics. In the long term, sound insurance schemes would allow for a more stable policy framework and can reduce the need for one-off support payments to farmers. These pilot programmes should be assessed before being extended across a wider range of provinces and commodities. Such evaluation would need to include the cost of the programmes, the extent to which benefits reached intended beneficiaries, the actuarial soundness of the system, and their cost-effectiveness relative to other policy alternatives. In the short-term, a subsidy on the insurance premium can demonstrate to farmers the value of insurance and can help create a relevant database for developing viable insurance schemes. However, in the long run, a wider package of policies can serve to equip farmers better with the information and tools needed to manage a wide range of risks normally associated with farming.

2. Enhance farm restructuring

- ***Encourage farm consolidation.*** For most commodities, there are economies of size that help reduce some categories of farm costs. Larger land holdings become more valuable when farm labour becomes expensive and when there are options to mechanise to save labour. Even if raising farm size is not yet an economic imperative, the process should not be discouraged. A useful initial step would be to remove any barriers to growth in farm size: a) removal of the land size upper limits and the restrictions to land transfers under the Land Law, b) improving the availability of farm credit, including to smaller and medium sized farms, c) improving rural education, training and extension so that farmers can learn about and operate more efficient production technologies that involve larger scale, and d) avoid policy distortions that alter factor price ratios.
- ***Limit the scope of compulsory land conversions.*** Most land conflicts could have been avoided if the legal framework did not allow for compulsory land conversions for so called "socio-economic development" uses of land. If instead voluntary conversions or transactions between the farmer and the investor were allowed, corruption would be reduced, the need for costly support mechanisms such as resettlement would be smaller and social unrest would almost certainly decline. This would not preclude state designation of certain land areas for specific uses, such as for public investment and military uses, or land areas where defined uses would be prohibited. A specific area might be restricted to agricultural uses. But within the allowed uses as defined by approved land use plans, land tenancy transactions would be voluntary between buyer and seller.

- ***Base compensation for land on open market land prices.*** The Land Law of 2013 refers to the principle of compensation at market prices, but how district or provincial party committees do this is left open, and compensation is still based on the agricultural use value, thus much below market prices for alternative uses of land. The negotiation over the price of land should be left to the buyer and seller, so farmers could negotiate a higher price if they chose and could do so. It would then be less critical to alter the procedures for compulsory takings and price arbitration for truly state uses of land such as for a highway, which account for a small minority of current land conflicts.
- ***Remove restrictions on agricultural land use.*** Designating 3.8 million ha for rice production exclusively is unlikely to be the best policy approach in a country exporting large quantities of rice. If the main objective is food security, there are more effective means of achieving it. Indeed, diversification to achieve lower risk is a measure that adds to food security and is a separate (and commendable) objective of the government. The restrictions on crop choice work against diversification. And if the goal is increased exports, farmers should not be prevented from producing higher-value crops.
- ***Enhance transparency in land management.*** Bribery and the lack of transparency constitute significant impediments to investment. Social conflicts and corruption in the land administration may be reduced by developing participatory land use plans to clarify land allocation, limiting compulsory land conversions, and allowing direct transactions between land users without state involvement. Participatory land use plans would define land preserved for agricultural use and would guide farmland conversion to non-agricultural use in designated areas. Simplifying the procedures to obtain land use right certificates and publicising the various related fees would also enhance transparency.
- ***Enhance various forms of co-operation between farmers.*** Producers lack trust in the large co-operatives that existed prior to 1986 even though they have been transformed and restructured. Smaller co-operatives created around specific commodities such as milk, vegetables, and horticulture, can function well and provide input and marketing services. If supported by extension services they could more effectively help farmers access agricultural inputs, training, technology, and market information.

3. *Improve the efficiency of resource use to minimise negative impacts on the environment*

- ***Reintroduce the water fee for farmers to cover operation and maintenance costs.*** While the waiver of irrigation service fees has increased farmer income, it has reduced the incentive for farmers to save water, made the national budget fully responsible not only for capital investment, but also for financing operation and maintenance costs, and diminished incentives for irrigation and drainage management companies to provide quality irrigation services. While the government could remain responsible for all capital investment in the irrigation systems, farmers should cover all operation and maintenance costs. Re-establishing a water fee based on a per unit of water charge rather than a per hectare charge as previously used would encourage greater water use efficiency.
- ***Reinforce monitoring, compliance and enforcement of environmental legislation.*** Viet Nam has undertaken efforts to enhance environmental protection, promote sustainable water use and forest management, reduce GHG emissions, and respond to climate change, but enforcement mechanisms are weak. Education and extension services should better

demonstrate to farmers the short- and long-term benefits from implementing environmental legislation, e.g. lower production costs through reduced use of chemicals, particularly in areas characterised by overuse of such inputs.

4. *Reinforce agricultural innovation systems*

- ***Improve the institutional design of agricultural research and development.*** Despite increasing by an average rate of 11% per annum between 2000 and 2012, government funding for R&D remains relatively low. Improving Viet Nam's domestic capacity to develop and improve plant varieties, improve animal breeding and develop technological solutions for farmers should be complemented by much greater efforts on more effective adoption of technologies developed by technology leaders. Good co-ordination with international, regional and sub-regional research networks would be important to improve Viet Nam's absorption capacity and to up-grade the national research system. To increase the available funding, the government should explore ways to harness the considerable R&D capacity in the private sector, for example through designing effective public-private partnerships. However, any increase in funding should be linked with a stronger focus on research that meets the practical needs of farmers and on areas going beyond primary production, such as post-harvest, processing, product hygiene and safety and environmental protection.
- ***Re-orient the focus of agricultural education and extension services to improve farm management skills.*** The current focus on primary agricultural production needs to be re-oriented to areas such as: marketing skills, preparation of business plans, co-operation arrangements between farmers, and use of more environment-friendly methods of production. The current top-down approach, with the government deciding what extension advice is to be provided to farmers, should be re-oriented towards a greater role given to farmers who could guide extension services according to their needs. While the current use of a competitive bidding process for the selection of extension projects creates the possibility for more efficient allocation of resources, there appears to be potential for overlap in projects awarded at the central and local government levels.

5. *Further integrate into international agro-food markets*

- ***Improve the transparency of non-tariff measures affecting agro-food imports.*** Viet Nam has made significant steps in removing quantitative restrictions on trade. However, less transparent forms of licensing for the purpose of quality control, the collection of data, or the issuing of government guidance about what should or should not be imported have been introduced. While the policy objective of ensuring quality control is legitimate, the licensing system should not be used as a non-tariff barrier to trade. The import tariff quotas that exist for eggs, sugar and unmanufactured tobacco should be auctioned off to increase competition rather than given to existing end users. If there is no demand for product, then consideration should be given to removing the tariff quota altogether.
- ***Strengthen the capacity of policy-making and implementation in quarantine and food safety.*** This action is needed to ensure the protection of human, plant and animal health, improve Viet Nam's regulatory reputation and support the export of value-added agricultural goods. It is important that import requirements for food safety, quarantine, and standards and labelling purposes are implemented in a transparent manner, consistent with international guidelines and practice. This would help to facilitate the achievement of Viet Nam's ambitious goals on both trade and food security.

- ***Overhaul the current system for controlling rice exports.*** The current system creates a conflict between the objectives of improving the market orientation of the sector and ensuring food security. It limits competition, creates market uncertainty and reduces the incentive to develop long-term marketing arrangements. The result is a continued focus on supplying low-quality rice. The failure of the system to prevent the transmission of rising world prices onto the domestic market in 2008 suggests that the rational for maintaining the policy in place is not sound.

References

Arita, S.S. and J. Dyck (2014), "Vietnam's Agri-Food Sector and the Trans-Pacific Partnership", *Economic Information Bulletin* No. 30, ERS/USDA, October.

FAO-IFAD (International Fund for Agricultural Development) – WFP (World Food Programme) (2014), *The State of Food Insecurity in the World 2014, Strengthening the Enabling Environment for Food Security and Nutrition*, Rome, FAO.

Fuglie, K. and N. Rada (2013), *International Agricultural Productivity Dataset*, USDA Economic Research Service: *www.ers.usda.gov/data-products/international-agricultural-productivity*.

GSO (General Statistical Office) (2013a), *Statistical Yearbook of Vietnam*, Statistical Publishing House, Hanoi, *www.gso.gov.vn*.

ILO (International Labour Organisation) (2011), *Vietnam Employment Trends 2010*, National Centre for Labour Market Forecast and Information, Bureau of Employment, ILO Office in Vietnam.

ISPONRE (Institute of Strategy and Policy on Natural Resources and Environment) (2009), *Vietnam Assessment Report on Climate Change*, With support of the United Nations Environment Programme (UNEP).

OECD (2013), *Southeast Asian Economic Outlook 2013: With Perspectives on China and India*, OECD Publishing, Paris, *http://dx.doi.org/10.1787/saeo-2013-en*.

OECD-FAO (2014), *OECD-FAO Agricultural Outlook 2014*, OECD Publishing, Paris, *http://dx.doi.org/ 10.1787/ agr_outlook-2014-en*.

WB WDI (2014), *World Development Indicators Database*, *http://data.worldbank.org/data-catalog/world-development-indicators*.

Chapter 1

The agricultural policy context in Viet Nam

This chapter provides the broad context in which the Vietnamese agricultural sector developed over the last two decades, including political, demographic, macroeconomic and social factors. It evaluates agriculture's performance in terms of production, productivity and trade; outlines social impacts in terms of employment, incomes, poverty and food consumption; discusses environmental consequences; and finally, analyses structural issues both in agriculture and in its upstream and downstream sectors.

1.1. Introduction

Viet Nam's *Doi Moi* reforms in the mid-1980s triggered the modernisation process of the economy. High rates of economic growth contributed to three-fold real growth in GDP per capita over the two decades of 1990-2010. The growth was particularly inclusive such that the poverty rate fell faster than in any other country in the world with the exception of China. Within this favourable macroeconomic context, further enhanced by agriculture-specific reforms, such as the de-collectivisation of farms and the issuance of land use rights, Vietnamese agriculture has performed exceptionally strongly across a wide range of agricultural subsectors. As a result, Viet Nam has become one of the leading producers and exporters of such agricultural commodities such as rice, coffee, natural rubber, cashews, cassava, and black pepper, but also of aquaculture products such as catfish and shrimps.

However, these trends are unlikely to continue. Already many growth rates are declining. A good part of the past growth in revenues or export values was due to commodity price growth in the 2000s. Prices of many commodities declined over the last 2-3 years and are projected to fall further in real terms over the next decade. Ongoing real appreciation of the Vietnamese Dong puts additional pressure on exporters and makes imports more competitive. In addition, most resources for expansion are already used (e.g. arable land) and there is increasing evidence of negative environmental impacts of both overexploitation of resources and overuse of chemical inputs. Moreover, the agricultural sector is steadily moving up its supply curve. Labour costs will increase if GDP and job growth continue along their recent paths. The rural-to-urban migration process that has started will continue as long as the Vietnamese economy continues to grow, just as it has for China and some other Asian countries. This will open opportunities to adopt productive new technologies and encourage larger farms. However, the higher labour costs from rising wage rates will also dampen expansion of labour-intensive subsectors, in particular if newer labour-saving techniques are not so readily available, too costly or not adaptable to the dominant small-scale farming.

1.2. General aspects

Viet Nam's total land area is 330 951 km^2 of which 72% is mountain or hill and only 28% is plain. Its S-shaped territory spreads over the distance of about 1 650 km and its coastline is 3 444 km long. In the north it is characterised by a subtropical climate with four separate seasons – spring, summer, autumn and winter and in the south by a tropical climate with only two seasons – dry and wet. Viet Nam's population at 89.7 million in 2013 makes it the 13th most populous country in the world. Its population density is high at 270 persons/km^2, which is just above the level of the United Kingdom and slightly below that of the Philippines.

Viet Nam's Gross Domestic Product (GDP) is currently (2013) measured at USD 171.4 billion, which translates into USD 5 293 per capita at purchasing power parity (PPP). Its Gross National Income at USD 1 740 per capita at the World Bank Atlas conversion method places

Viet Nam in the World Bank category of **lower middle-income** countries, which is an impressive accomplishment for a country that in the mid-1980s was one of the poorest in the world (WB WDI, 2014).

Basic political characteristics

The political environment in Viet Nam is dominated by the role of the **Communist Party of Viet Nam**. Its most senior institution is the National Congress which meets every five years to elect the Central Committee. This is the most important decision-making body for Party affairs and government policy. The Central Committee adopts key policies and elects the Politburo and its Secretariat. Between the five-year meetings of the National Congress, the Politburo takes over as the highest-ranking body, but only to implement policies decided by the National Congress or Central Committee. The legislative body for the country is the National Assembly, composed of delegates elected for five-year terms. It agrees upon new laws, and elects the President, who is the Head of State. In addition to appointing or dismissing the Prime Minister and acting as Chairman of the Council of Defence and Security, the President appoints many other key officials, ambassadors, and signs international agreements. The Prime Minister, as leader of the Government, is responsible for implementing Government operations, including the work of the various Ministries.

Beneath the central government is local or regional government: 58 provinces and 5 cities which all have smaller subdivisions such as districts, municipalities, precincts, and in the rural areas, communes. They come under the authority of the central government, but they have **People's Councils** that are elected locally. They supervise the implementation of central government laws and regulations at the local level, and have executive bodies under them called People's Committees. The leadership and members of these committees are elected by the People's Council.

With this centralised structure, the government has a wide set of levers and a limited level of opposition in its choice of policy directions and decisions. These have been chosen quite strategically (and decisively) which was particularly evident in the country's choice of economic policies. The most dramatic policy change was the shift, beginning in the mid-1980s, away from a central planning framework and toward a greater market orientation through a variety of economic reforms known widely as **Doi Moi** or "Renovation". Since that time a long series of policy changes have moved the economy, including the agricultural sector, in the direction of open markets, private decision-making, private ownership of land use rights, the acceptance of private firms, and measures that have embraced foreign trade and investment.

Demographic factors

Demographic factors have been closely linked with Viet Nam's rapid economic progress. The annual **population growth** rate has almost halved in the period from 1990 to 2012, falling from 1.90% per year to only 1.05% per year. This drop is mirrored by a full reduction by half in fertility, observed in the number of live births per woman, which has declined from 3.6 in 1990 to 1.8 in 2012 (WB WDI, 2014).

In 1990, 5.7% of the population was aged 65 and over, by 2012 the percentage had only reached 6.6% which indicates a slow ageing. The population is also living longer, which contributes to population growth. **Life expectancy** has increased from 70.5 years in 1990 to 75.6 in 2012 (WB WDI, 2014).

Urbanisation has been progressing as people migrate away from the agricultural sector to better paid jobs, which are mostly in urban areas. In 1990, the population was 80% rural and only 20% urban.[1] By 2013, the ratios had become 68% rural and 32% urban (GSO, 2014). This means that the population share of urban regions increased by half in only 20 years, which reflects the rapid rate of Vietnamese economic growth and creates a major challenge for a more integrated development of rural and urban areas.

Natural resource endowments

A considerable proportion of Viet Nam's economic growth in the past two decades has been the result of **exploiting natural resources**, especially the intensified use of both land and water, and a large degree of deforestation to plant export crops.

Viet Nam is relatively rich in **water resources**, but regional and seasonal differences are significant and local shortages occur during the dry season, in particular in Southeast provinces. Moreover, almost 60% of Viet Nam's total water resources are generated outside its borders, making the country vulnerable to decisions made about water resources in upstream countries (FAO AQUASTAT, 2013).

The economic **scarcity of land** is significant, with just 0.12 ha of agricultural land per capita, one-sixth of the world average, on par with Belgium, just below the Netherlands, but less than China or Indonesia and just above the Philippines and India (FAOSTAT, 2015). There are also growing pressures to convert agricultural land into higher-value non-farm uses (both urban and industrial). This has created a strong incentive to increase land intensity given the availability of relatively cheap labour, high soil fertility in some regions, and relatively good climate conditions. The key issue is the management of these resources, in view of their degradation and the potential impacts of climate change (Section 1.7).

Deforestation, from increased planting to profitable agricultural crops, notably coffee, occurred heavily until the early 1990s. While the area of natural forest continues to decline, re-forestation efforts in the last 15 years have increased total forested areas, in particular of planted and naturally regenerated forests (Section 1.7).

General features of the Vietnamese economy

The most striking feature of the Vietnamese economy is its overall **rapid rate of economic growth** since the mid-1980s when the *Doi Moi* reforms began. Annual growth rate averaged 7.4% in the 1990s and 6.8% in the 2000s, contributing to three-fold real growth in GDP per capita (calculated at constant 2005 USD) over these two decades (Figure 1.1. and WB WDI, 2014). It has slowed to 5.7% in 2010-14, but still compared favourably with most Asian countries. Not only did the reforms generate rapid overall growth but the growth was particularly inclusive, such that poverty was alleviated as much as in any country in the world in the 1990s except for China.

Changing structure

The Vietnamese economy is moving from being heavily agricultural to a diverse mix of agriculture, services and industry. The **agricultural share in GDP halved** from 39% in 1990 to 20% in 2012.[2] Services rose from 39% to 42%, and industrial production rose from 23% to 39% over the same period. In turn, agriculture's share in employment fell from 70% in 1996 (earliest year available for Viet Nam) to 47% in 2012, paralleled by significant increases for industry from 11% to 21% and for services from 21% to 32% (WB WDI, 2014).

The importance of the **shift of labour from agriculture** to the non-agricultural sector in maintaining economic growth and in reducing poverty cannot be overstated. Migration raises incomes of migrants, contributes to higher incomes of migrants' families through remittances in various forms, raises the wage rates of agricultural labour remaining in the countryside as its supply shrinks, and improves land and water availability for those who remain dependent on farming. In early years, Viet Nam restricted this movement by a variant of the household registration system *Hộ Khẩu*. This had the effect of inhibiting labour migration out of agriculture and rural areas, despite the existence and gradual growth of "grey" labour markets on the outskirts of Hanoi and Ho Chi Minh City as from the early 1990s. The reduced enforcement of those restrictions has been an important structural change in the labour market, but still there is evidence that government-provided services for health, schooling, and social protection are tied to the registration system, which restricts or privileges access to those permanently registered (Coxhead et al., 2015).

Competitiveness

Overall, Viet Nam's **competitiveness is ranked 70th** out of 148 countries classified by the World Economic Forum. It compares rather favourably with other countries classified as factor-driven economies, but rather poorly when compared with other members of the Association of Southeast Asian Nations (ASEAN). Singapore, Malaysia, Brunei Darussalam, Thailand, Indonesia and the Philippines are all classified above Viet Nam's rank and only Lao PDR, Cambodia and Myanmar are positioned below. While large market size works to Viet Nam's advantage, there are a large number of concerns including weak institutions, low quality of infrastructure, underperforming higher education and training, weak financial market development, and slow adoption of the latest technologies by Vietnamese businesses (WEF, 2014).

Within the institutional pillar, the **lack of transparency** concerning policy decisions, corruption, weak property rights, weak intellectual property protection and weak auditing and reporting standards are areas of particular concern (WEF, 2014). The problem of low quality or insufficient provision of **public infrastructure** is most evident in terms of poor quality of transport infrastructure, including roads, ports and airports, and electric power transmission and distribution (Chapter 3). Viet Nam performs very well in terms of primary education enrolment. However, the high skill end of the labour market is weak and the level of innovation is poor, which undermines competitiveness of firms and industries (WEF, 2014; OECD/WB, 2014).

One widely noted problem area across many sectors of the economy is the often low productivity and international competitiveness of the **state-owned enterprise (SOE)** sector (IMF, 2012). Despite on-going privatisation process, SOEs still contributed more than one-third to GDP, about 50% to exports, 28% of total domestic government revenue (excluding revenue from crude oil and the import-export tax) and almost 40% of the value of industrial production in 2011 (OECD, 2013). The government's reform effort includes transforming SOEs into joint-stock companies through so called "equitisation", it means privatisation of a wholly-state-owned enterprise by selling a part or all of the assets and liabilities of the SOE to the private sector. However, in 2011 the government still retained an ownership stake of 57% in the equitised firms with the rest divided between employees (14%) and other shareholders (29%). Thus, changes in the SOE's ownership structure have been much slower than originally expected (OECD, 2013). It is argued that the protection of SOEs brought a misallocation of resources and slowed the flows of resources to sectors that would be more in line with Viet Nam's comparative advantage (Tran Van Tho, 2013).

At the industry-specific level, agriculture and its export subsector have been real bright spots. While exports of wood products, cat fish or textiles were successful, the export performance of rice, coffee, natural rubber, cashew nuts, cassava, and pepper could be considered remarkable. Viet Nam not only became **export-competitive** but also became one of the world's largest exporters for each of the six commodities.

Openness

Viet Nam has been strikingly successful in opening its economy to international trade. When measured by the ratio of traded goods (imports plus exports) to GDP, Viet Nam's **trade openness** increased from 30% in 1990 to 79% in 2000 and then to 161% in 2012 (WB WDI, 2014). Another measure of openness is the weighted average applied tariff across all imports and it shows the same striking increase in openness, especially since 2001. From a high of 21.1% in 1994, the average tariff fell to 17.4% in 2001, and 5.7% in 2010. This rate was marginally higher in primary products (6.0%) compared to manufactured goods (5.6%) (WB WDI, 2014). This partially reflects WTO commitments, but it is consistent with an overall policy of increasing openness.

Macroeconomic performance

Aside from the strong overall GDP and export growth performance, the macroeconomic environment in Viet Nam has historically been challenging, particularly the difficulty in keeping inflation at moderate levels. In the first five years of the 1990s, **inflation** ranged from 17 to 73% and in the last seventeen years (1996-2013) it averaged 9%. It peaked at 23% in 2008 and again at 21% in 2011, but in 2013 it has been brought down to 5% (GDP deflator; Figure 1.1.), an indication of considerable progress in macro stabilisation. This reduction in inflation is partly due to credit tightness, which is also part of the reason why GDP growth has moderated in recent years.

At the same time, Government **budget revenue** growth has been sluggish, pointing to another longstanding challenge, of generating sufficient government (especially tax) revenue to deal with potential SOE and bank liabilities (due to the tighter credit conditions since 2011 and the weak condition of some of these firms), social support expenditures, and public investment priorities.

The external environment has improved with the **external debt** falling from 384% of GNI in 1990 to a range of 30-45% in more recent years (Figure 1.1.). Viet Nam applies a crawling-peg system, allowing the dong's value against the USD to adjust to changing market conditions. The nominal exchange rate has been gradually depreciating from VND 16 000 per USD in early 2008 to VND 21 300 per USD in early 2015. This depreciation has been offset by bouts of inflation, which has made the real exchange rate quite variable, depending upon the particular period.[3] IMF data show an overall **appreciation** from April 2008 to early 2012 of 12% (IMF, 2012). Thus, the apparent depreciation in the Vietnamese dong has not been sufficient to offset domestic inflation, and to 2012 domestic firms' competitiveness has been eroded. More recently, the nominal exchange rate has stabilised around VND 21 000 per USD, even as inflation has continued (admittedly slowed to 4-7%), appreciating the dong further in real terms, hence exerting pressures on export-oriented sectors, including agriculture.

Even though the real exchange rate has appreciated, the **current account** has improved strongly from a deficit of USD 4.2 billion in 2010 to an annual surplus of USD 8-9 billion in 2012-14. This reflects continued strong export performance and some slowdown in

Figure 1.1. **Viet Nam: Selected macroeconomic indicators, 1990-2013**

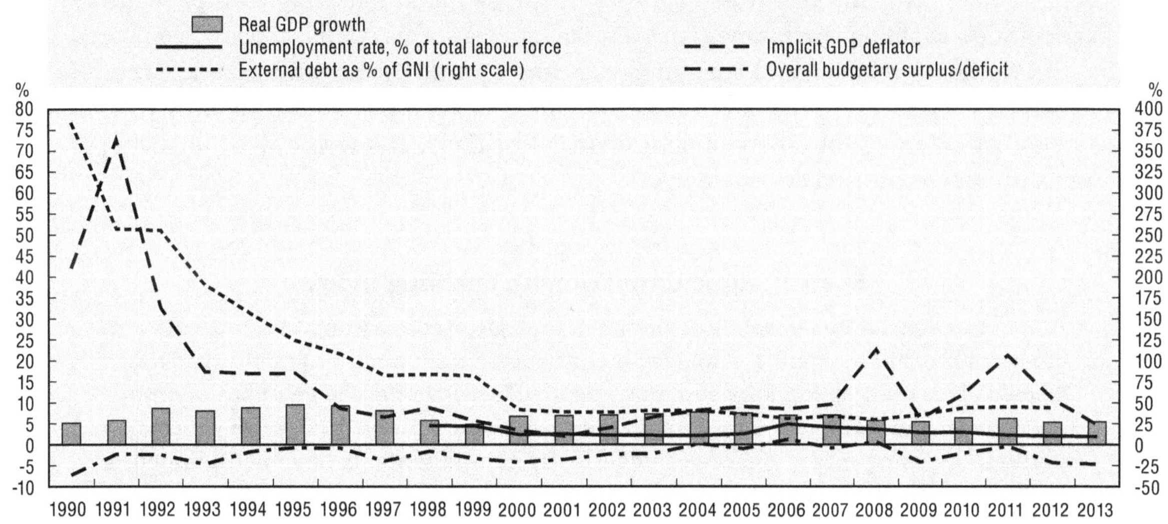

Note: Overall budget surplus/deficit in 1990-95, excluding grants.

Source: ADB (2014), *Key Indicators for Asia and the Pacific 2014*; ADB (2005), *Key Indicators 2005*.

StatLink http://dx.doi.org/10.1787/888933223200

imports, but also an increase in overseas remittances. Gross international reserves have tripled between late 2010 and late 2014 to about 3.3 months of imports (EIU, 2015).

Social situation

The robust economic growth over the last two decades has been accompanied by an impressive fall in the **incidence of poverty**. Using the World Bank poverty definition of USD 2/day (PPP) or below, the proportion of the population at this low income level halved in Viet Nam between 1993 and 2008 from 85.7% to 43.4%. In the early 1990s, the poverty rate was higher than in China, Indonesia, Thailand and Malaysia. By 2008 (last year available for Viet Nam), its poverty rate had been marginally below Indonesia and the Philippines, marginally above China, and still higher than in Malaysia. Over this fifteen year period, Viet Nam's poverty reduction performance is arguably equal to that of China, despite China's advantage of faster overall growth rates, and better than Indonesia's, especially since 2002 (WB WDI, 2014). Poverty is far more prevalent in rural areas, as is the case in most countries, but even there it is falling quickly (Section 1.5). In fact, by 2002 Viet Nam had already met its Millennium Development Goal of reducing poverty defined as USD 1/day by half.

Indicators of **income inequality**, in particular the Gini coefficient, show relative stability over the period 1993-2012, while in some other countries in the region the inequality tended to increase. For Viet Nam, for all years for which data are available, the Gini coefficient is in a range of 35.5 to 39.3. It increased by almost 4 points at the end of the 2000s, but then fell by those 4 points from 2010 to 2012. The most recent data available for the Southeast Asian countries would indicate that income inequality in Viet Nam is the lowest, with the exception of Cambodia (WB WDI, 2014).

Agriculture's enabling environment

The above discussion provides a brief overview of conditions within which Vietnamese agriculture has been functioning over the last two decades. Box 1.1 presents a tool which quantifies performance of selected components and synthesises them into one index which allows comparisons across countries. It can be found that despite improvements agriculture's **enabling environment** in Viet Nam still performs relatively poorly compared with 19 other developing and emerging countries covered by this assessment. Areas of particular concern are: weak governance, underdeveloped infrastructure, inefficient food safety institutions, poor functioning of financial markets, and low level of financing of agricultural research and development.

Box 1.1. Agricultural Growth Enabling Index

To assess agriculture's enabling environment in a given country and to compare it with other countries Diaz-Bonilla et al. (2014) constructed preliminary Agricultural Growth Enabling Index (AGEI). It allows summarising a wide array of available information in a structured manner and can be used to provide cross-country comparisons or single-country evaluations using either the index itself or its components. It has been applied to 20 developing and emerging economies, including Viet Nam.

Relative scores on the AGEI overall and its four main blocks are shown in Figure 1.2. It can be seen that Viet Nam's AGEI overall score is negative (below average) and it ranks 14 out of 20 countries covered. Looking across the components of the AGEI, Viet Nam performs

Figure 1.2. **Agricultural Growth Enabling Index and its sub-component blocks, early 2010s**

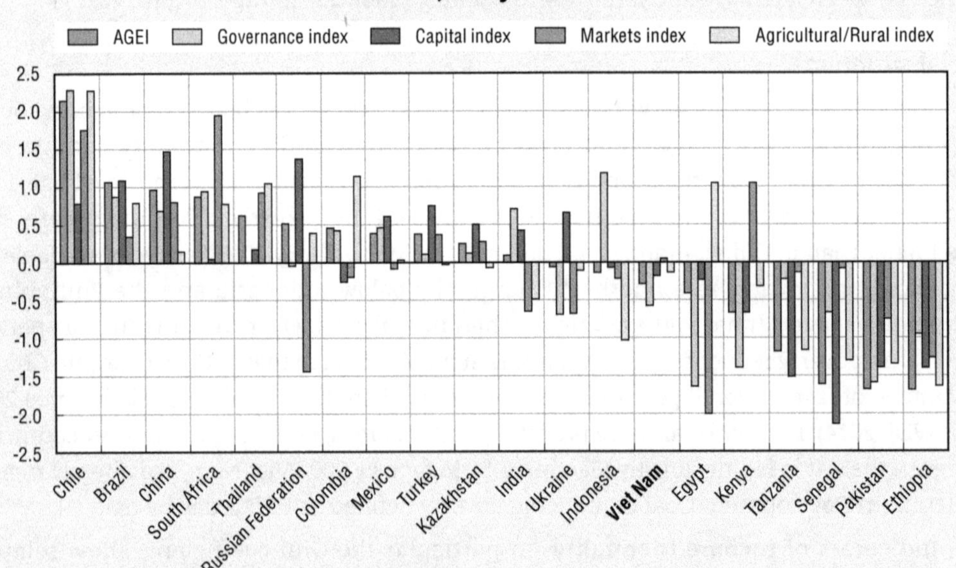

Note: The index is comprised of four blocks with 40% of the weight on agriculture/rural factors and 20% each on broader economy-wide governance, capital and market operation. The indicators selected measure circumstances within each country around the early 2010s. To account for the differences in averages of scores of the 20 countries and the variances of these scores across the index and its blocks, this figure shows the normalised score of each country on the AGEI index and on each component. Specifically, for the AGEI and each of its four blocks the average for the 20 countries has been subtracted from each country value and the resulting country value divided by the standard deviation for the series, to create series with zero mean and unit standard error. For example, a value of 2 means that the observation for a given country is 2 standard deviations above the average (which is zero) for the 20 countries.

Source: Diaz-Bonilla et al. (2014).

StatLink http://dx.doi.org/10.1787/888933223217

Box 1.1. **Agricultural Growth Enabling Index** *(cont.)*

particularly poorly on governance, somewhat better, but still below average, on capital and agricultural/rural index, and slightly above average on markets index.

Similar decomposition can be made for the indicators within each main block of the AGEI. Within the governance block Viet Nam performs relatively poorly on each subcomponent, it means on macro stabilisation, institution and political stability. Within the capital block, Viet Nam performs well (above average) on human capital captured by health/education indicators, but poorly on infrastructure and food-safety net. Within the markets block, Viet Nam scores well on labour market, around average on goods markets, but poorly on financial markets. On agricultural/rural block, Viet Nam scores slightly below average, but results differ quite strongly between various sub-components with particularly low on financing of agricultural research and development and high on the intensification of land use (Diaz-Bonilla et al., 2014).

1.3. Agricultural situation

Agriculture and the food sector within the economy

Despite high agricultural GDP growth (subsection below), the non-agricultural economy has grown substantially faster, pushing down the agricultural sector's share in GDP, as well as its share in total employment. However, agriculture (including fisheries and forestry) continues to be a **key source of income** for almost half of the population. Its share in GDP fell from 39% in 1990 to 19% by 2005 and has remained at this high level up to 2013. Its share of employment fell from 70% in 1996 to 47% in 2013, but it still remained 2.5 times higher than the sector's share in GDP. This indicates relatively low labour productivity, which is one of the reasons of the low incomes of households dependent on farming.

The share of agro-food exports (including fisheries) in total exports had fallen from 27% in 2000 to 17% by 2013. Over the same period, the share of agro-food imports in total

Figure 1.3. **The share of agriculture in GDP, employment, total exports and imports, 2000-13**

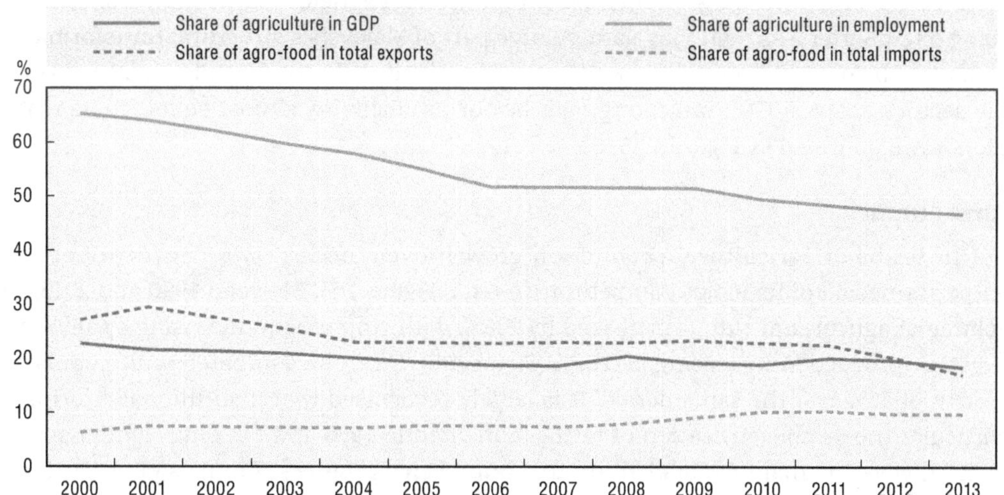

Source: WB (2015), *World Development Indicators*; UN (2015), *UN Comtrade Database*; GSO (2014), *Statistical Yearbook of Vietnam 2013*.

StatLink http://dx.doi.org/10.1787/888933223228

Figure 1.4. **Evolution of agriculture's share in GDP and in employment in selected Asian countries, 1990-2012**

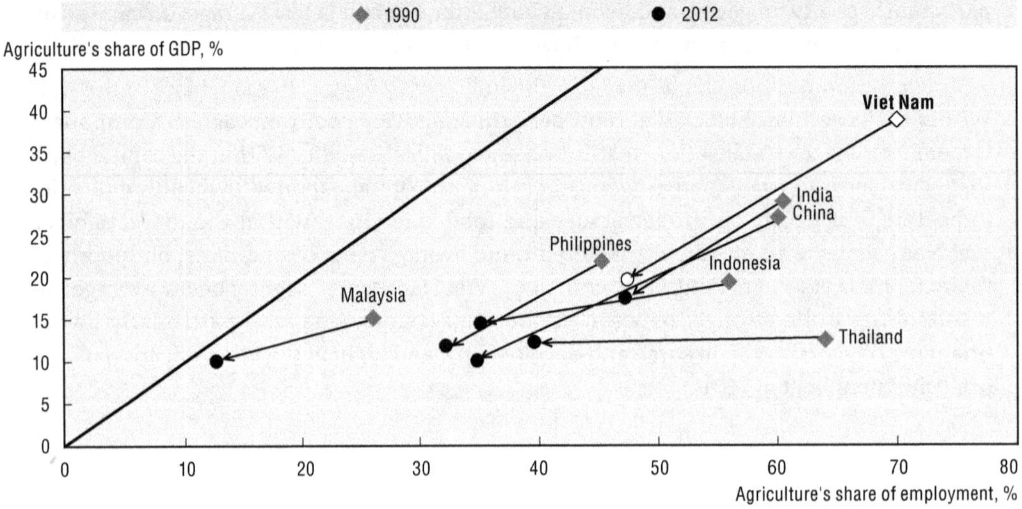

Note: The share of agriculture in total employment is 1996 instead of 1990 for Viet Nam, 1994 instead of 1990 for India and 2011 instead of 2012 for China.
Source: WB (2015), *World Development Indicators*.

StatLink http://dx.doi.org/10.1787/888933223231

imports increased from 6% to 10% (Figure 1.3). When compared with agriculture's share in GDP at 18-20% in recent years, these shares show that the sector's **openness to international trade** is strong on the export side, almost equal to the openness of the rest of the economy, but weaker on the import side (Section 1.6). This might also indicate Viet Nam's comparative advantage in agricultural production.

The recent stability of agriculture's share in GDP despite a continued fall of the share of agriculture in total employment indicates growing labour productivity of the sector. The comparable employment data for Viet Nam are available as from 1996, but the pattern is unambiguous (Figure 1.4). Agriculture's employment share is dropping fast but is still highest, getting close to that in India. China, Thailand and Indonesia have been successful in moving a large percentage of their labour force out of agriculture, with their shares now being one-quarter less than Viet Nam. A large part of Malaysia's **structural transformation** took place before 1990 and the current share of agriculture in total employment is close to the sector's share in GDP, indicating high labour productivity, almost equal to the rest of economy (confirmed by Figure 1.15, Section 1.4).

Farm output

In terms of agricultural production growth over the last two decades Viet Nam outperformed all of its major competitors in Asia (Figure 1.5). Between 1990 and 2013, the volume of agricultural output increased by 206%, with crop production rising by 189% and livestock production increasing by 282% (Figure 1.6). It can be compared with population growth of 36% over the same period. It is largely recognised that institutional reforms, in particular the de-collectivisation of farms mandated in 1988 and the land rights issuance in 1993, were the main factors behind this **impressive progress** (Kompas et al., 2012 and Nguyen and Goletti, 2001; see also Chapter 2).

Figure 1.5. **Growth in gross agricultural output in selected Asian countries, 1990-2013**

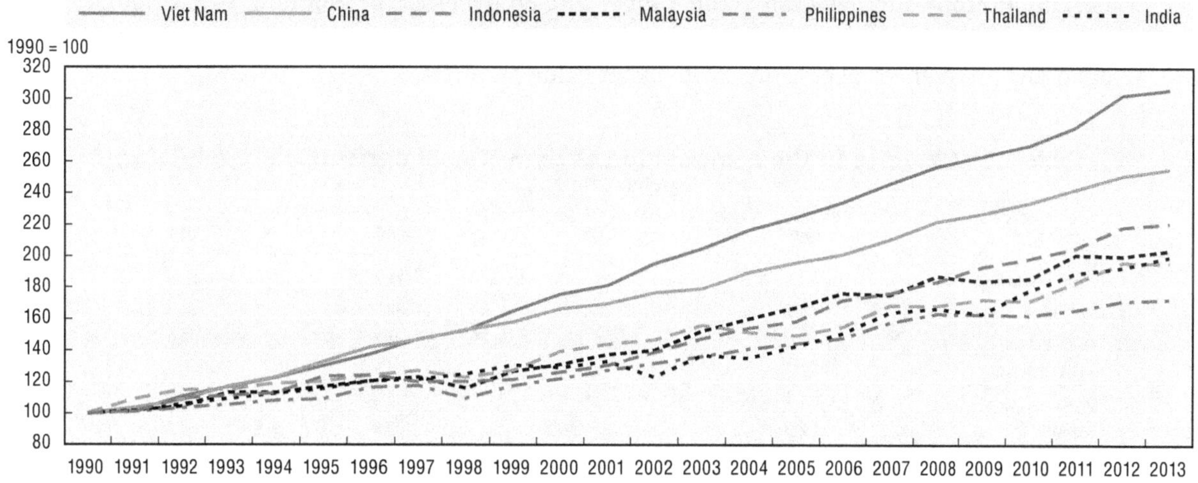

Note: The FAO indices of agricultural production show the relative level of the aggregate volume of agricultural production for each year in comparison with the base period 2004-06. They are based on the sum of price-weighted quantities of different agricultural commodities produced after deductions of quantities used as seed and feed weighted in a similar manner. In this figure, indices based on the 2004-06 period have been recalculated taking indices for 1990 as 100.

Source: FAOSTAT (2015).

StatLink http://dx.doi.org/10.1787/888933223243

Figure 1.6. **Growth in agricultural output in Viet Nam, 1990-2013**

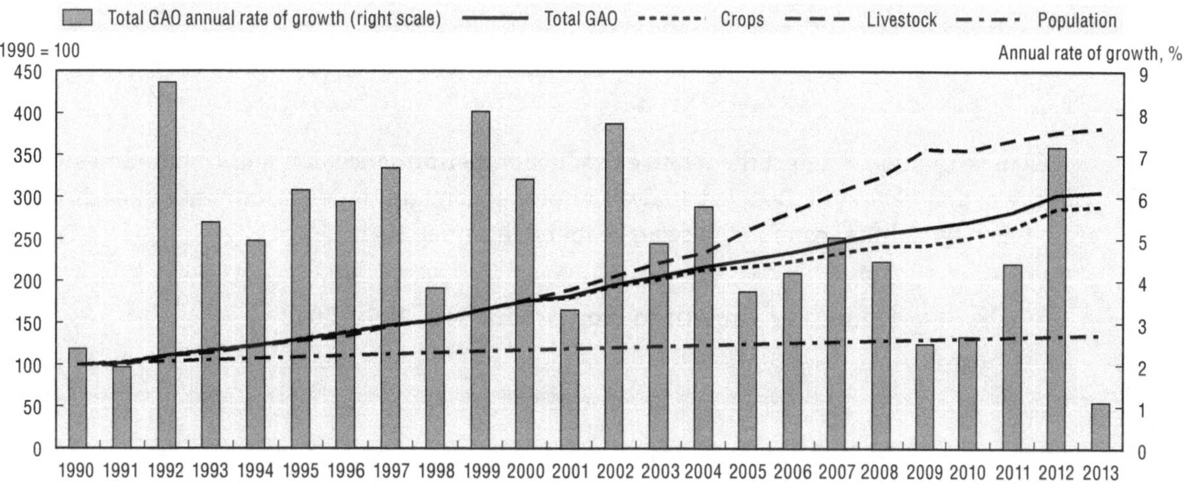

Note: FAO indices based on the 2004-06 period have been recalculated taking indices for 1990 as 100.

Source: FAOSTAT (2015); WB (2015), *World Development Indicators*.

StatLink http://dx.doi.org/10.1787/888933223250

The annual rate of agricultural production in Viet Nam slowed from an impressive average of 5.7% in 1990-2002 to an average of 4.2% in 2002-13 (Figure 1.6). Compared to other countries in the region, it maintained rates higher than most countries, but some **slowdown** in more recent years, with the exception of 2012, is noticeable. Most likely, the rates would have declined still further in the last period had there not been the agricultural price boom that elevated many world prices by a factor of two. This might be taken as a warning signal that the earlier sources of the sector's boom might not be sustainable.

While rice remains by far the most important commodity, accounting for 35% of total value of agricultural production in 2012, there has been an important **change in the**

composition of production away from staple foods to other commodities, in particular perennial crops such as coffee and rubber and to livestock production, in particular pigmeat (Table 1.1.). This reflects the strong export orientation of perennial crops and changing preferences of consumers to higher value products.

Table 1.1. **Changes in the composition of the value of agricultural production, 1991-2012, %**

	1991	2000	2010	2012
Crops, including:	**77.3**	**79.6**	**74.4**	**73.0**
Rice paddy	39.7	45.7	39.7	34.9
Coffee green	1.2	4.5	4.7	7.0
Rubber natural	1.2	2.5	4.7	5.4
Maize	1.4	2.9	4.2	4.1
Cassava	2.9	1.1	3.5	3.2
Cashew nuts	2.6	2.1	3.3	3.2
Sugar cane	4.7	5.4	2.1	2.6
Pepper	0.9	1.8	1.3	2.5
Other	22.8	13.6	10.9	10.1
Livestock, including:	**22.7**	**20.4**	**25.6**	**27.0**
Meat pig	13.4	12.5	16.3	17.6
Meat chicken	5.3	5.7	6.3	6.5
Meat cattle	1.8	1.0	1.8	1.7
Eggs	0.3	0.2	0.3	0.3
Total	**100.0**	**100.0**	**100.0**	**100.0**

Source: FAOSTAT (2015).

StatLink http://dx.doi.org/10.1787/888933223262

Figures 1.7 and 1.8 depict the **relative production performance** of major commodities over the period 1990-2013. Detailed analysis of production and trade performance of each of 13 major agricultural commodities can be found in Annex 1.A1.

Figure 1.7. **Growth in crop production, 1990-2013**

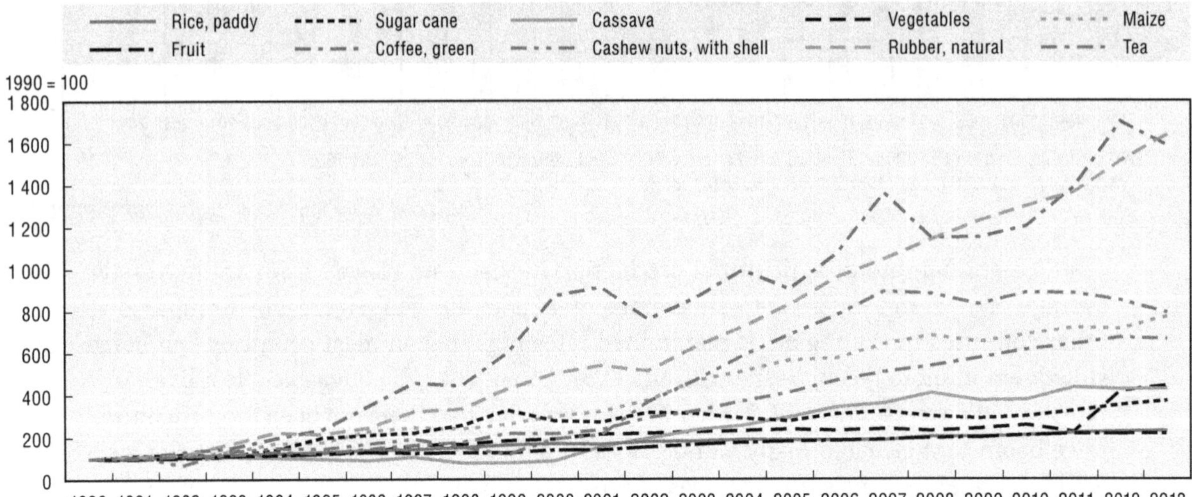

Source: FAOSTAT (2015).

StatLink http://dx.doi.org/10.1787/888933223274

Figure 1.8. **Growth in livestock production, 1990-2013**

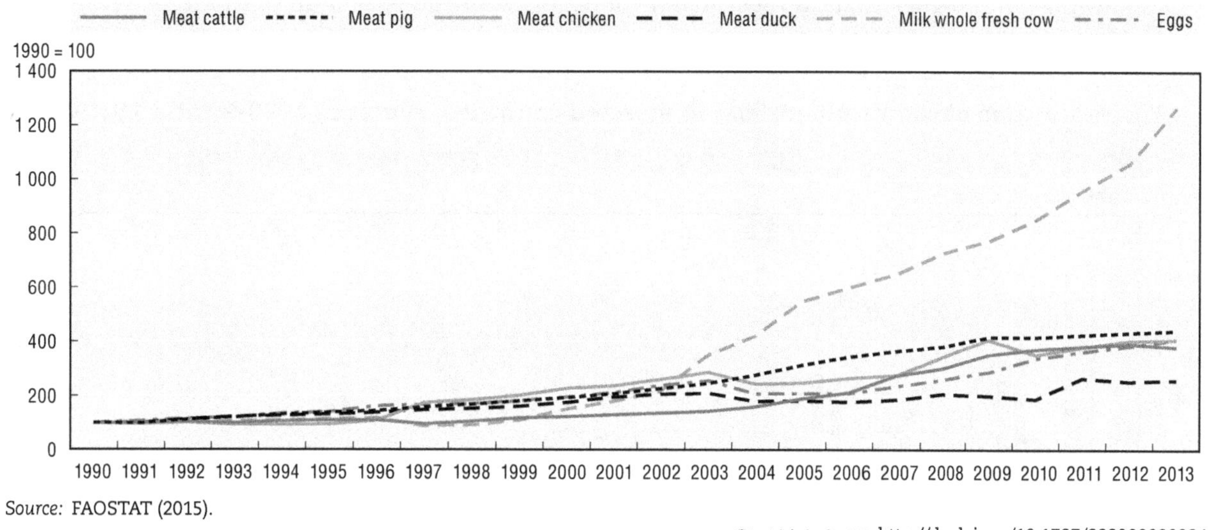

Source: FAOSTAT (2015).

StatLink http://dx.doi.org/10.1787/888933223284

Figure 1.7 shows that among **crops**, coffee and natural rubber were by far the **best performers** in terms of production growth over the last two decades, both increasing about 16 times since 1990. In turn, production of staple foods, in particular rice and fruits increased the least, but still more than doubled over this period.

Among **livestock products**, milk production increased the most, almost 13 times, but from a very low level and it still accounts for less than 1% of the total value of agricultural production in Viet Nam (Figure 1.8). Production of pigmeat, chicken meat, beef and eggs increased about four times each in 1990-2013. While production of the best performing crops is driven by exports, livestock products are destined almost exclusively for the domestic market and their production growth reflects growing demand from domestic consumers.

1.4. Factors of production and productivity

Farm input use

Capital inputs remain relatively small in Viet Nam (Section 1.9 discusses machinery use and Chapter 3 provides an overview of capital investment). Given the low wage rates, high labour intensity relative to capital is widely observed in South East Asia, changing only when wage rates rise sufficiently. Only recently growing hired-labour costs have encouraged Vietnamese farmers to apply more mechanisation in selected segments of production.

Fertilisers and seeds are the key purchased input applied by Vietnamese farmers. In 2013 total domestic supply of **fertilisers** in Viet Nam was 8.3 million tonnes. Total demand was 10.3 million tonnes; hence total imports were about 2.0 million tonnes, about 50% less than in 2011.[4] Viet Nam imports fertilisers mainly from China (about half of total imports) as they are cheaper than from other sources and cheaper than those produced domestically (Nguyen Hang T., 2013; Ken Research, 2014a).

About two-thirds of fertilisers are used for rice production, especially for new rice hybrids that require more nitrogen and phosphate; followed by maize (9%), rubber (8%) and coffee (5%) (Nguyen Hang T., 2013). The rate of fertilisers applied per hectare in 2010-12 was

80% higher than in 1990-92 and, at almost 200 kg/ha, was 90% higher than the regional average and higher than in most Asian countries, with the exception of China, Korea, Malaysia and Japan (Figure 1.9).

Figure 1.9. **Use of chemical fertiliser in selected countries, averages 1990-92 and 2010-12**

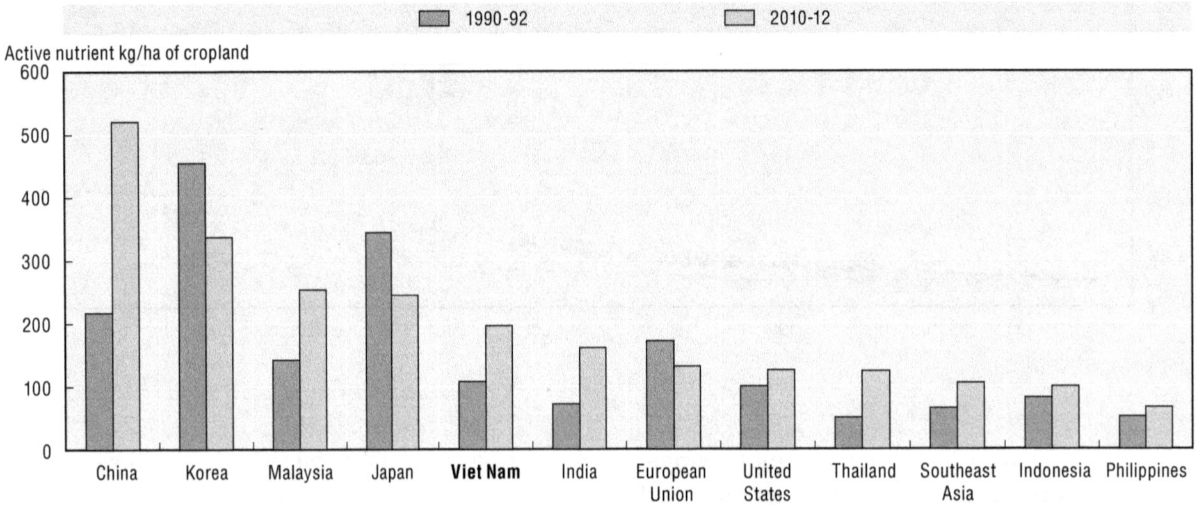

Note: Use of fertiliser includes nitrogenous, phosphate and potash fertilisers in nutrient terms. Cropland includes arable land and perennial crops.
Source: FAOSTAT (2015).

StatLink http://dx.doi.org/10.1787/888933223294

However, annual data suggest that while the **application rate** was increasing very fast in the 1980s and 1990s, it stabilised over the last decade (FAOSTAT, 2015; Coxhead et al., 2010). One explanation for this stability is that arable land is becoming saturated with fertiliser applications. With the exception of maize and rubber, where future fertiliser use may rise, overall increases in agricultural productivity are unlikely to be generated by more fertiliser use, given current high application levels (Technoserve, 2013 for coffee; Pham Quang Ha et al., 2006 for nutrient imbalances).

In volume terms, market demand for **seed** is dominated by the rice sector (89%), followed by groundnuts (5%), corn (2%), soybeans (2%) and vegetables (1%) (Nguyen Trung Kien, 2012). Rice seed demand in 2011 was 1.0-1.2 million tonnes, maize was 40 000 tonnes, and potato seed 25 000 tonnes (Nguyen Mau Dung, 2013). Over the five years from 2008 to 2013, the estimated total value of the seed market is estimated to have grown at 1.7% per year (Ken Research, 2014b), but the non-hybrid market grew at only 0.7% per year. However, the **formal seed sector** only supplies 16% of rice seed demand, whereas in maize it supplies 80-90%, vegetables 49%, soybeans 7-8%, and groundnuts 3% (Nguyen Trung Kien, 2012).[5] The balance of the seed supply for each commodity comes from farm-saved seed.

Seed imports are significant, amounting to about USD 200 million in 2011, especially of hybrid seeds, including 70-80% of hybrid seeds for rice, vegetables, and maize (Nguyen Mau Dung, 2013). Hybrid rice seeds come mainly from China, whereas hybrid corn seeds are imported from Thailand and India. Vegetable seeds originate in Thailand, China, Japan, Korea and France. The use of improved seed varies considerably by crop. It is estimated that the shares of total planted area under improved seeds are: rice 67%, maize 83%, soybeans 68%, peanuts 55%, rubber 98%, cashews 29%, and tea 20% (Nguyen Mau Dung, 2013).

Land use and allocation

At just 0.12 ha per capita, **agricultural land is very scarce** in Viet Nam (Section 1.2). This is despite a 61% increase in total agricultural land in 1990-2012 (Figure 1.10). However, given the population increased by 34% in this period, per capita availability increased by only one-fifth or just 0.02 ha.

Figure 1.10. **Agricultural land, 1990-2012**

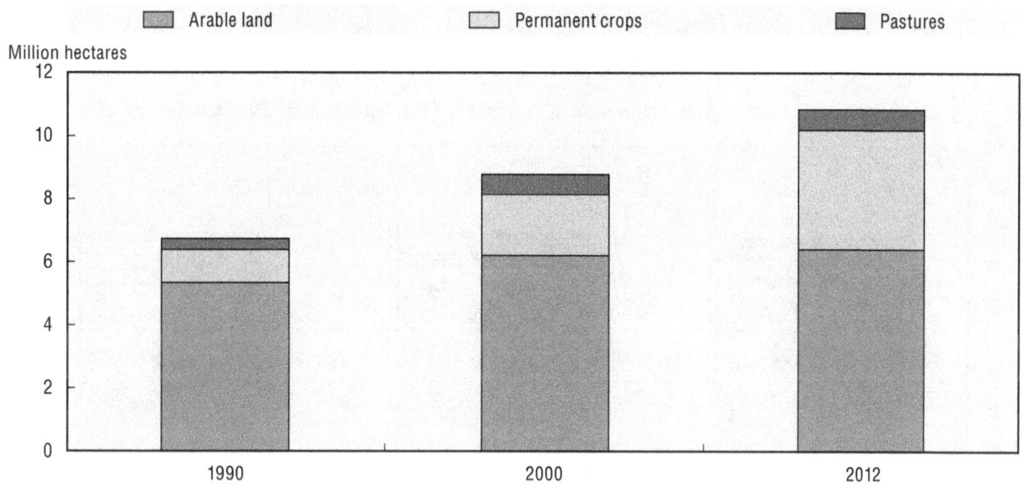

Source: FAOSTAT (2015).

StatLink http://dx.doi.org/10.1787/888933223302

Two-thirds of the increase in total area took place in the 1990s, followed by a more moderate increase in the 2000s (Figure 1.10). In fact, until the early 1990s large tracts of forest were converted and adapted to allow for intensive agricultural operations. Compared to other countries in the region, Viet Nam's total agricultural land has grown more than three times as quickly as anyone else in the last 20 years (Fuglie and Rada, 2013). As of 2012, agricultural land covers **10.8 million ha** representing 35% of the land area and consists of 6.5 million ha of arable land (60%), 3.7 million ha of perennial crops (34%) and just 0.6 million ha (6%) of permanent pastures and meadows (Figure 1.10).

The **arable land** component increased by almost one-fourth in 1990-2000, but since then remained relatively stable at around 6.5 million ha (FAOSTAT, 2015). Land allocated to perennial crops (e.g. coffee, rubber, and cashews) more than doubled in the 1990s and almost doubled in the 2000s, thus on average increasing 6% annually in 1990-2012. This is consistent with the explosion in production of those largely exported perennial crops as described in Annex 1.A1. Area allocated to pastures and meadows doubled in the 1990s, but remained stable since then (Figure 1.10).

The stabilisation in arable land might indicate that almost all accessible arable land is already in cultivation and small increases in some areas is compensated by small losses each year due to urbanisation and related conversions. While Viet Nam climate pattern allows for multiple cropping during the year, future crop productivity likely rests almost entirely on yield growth. With the current emphasis on re-forestation (Section 1.7), the growth in land allocated to perennial crops is likely to decrease, again suggesting a focus on yield growth over the next decade.

The allocation of **harvested area** by commodity shows that the share of rice has been falling but it still occupies the top position with 55% of total area in 2013. In absolute terms, the area allocated to rice has stabilised since 2000. However, if it were not for land use restrictions (Section 1.8), rice area would probably decline. A large part of the remaining cropland is planted in maize (8% of the total) in 2013 and it is growing quickly. The area sown to maize almost tripled in 1990-2013. Coffee, cassava, natural rubber and cashew nuts are other crops which have seen area expansions, but their shares are still relatively small at 2-4% each in 2013 (Figure 1.11).

Figure 1.11. **Composition of harvested area, 1990-2013**

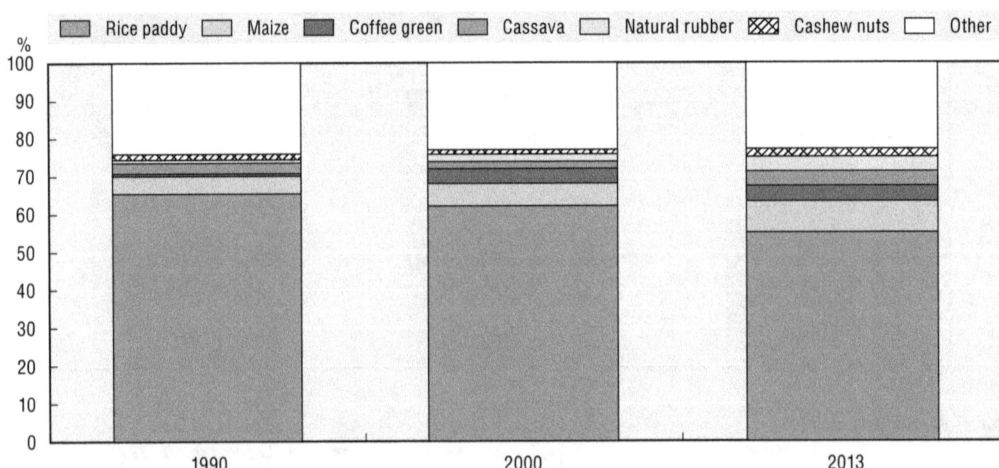

Note: Area harvested refers to the area from which a crop is gathered. It excludes the area from which there was no harvest due to e.g. damage, but if a given crop is harvested more than once during the year as a consequence of successive cropping the area is counted as many times as harvested.
Source: FAOSTAT (2015).

StatLink http://dx.doi.org/10.1787/888933223313

Farm employment

Historically, Vietnamese agriculture has been **labour-intensive**, though the weight that labour represents in agricultural production has been declining in recent years. Still, while the agricultural share in total employment has fallen over the last two decades to 47% in 2013, the total number of persons employed in agriculture increased up to 2009[6] and since then stabilised at around 24.4 million (including forestry and fisheries) (GSO, 2014) indicating still relatively low labour absorption by industry and services. Thus, Viet Nam's agriculture is not yet at the stage of **shedding labour** in absolute terms, but it might be at the turning point and, according to some projections, farm employment might fall by 9% in the 2010s (ILO, 2011).

Within the economically active rural population (EAP), those working on farms still accounted for almost 59% of the total in 2012, including 53% self-employed and 6% farm wage earners (GSO, 2013). The percentage of the population working on farms (self-employed or as hired labour) has decreased systematically from 2002 to 2012, in favour of wage employment in the nonfarm sector which almost doubled its share over the same period. This indicates a positive trend of economic diversification in rural areas.

While the majority of farm population belongs to rural households, a surprisingly large percentage of urban households still rely on agriculture as a key source of income.

According to the recent Household Living Standards Survey as much as 14.5% of the urban EAP was employed in agriculture in 2012, including 12.4% as self-employed and 2.1% as wage earners. Both percentages tend to fall (GSO, 2013).

The last three agricultural censuses (2001, 2006, 2011) confirm the general trend of the falling importance of agriculture (including fishery and forestry) in **rural employment** (Figure 1.12). There has been a steady shift of labour shares out of agriculture by 10 percentage points each five years, into both industry and services.

Figure 1.12. **Composition of rural employment by sector, 2001-11**

Source: GSO (2012), *Results of the 2011 Rural, Agriculture and Fishery Census.*

StatLink http://dx.doi.org/10.1787/888933223326

Figure 1.13 breaks this down for the three most important **agricultural regions**, Red River Delta, Mekong River Delta, and Central Highlands. The three regions are quite different. Red River Delta shows the most significant shift of labour out of agriculture, falling from 77% in 2001 to 43% in 2011. This may reflect the diversity of the Greater Hanoi economy and the large number of off-farm employment options. The major rice bowl,

Figure 1.13. **The share of agriculture in rural employment, selected regions, 2001-11**

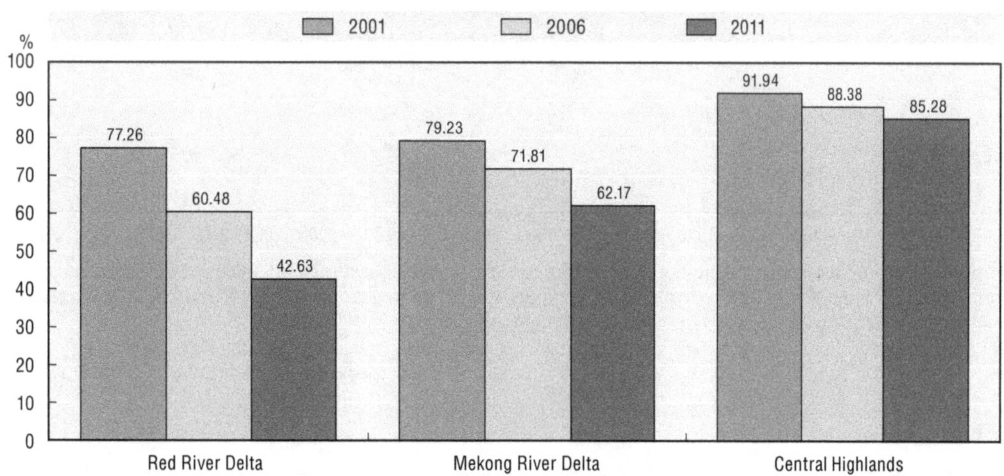

Source: GSO (2012), *Results of the 2011 Rural, Agriculture and Fishery Census.*

StatLink http://dx.doi.org/10.1787/888933223338

Mekong River Delta, shows a more gradual decline, from 79% in 2001 to 62% in 2011. This is very similar to the country-wide data that show declines from 80% to 60%. The Central Highlands has the largest labour force share in agriculture, and shows the least decline, starting at 92% in 2001 and falling only to 85% in 2011. This region is in contrast to the Red River delta in that there is no large urban centre that is either within or close to the region.

Productivity

This section compares the evolution in land, labour and total factor productivities in Viet Nam with those in selected Asian countries: China, Indonesia, Malaysia, the Philippines, Thailand and India. The comparison is based on a comprehensive database constructed by Fuglie and Rada (2013).

Land

In Viet Nam **land productivity** increased by 67% in 1990-2010, less than in China where it more than doubled, slightly less than in Malaysia, but more than in Indonesia, Thailand, the Philippines and India (Figure 1.14). For all major crops yield growth was strong ranging from 2% per year for coffee and sugar cane to above 6% for cashew nuts in 1990-2012 (Annex 1.A1). In absolute terms, Viet Nam's yields per hectare are higher than the South East Asia average for all major crops with the exception for sugar cane and one of the highest in the world for coffee, rubber and cashews. If compared with China, Viet Nam's yields are lower for rice, maize and sugar cane, but higher for e.g. coffee, rubber and cashews (FAOSTAT, 2015). Thus, it might be concluded that in quantity terms, Viet Nam's land productivity is high and that major effort should be put on improved quality and on diversification to higher value crops to create more value from given amount of land.

Figure 1.14. **Growth in land productivity in selected Asian countries, 1990-2010**

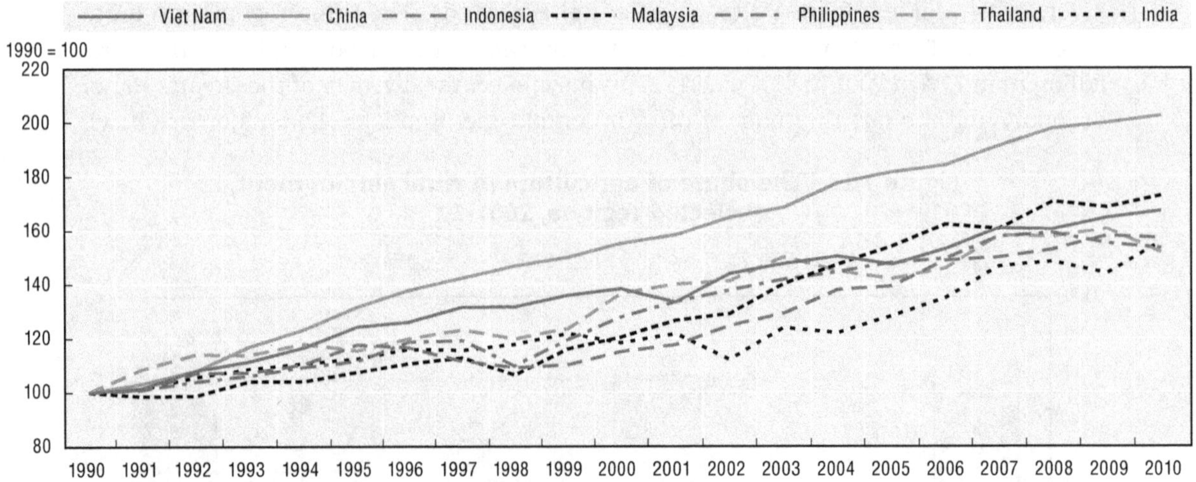

Note: Agricultural land productivity is calculated as total agricultural output (constant 2005 USD) divided by total agricultural land, expressed in hectares of "rainfed cropland equivalents". This is the sum of rainfed cropland (weight equals 1.00), irrigated cropland (for Asia weight equals 2.9933) and permanent pasture (for Asia weight equals 0.0566).

Source: Own tabulation based on Fuglie and Rada (2013), *International Agricultural Productivity Dataset*, ERS, USDA.

StatLink http://dx.doi.org/10.1787/888933223344

Labour

Figure 1.15 shows that farm sector's **labour productivity** growth in Viet Nam has been third highest among the countries listed, after China and Malaysia. In the latter two

Figure 1.15. **Growth in labour productivity in selected Asian countries, 1990-2010**

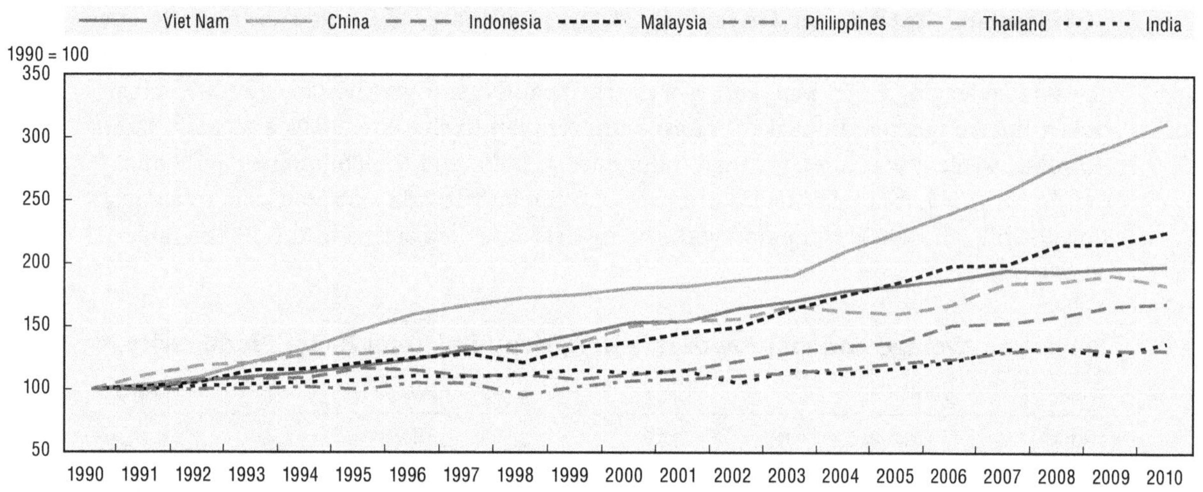

Note: Labour productivity is measured as total agricultural output (constant 2005 USD) divided by the total number of economically active persons in the sector in a given year.
Source: Own tabulation based on Fuglie and Rada (2013), *International Agricultural Productivity Dataset*, ERS, USDA.
StatLink http://dx.doi.org/10.1787/888933223354

countries, growth in farm labour productivity is driven both by a fall in the growth of agricultural production and a fall in farm employment. As discussed above, this is not yet the case in Viet Nam where employment in agriculture increased until 2010 and has only recently stabilised.

The actual levels of real agricultural value-added per worker, across countries and by year are shown in Figure 1.16. Despite the quite rapid growth in Viet Nam's labour productivity, the level of that productivity is still considerably lower compared with most other countries in region, much lower than in Malaysia[7] and only slightly higher than in India.

Figure 1.16. **Agriculture value added per worker in selected Asian countries, USD 2005, 1990-2010**

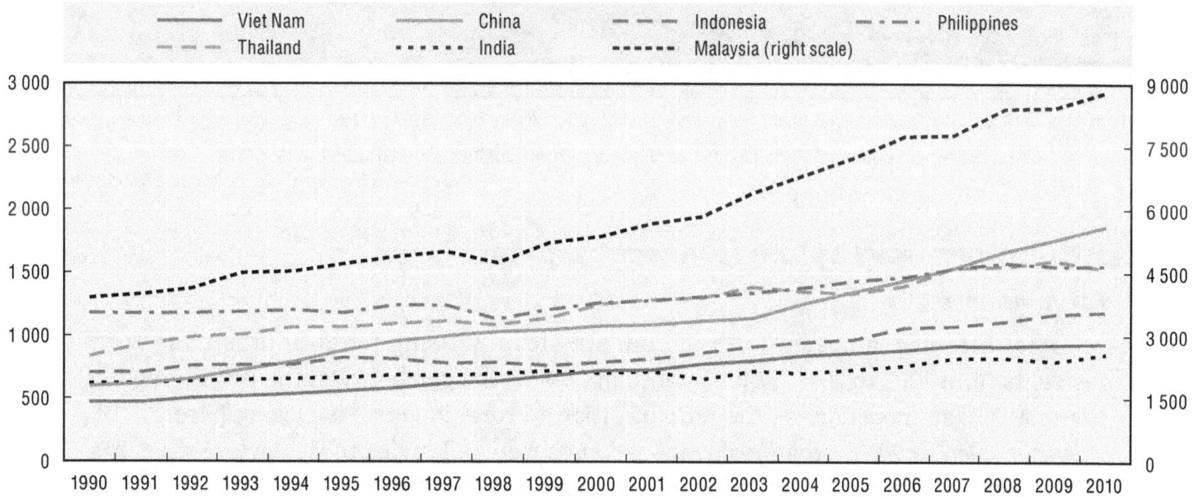

Note: Labour productivity is measured as total agricultural output (constant 2005 USD) divided by the total number of economically active persons in the sector in a given year.
Source: Own tabulation based on Fuglie and Rada (2013), *International Agricultural Productivity Dataset*, ERS, USDA.
StatLink http://dx.doi.org/10.1787/888933223367

Total Factor Productivity

Viet Nam's **Total Factor Productivity** (TFP) growth rate, calculated as the difference between output and input growth rates, has been strong and sustained over the last 20 years (averaging 2.65% per year). It was significantly stronger than in the 1980s, clearly reflecting the positive impact of reforms undertaken in the late 1980s and early 1990s. However, while it was stronger than in Indonesia, India and the Philippines and equal to that in Thailand, it lagged behind China and in the last decade also Malaysia, reflecting a slowdown in the 2000s compared with the highest rates registered in the 1990s (Table 1.2 and Figure 1.17)

Table 1.2. **Average annual growth rate in Agricultural Total Factor Productivity, %**

Years	Viet Nam	China	India	Indonesia	Malaysia	Philippines	Thailand
1981-90	1.03	1.69	1.32	0.52	3.32	0.30	0.47
1991-00	2.86	4.13	1.12	1.23	1.87	0.46	3.27
2001-05	2.52	2.39	1.11	3.36	3.73	2.64	2.18
2006-10	2.18	3.25	2.36	2.62	2.94	1.68	1.60
1991-10	2.65	3.10	1.25	2.26	2.92	1.67	2.73

Source: Fuglie and Rada (2013), *International Agricultural Productivity Dataset*, ERS, USDA.

Figure 1.17. **Growth in total factor productivity in selected Asian countries, 1990-2010**

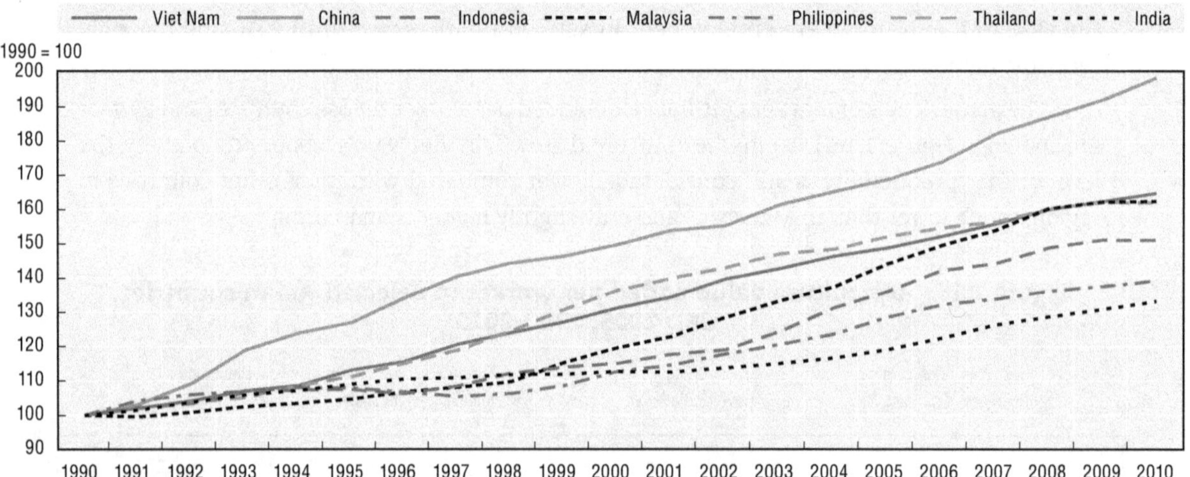

Source: Own tabulation based on Fuglie and Rada (2013), *International Agricultural Productivity Dataset*, ERS, USDA.

StatLink http://dx.doi.org/10.1787/888933223376

1.5. Farm incomes, poverty and food consumption

Farm incomes

Real incomes, adjusted for inflation, are steadily rising for both urban and rural residents from 2002 to 2012. While in absolute terms the gap between the two is growing, the relative gap measured as the ratio of urban to rural incomes is closing (Figure 1.18). However, even by 2012, urban residents' incomes were still twice those of rural residents. Taking into account higher cost of living in urban areas, the gap in purchasing power parity incomes would be smaller.

The **share of agricultural income** in total rural income is falling in most years, with the slight exception of the agricultural price boom in 2008. It falls from 37% in 2002 to 28% in

Figure 1.18. **Monthly income per capita by residence, in 2005 prices, 2002-12**

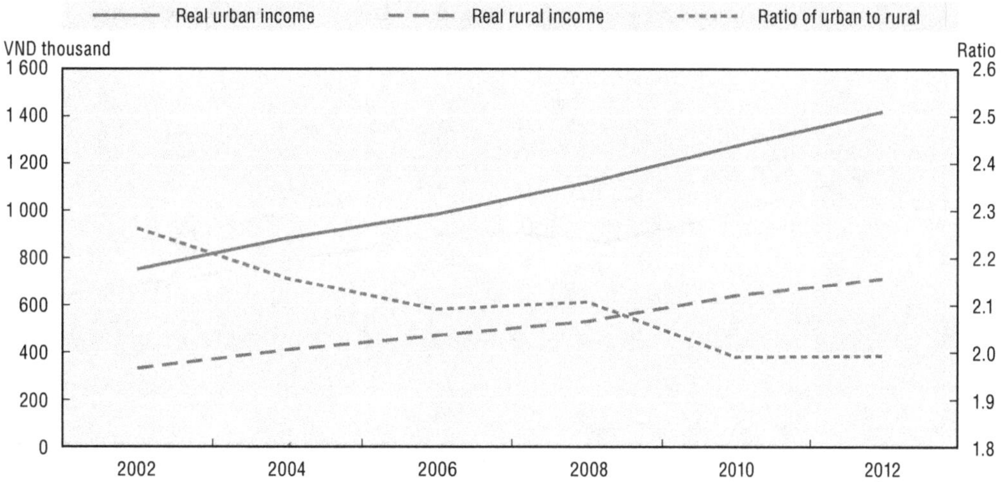

Note: Nominal incomes have been deflated by the Consumer Price Index (CPI), base year 2005.
Source: GSO (2013), *Household Living Standards Survey 2012.*

StatLink http://dx.doi.org/10.1787/888933223383

2012 (Figure 1.19). The decline is more pronounced in recent years if forestry and fishery incomes are added to agricultural incomes. By contrast, salary and wage income rises steadily from 24% in 2002 to almost 40% in 2012, becoming the largest component by 2010. This reflects changes in the structure of rural employment as shown earlier by Figure 1.12.

Figure 1.19. **Monthly income per capita for rural residents by source, 2002-12**

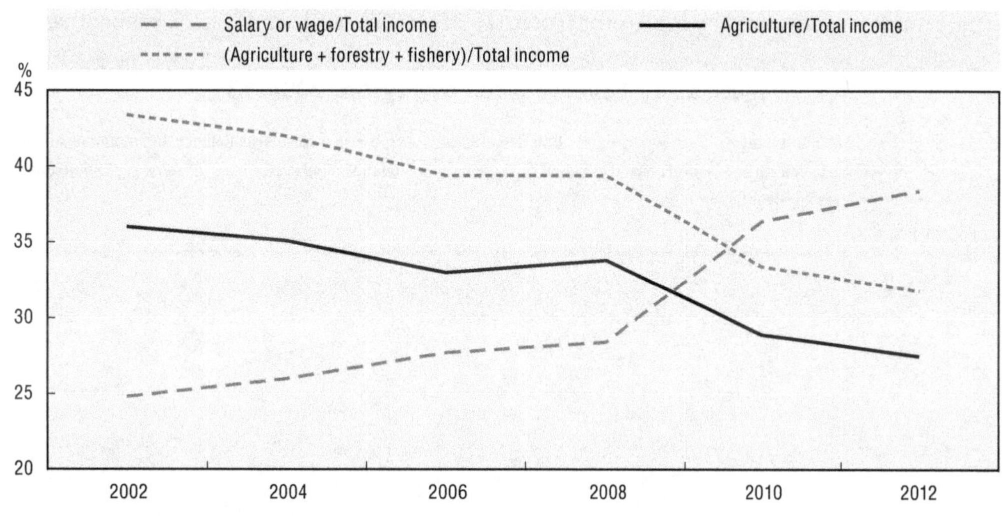

Source: GSO (2013), *Household Living Standards Survey 2012.*

StatLink http://dx.doi.org/10.1787/888933223394

Poverty

According to the national definition of the poverty line, rural **poverty rates** are much higher than those in urban areas.[8] The gap tends to decline, but remains large (Figure 1.20). This decline in rural poverty rates from 21% in 2004 to 13% in 2013 reflects Viet Nam's success in increasing agricultural productivity for many farm commodities and in diversification of sources of rural incomes.

Figure 1.20. **Urban versus rural poverty rates, 2004-13**

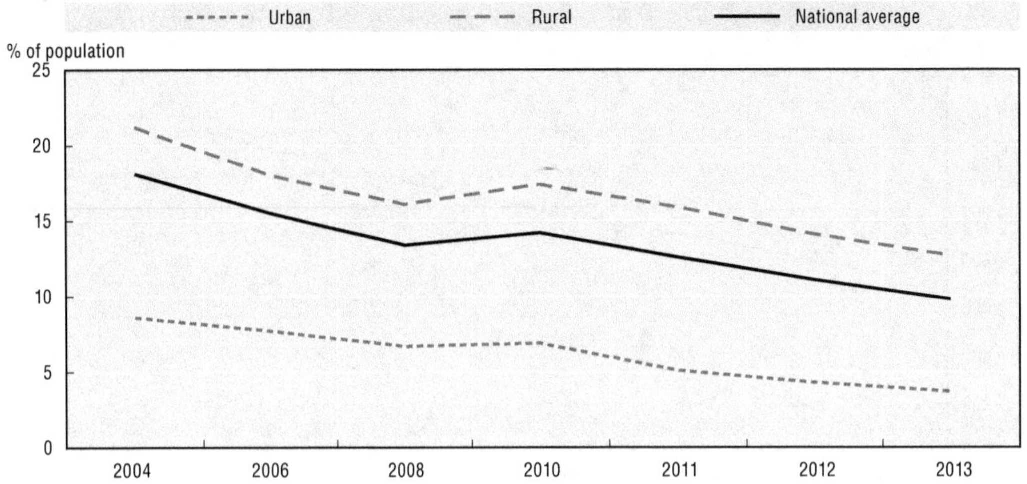

Note: The poverty rate is based on the GSO definition of the poverty line, different for rural and urban areas to take into account differences in cost of living, and adjusted for inflation. For 2013 the line was VND 570 000/month/capita for rural areas, and VND 710 000/month/capita for urban areas.

Source: GSO (2013), *Household Living Standards Survey 2012*; GSO (2014), *Statistical Yearbook of Vietnam 2013*.

StatLink http://dx.doi.org/10.1787/888933223401

Poverty rates fell across all regions over the same 2004 to 2013 period, but remained stubbornly high in the poorest region of the Northern Midlands and Mountain areas (Figure 1.21). Progress in poverty alleviation in other poor areas, such as Central Highlands as well as North Central and Central Coast, has been much stronger. The lowest poverty rates are always in the Southeast and Red River Delta, both benefiting from jobs, incomes and output markets generated by large urban centres of Ho Chi Minh and Hanoi, respectively.

Figure 1.21. **Poverty rates by region, 2004-13**

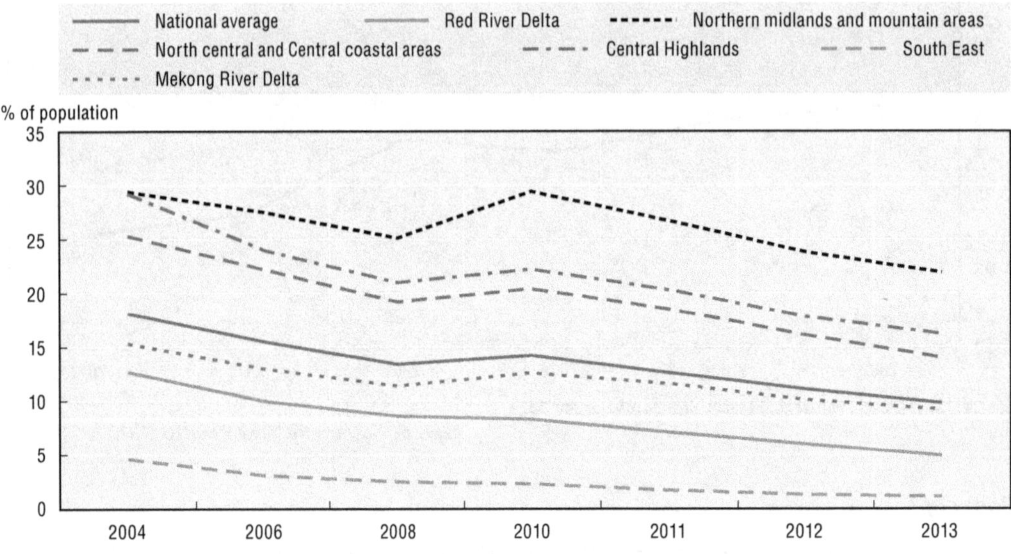

1. Poverty rates are based on national definition of poverty line.
2. This figure applies a standard division of the Vietnamese territory into 6 large regions, each consisting from 5 (Central Highlights) to 14 (Northern Midlands and Mountain Areas as well as North Central Area and Central Coastal Areas) provinces. Detailed information on allocation of provinces to regions can be found e.g. in GSO (2014).

Source: GSO (2013), *Household Living Standards Survey 2012*; GSO (2014), *Statistical Yearbook of Vietnam 2013*.

StatLink http://dx.doi.org/10.1787/888933223416

Food consumption

Viet Nam has made astonishing progress in combatting **undernourishment**. The proportion of undernourished in the total population fell from 45.6% in 1990-92 to 12.9% in 2012-14. This represents a decrease of 71.7%, which is one of highest rates for all countries, just after Thailand (80.9%), and larger than in China (55.4%) (FAO-IFAD-WFP, 2014).

Nevertheless, 11.9 million Vietnamese suffered from undernourishment in 2012-14 (FAO-IFAD-WFP, 2014). Moreover, the prevalence of **stunting** (moderate and severe) among children under 5 is still high at 23% in 2007-11 (WHO, 2015). Most food insecure people live in rural areas, in particular those affected by poverty. It can also be noted that the Mekong Delta, the country's rice bowl and dominant source of exported food of various forms, ranked last or next to last in nutrition progress last decade (Le Canh Dung et al., 2011).

On average, there has been a steady increase (1.8% per annum) in the **daily per capita energy intake** from 2 311 kcal in the mid-90s to 2 769 kcal one decade later (Figure 1.22). Daily per capita energy intake is higher than in India (2 300 kcal) and most Southeast Asian countries, including Indonesia (2 538 kcal), Thailand (2 530 kcal) and the Philippines (2 518 kcal), but remains lower than in most OECD countries and in some Asian countries such as China (2 974 kcal) and Malaysia (2 908) (FAO, 2010a).

Figure 1.22. **Daily per capita energy intake, 1995-2007**

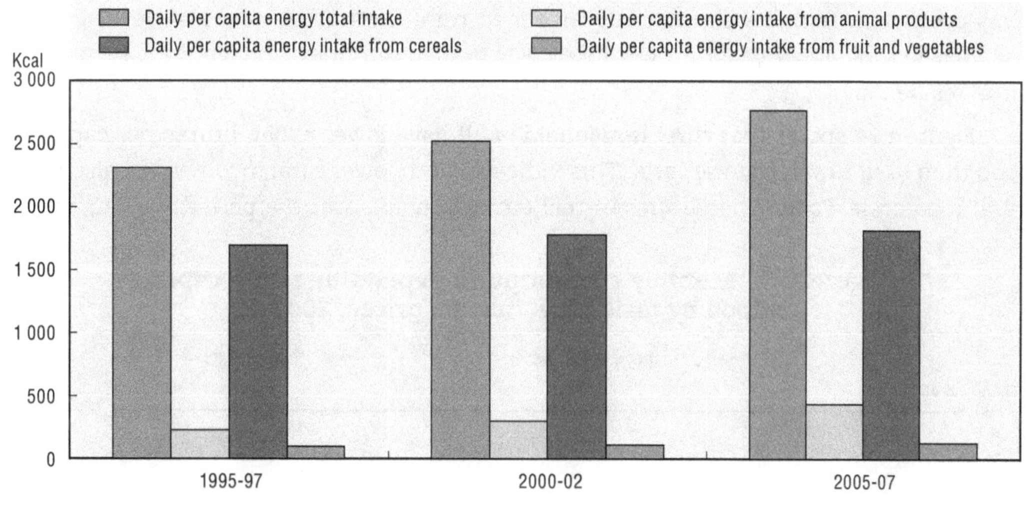

Source: FAO (2010a).

StatLink http://dx.doi.org/10.1787/888933223425

Daily energy intake from **animal products** has grown fairly rapidly (6.7% per annum) over this decade, from 230 kcal to 439 kcal, due to the high income elasticity for meat. By contrast, cereal energy intake rose only by 0.7% per year, and fruits and vegetable energy intake rose at the rate of 2.7% per year. Cereals accounted for 66% of total energy intake in 2005-07, animal products for 16%, and fruits and vegetables for just 5% (FAO, 2010a).

The **share of expenditures on food** in total household consumption expenditures, known as the Engel coefficient, provides an indication of food security: the lower the share, the greater the food security (Figure 1.23). Aggregate data show a gradual decline from 52.0% in 2002 to 47.1% in 2008. However, it rose to 52.5% in 2012. This partly reflects the impact of a relatively large increase in food prices compared to non-food prices in 2007-11.[9]

Figure 1.23. **Share of expenditures on food consumption in total expenditures, 2002-12**

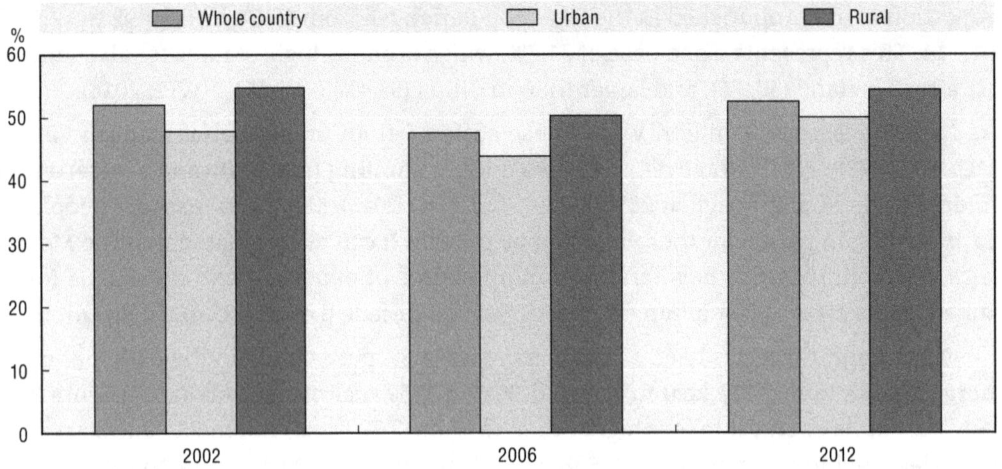

Whole country
Urban
Rural
%
60
50
40
30
20
10
0
2002
2006
2012

Note: Food consumption expenditure refers to the monetary value of acquired food, purchased and non-purchased, including non-alcoholic and alcoholic beverages as well as food expenses on away from home consumptions in bars, restaurants, food courts, work canteens, and street vendors.

Source: GSO (2013), *Household Living Standards Survey 2012.*

StatLink http://dx.doi.org/10.1787/888933223430

This coefficient remains consistently higher in rural than urban areas, reflecting lower incomes in rural areas, despite lower food costs than in the cities, and less food security for rural residents.

Figure 1.24 shows that **rural households** still have lower expenditures per capita on food than their urban counterparts. This indicates both lower rural incomes but also lower rural food costs. Food expenditures in real terms doubled over the period 2002-12, with a

Figure 1.24. **Monthly consumption expenditures per capita on food by residence, in 2005 prices, 2002-12**

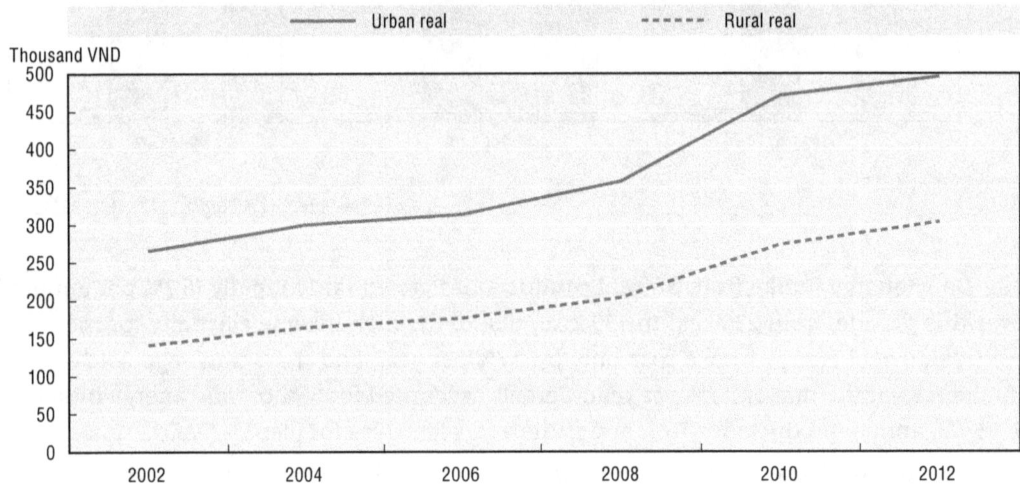

1. Food consumption expenditure refers to the monetary value of acquired food, purchased and non-purchased, including non-alcoholic and alcoholic beverages as well as food expenses on away from home consumptions in bars, restaurants, food courts, work canteens, and street vendors.
2. Nominal values of consumption expenditures have been deflated by the Consumer Price Index (CPI), base year 2005.

Source: GSO (2013), *Household Living Standards Survey 2012.*

StatLink http://dx.doi.org/10.1787/888933223449

strong acceleration in 2008-10. On average, rural residents spent only 57% of what urban residents did, but this ratio increased from 53% in 2002 to 61% in 2012.

Rice consumption per capita fell on average from 144 kg in 2002 to 115 kg in 2012.[10] It remains higher in rural households, at 126 kg compared to 92 kg for urban residents. Expenditure on rice as a share of all food expenditure fell for both rural and urban residents. For rural residents it fell from 31% in 2002 to 17% in 2012, larger than the fall from 15% to 9.5% for urban residents. While meat consumption increased for both types of households, its share in total expenditure on food fluctuated between 20-22% for urban residents and increased from 19 to 24% for rural residents. Fruit and vegetable expenditure as a share of total food increased for both groups, from 4-5% in 2002 to 8-9% in 2012 (GSO, 2013).

All income quintiles show increased food expenditure per capita in real terms over the 2002-12 decade, and all appear to roughly double, showing that income elasticity for food remains relatively high for all income groups (Figure 1.25).

Figure 1.25. **Monthly per capita expenditures on food by income quintile, in 2005 prices, 2002-12**

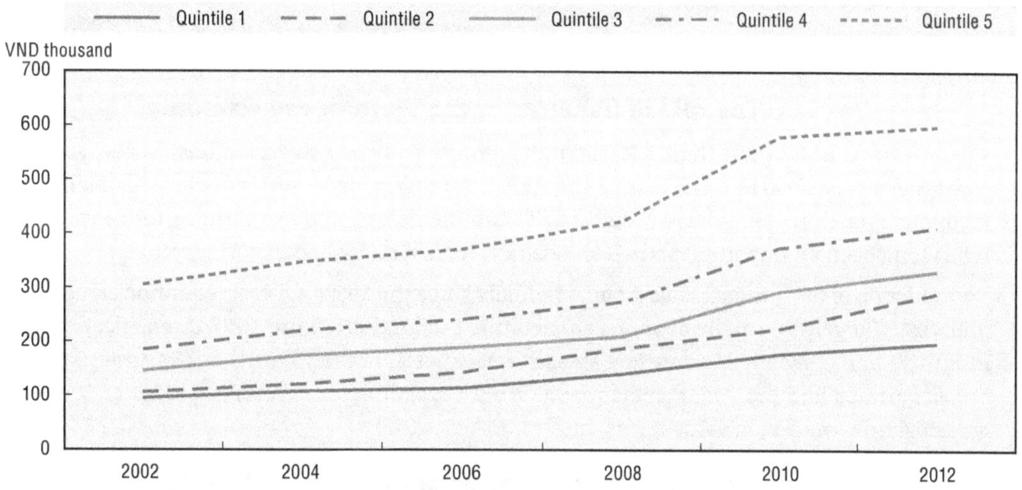

Note: Nominal values of expenditures on food have been deflated by the CPI, base year 2005.
Source: GSO (2013), *Household Living Standards Survey 2012*.

StatLink http://dx.doi.org/10.1787/888933223458

1.6. Agro-food trade flows

Prior to 1990, Viet Nam was not a significant player in world agricultural commodity markets. By 2011-13, Viet Nam had become the **world's largest exporter** of cashews and black pepper, the second largest exporter coffee and cassava, third largest exporter of rice and fisheries, and the fifth largest exporter of rubber. Annual exports for these commodities were well above or very near to USD 1 billion in recent years. Figure 1.26 shows a significant increase in Viet Nam's share in world exports for these commodities since 2000. Such trade performance for a relatively wide range of commodities for a country the size of Viet Nam starting from virtually no export market penetration and experience, and within two decades, is unmatched.

Viet Nam also benefits from advantageous conditions for the development of **fisheries** production. While Viet Nam's exports of fisheries and their share in the world's total have been increasing (Figure 1.26), a number of challenges may constrain further improvements in this respect (Box 1.2). Moreover, export performance of other commodities has been

Figure 1.26. **Share of Viet Nam in world's exports of selected commodities, 2000-13**

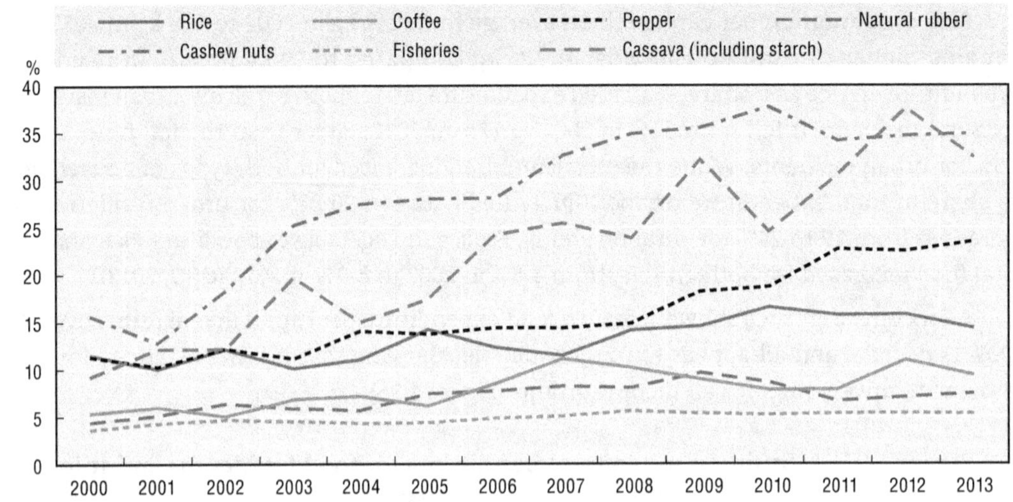

Rice | Coffee | Pepper | Natural rubber | Cashew nuts | Fisheries | Cassava (including starch)

%

Source: UN (2015), *UN Comtrade database*.

StatLink http://dx.doi.org/10.1787/888933223468

Box 1.2. **The role of fisheries in the Vietnamese economy**

Viet Nam counts more than 3 000 islands, almost as many rivers and estuaries, and a coastline of over 3 000 km. The country has about 1.7 million ha of water bodies. The Mekong River and Red River Deltas have been home to capture fishing and fish farming for centuries. It has long been an important source of incomes, food security and export revenue.

Production of fish products has grown gradually since the 1950s, with acceleration since the mid-1990s, largely due to the booming aquaculture production (Figure 1.27). Intensification of input use and a switch to a few key species for exports, notably catfishes (*Pangasius* spp.), carps, shrimps and prawns, made aquaculture a key subsector accounting for over 60% of gross fisheries output, in value terms, in 2011 (MARD, 2013).

Figure 1.27. **Viet Nam's fisheries production, 1990-2012**

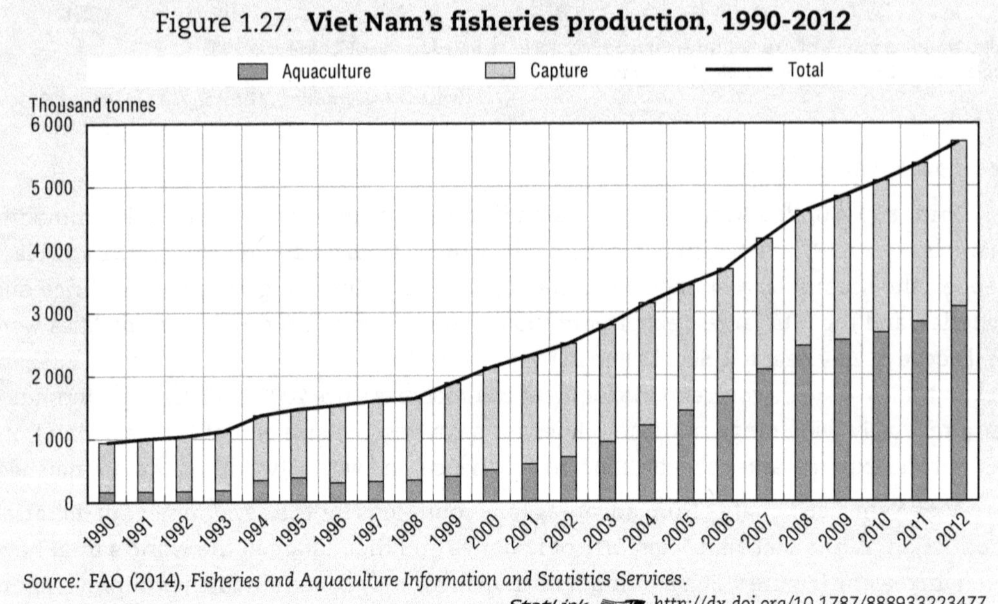

Source: FAO (2014), *Fisheries and Aquaculture Information and Statistics Services*.

StatLink http://dx.doi.org/10.1787/888933223477

Box 1.2. **The role of fisheries in the Vietnamese economy** *(cont.)*

Both the capture fisheries and aquaculture attract intensive, semi-intensive and small-scale producers. The latter two categories account for the largest part of production. Overall, the fisheries sector accounted for over 20% of the gross value of agriculture, forestry and fisheries production and for 3.7% of national GDP in 2012 (MARD, 2013). Around 4 million people, accounting for about 8% of total employment, derive their main income from the sector, including secondary activities such as transformation and transport. Inland fisheries (in floodplains and rice fields) play a particularly important role for seasonal income and nutrition of poor rural populations, including part-time fishers.

The booming aquaculture production made seafood one of the main export commodities of the country with fisheries accounting for about 6% of total export earnings in 2010-12. By 2011-13, Viet Nam had become the third largest seafood exporter worldwide with the European Union, the United States and Japan as the main markets for its exports (UN, *UN Comtrade Database*, 2014).

Consumption of fish is very high and still growing, with 34 kg per capita of annual consumption in 2011, almost double the world average (Figure 1.28 and FAO Food Balance Sheet, FAOSTAT, 2015). One of the reasons for growing domestic consumption in recent years was lower catfish export prices. This encouraged processors to develop products for the internal market, which is expanding due to both raising incomes and increasing demand for seafood as an important source of proteins.

Figure 1.28. **Fisheries production and utilisation in Viet Nam, 1990-2011**

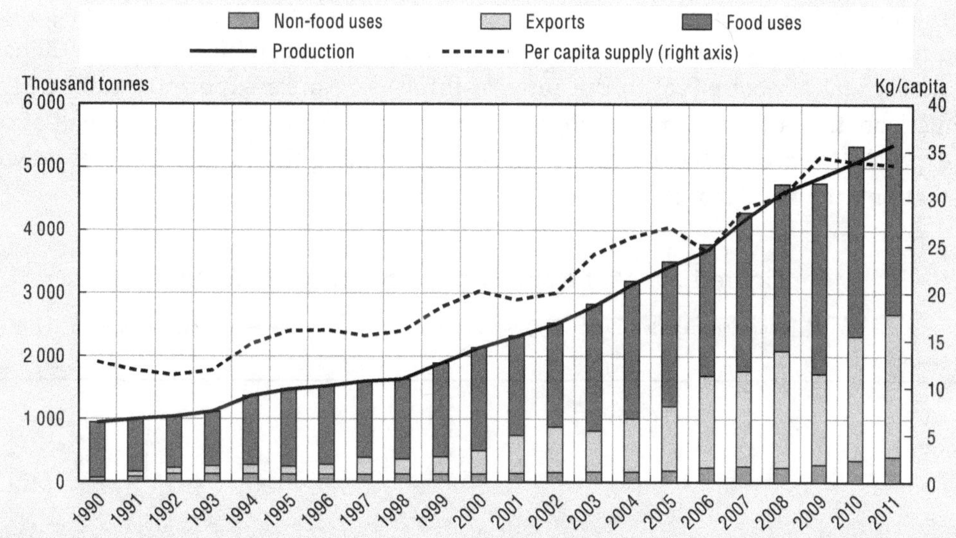

Note: All categories in live weight equivalent. In recent years, slightly higher total use (the sum of food and non-food uses and of exports) than production is due to small but growing imports.

Source: FAO (2014), *Fisheries and Aquaculture Information and Statistics Services*.

StatLink http://dx.doi.org/10.1787/888933223483

The rapid growth of the fisheries sector in Viet Nam poses a number of challenges. A lack of regulation of the capture sector and insufficient enforcement impacts stocks and ecosystems as well as long-term growth prospects. Concerns have been raised notably regarding overexploitation of inshore fisheries. But, even offshore fisheries are said to be fully exploited, following a rapid increase in capacity due to strong government support since

Box 1.2. **The role of fisheries in the Vietnamese economy** *(cont.)*

the late 1990s. Finding alternative livelihood options for coastal people will be key to easing the transition towards a more sustainable use of resources.

For aquaculture, the main challenges include disease control, water quality conservation, enforcement of a growing body of regulation, adaptation to food safety conditions and standards, including to increasingly complicated standards related to chemical and drug residues as well as certification imposed by importers. Indeed, Viet Nam's exports of fishery products face high relative rejection rate in most export markets (Henson, 2013).

Both the capture and aquaculture sub-sectors are also confronted with constraints that affect most of the economy: credit access difficulties, inefficient transportation and retailing facilities and underdeveloped marketing channels.

Source: FAO (2012a); Henson S. (2013); MARD (2013).

even stronger than that of fisheries leading to the significant fall in the share of fisheries in Viet Nam's total exports of agro-food products (Figure 1.30).

Led by expanding exports of the above mentioned commodities, the total value of Viet Nam's total **agro-food exports increased six-fold** between 2000 and 2012, but then declined by about 3% in 2013 (Figure 1.29 and Table 1.3). Exports are around double the value of agro-food imports, contributing to a positive balance of agro-food trade of about USD 10 billion in 2011-13. The agro-food sector shows on average a strong integration with international markets, particularly in terms of exports. The ratio of total value of agro-food exports to agricultural GDP was 70-80% in the early 2010s, much higher than in China or Indonesia and almost equal to the ratio of total Viet Nam's exports to total GDP. In comparison, the ratio of agro-food imports to agricultural GDP is only around half of that for exports. However, it has tripled since 1990 confirming the overall trend of growing integration with world markets.

Figure 1.29. **Viet Nam's agro-food trade, 2000-13**

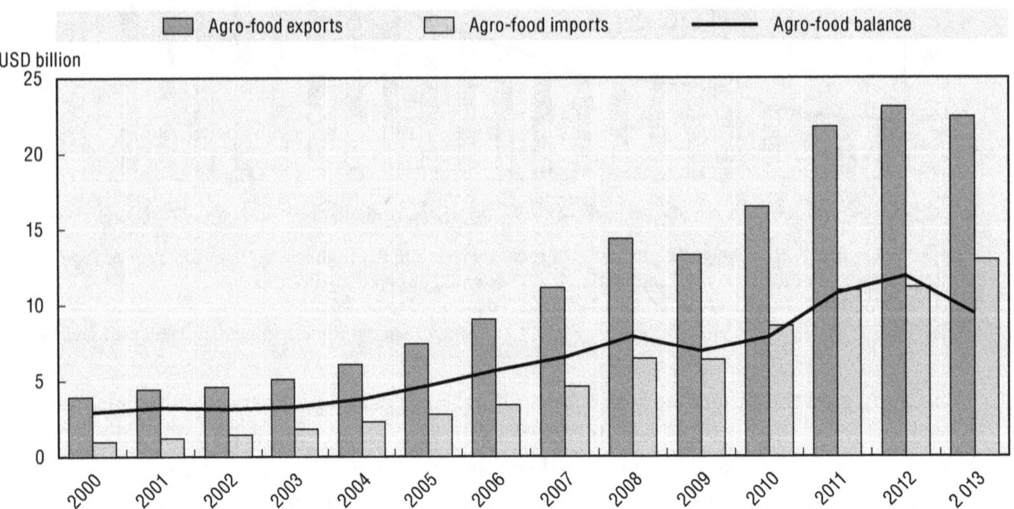

Note: Agro-food trade includes fisheries as well as natural rubber.
Source: UN (2015), *UN Comtrade database*.

StatLink http://dx.doi.org/10.1787/888933223492

Table 1.3. **Agro-food sector's integration with international markets, 2000-13**

		2000	2005	2010	2011	2012	2013
Agriculture, Gross Domestic Product (GDP), current prices	USD billion	7.6	11.1	21.9	27.2	30.6	31.5
Agro-food exports	USD billion	3.9	7.5	16.5	21.8	23.1	22.4
Agro-food imports	USD billion	1.0	2.8	8.6	11.0	11.2	13.0
Agro-food trade balance	USD billion	2.9	4.7	7.9	10.8	11.9	9.5
Coverage degree of imports by exports	%	396	268	191	198	206	173
Share of agro-food trade in total trade							
Exports	%	27	23	23	22	20	17
Imports	%	6	8	10	10	10	10
Ratio of agro-food exports to agricultural GDP	%	51	67	75	80	75	71
Ratio of agro-food imports to agricultural GDP	%	13	25	39	40	36	41
Ratio of total exports to total GDP	%	46	56	64	72	75	81
Ratio of total imports to total GDP	%	50	64	75	79	74	81

Note: Agro-food trade includes fisheries as well as natural rubber.
Source: Authors' calculations based on UN (2015), *UN Comtrade Database*; WB (2015), *World Development Indicators*.

StatLink http://dx.doi.org/10.1787/888933223508

The relative importance of various commodities in agro-food exports has evolved over time (Figure 1.30). In the early 2000s, fisheries were the **key agro-food export group** accounting for more than 40% of the total followed by rice and coffee. In total, these three commodities accounted for two-thirds of total agro-food exports. By the early 2010s, the share of fisheries had declined by one-third. The share of rice had remained at about 15%, but those of coffee, natural rubber, cashew nuts, cassava (including cassava starch) and pepper had increased. Viet Nam is also a growing exporter of various fruit and vegetables: fresh, dried, frozen and pre-processed. But, if cashew nuts and cassava are excluded, their exports remain relatively small at USD 558 million annually in 2011-13 accounting for just 2.5% of total agro-food exports in this period, half of that in the early 2000s. Considering their high labour intensity and revenue generation per unit of land, fruit and vegetables might become another export opportunity in the future.

Figure 1.30. **Composition of agro-food exports, averages 2000-02 and 2011-13**

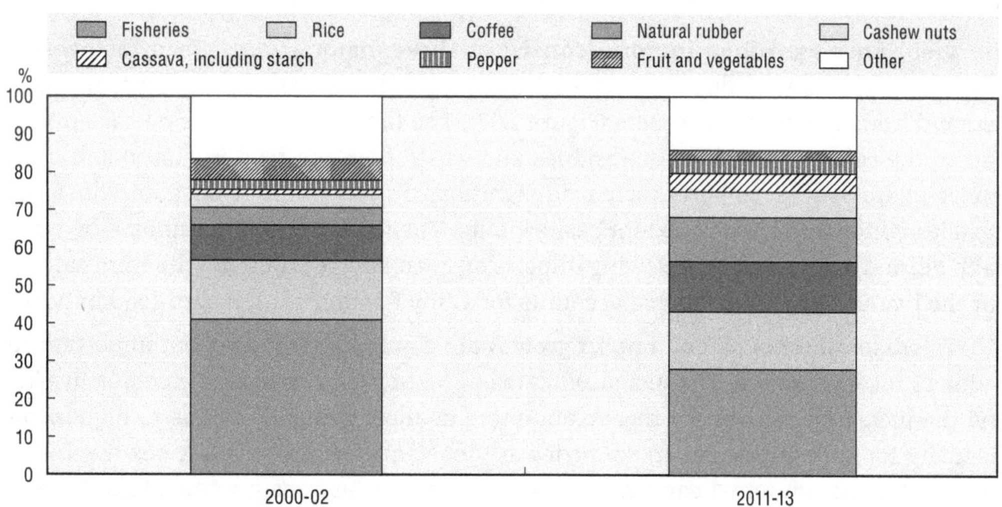

Note: Fruit and vegetables exclude cassava and cashew nuts, shown separately.
Source: UN (2015), *UN Comtrade Database*.

StatLink http://dx.doi.org/10.1787/888933223517

China is the main **destination for agro-food exports** accounting for 20% of the total in 2011-13 (Figure 1.31).[11] Main commodities exported to China include cassava starch, fresh and dried cassava, rice and cashew nuts. The United States is the second largest agro-food market, importing mostly cashew nuts, coffee, and fisheries. Japan imports mostly shrimps and prawns, coffee and frozen fish meat. Geographical proximity and trade liberalisation within ASEAN helped expand trade with such countries as the Malaysia, Indonesia, Philippines and Singapore. European Union countries are an important destination, but their share is relatively small compared with the size of the EU market.

Figure 1.31. **Main export markets for Viet Nam's agro-food products, 2011-13 average**

As per cent of total agro-food exports

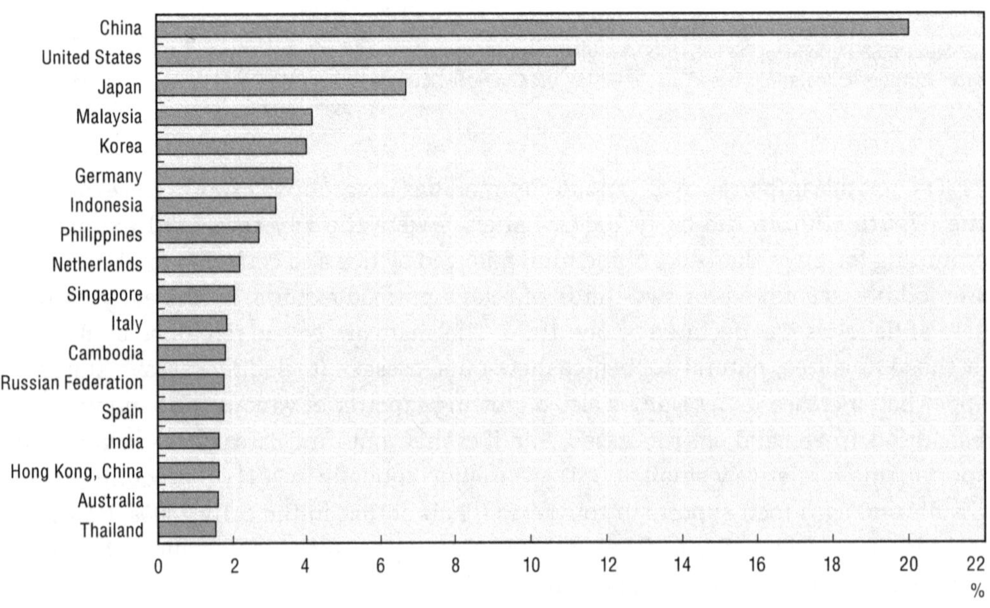

Source: UN (2015), *UN Comtrade database.*

StatLink http://dx.doi.org/10.1787/888933223525

Viet Nam's **agro-food imports** consist of three major groups: feed for livestock production; raw commodities for further processing and export; and products to meet food demand from domestic consumers (Figure 1.32). The first group includes oil cake (the key part of the category classified as residues and waste from the food industry) and maize, whose share in total imports increased significantly over the last decade. The second includes cotton and to some extent cashew nuts. The third group is dominated by wheat, palm oil, and dairy products. Growing imports of sugar and beef fall into the third category, but their value remains small and accounts for a tiny fraction of total agro-food imports.

The **determinants of food import growth** are commodity specific, but important role is due to income growth. The increased demand for soybean meal is for feeding livestock and the increased demand for meats, and proteins more generally, is due to high income elasticity for animal proteins in countries at Viet Nam's income level. A positive income effect is also known for wheat consumption within Asia, as consumption shifts from rice to wheat, and for cooking oils generally. Accordingly, increasing import demand for these commodities can be expected to continue.

Figure 1.32. **Composition of agro-food imports, averages 2000-02 and 2011-13**

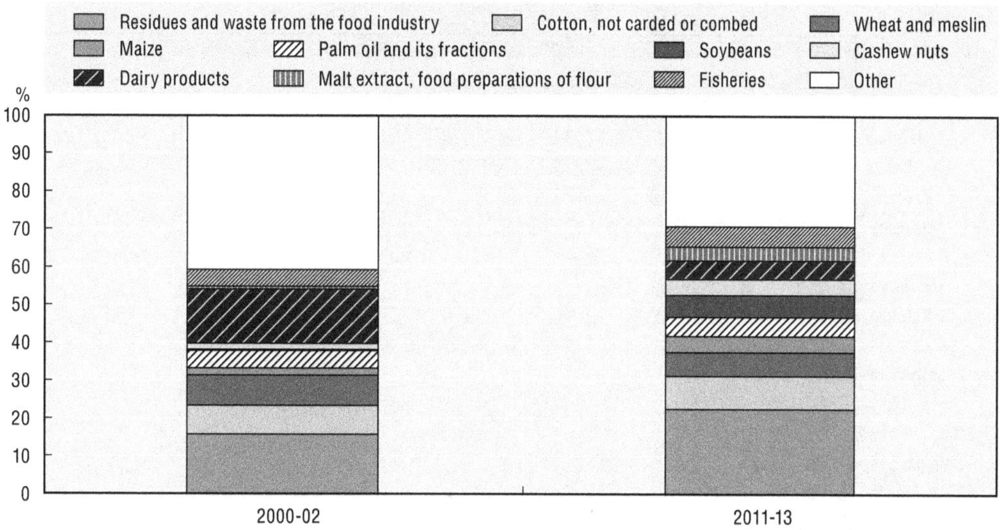

Residues and waste from the food industry
Cotton, not carded or combed
Wheat and meslin
Maize
Palm oil and its fractions
Soybeans
Cashew nuts
Dairy products
Malt extract, food preparations of flour
Fisheries
Other

%

2000-02
2011-13

Source: UN (2015), *UN Comtrade Database*.

StatLink http://dx.doi.org/10.1787/888933223531

As wheat is not produced in Viet Nam and cotton production is very small, imports of these two commodities satisfy around 100% of domestic use. In the case of wheat, large variations in the ratio of imports to total domestic supply from one year to the next are caused by changes in domestic stocks. Dependence on imports is also high for dairy products at around 80% and for soybeans at almost 50%. The shares for maize, sugar and bovine meat are smaller at 10-40%, but have increased strongly since 2000 (Figure 1.33).

Figure 1.33. **Share of imports in Viet Nam's domestic use of selected commodities, 2000-11**

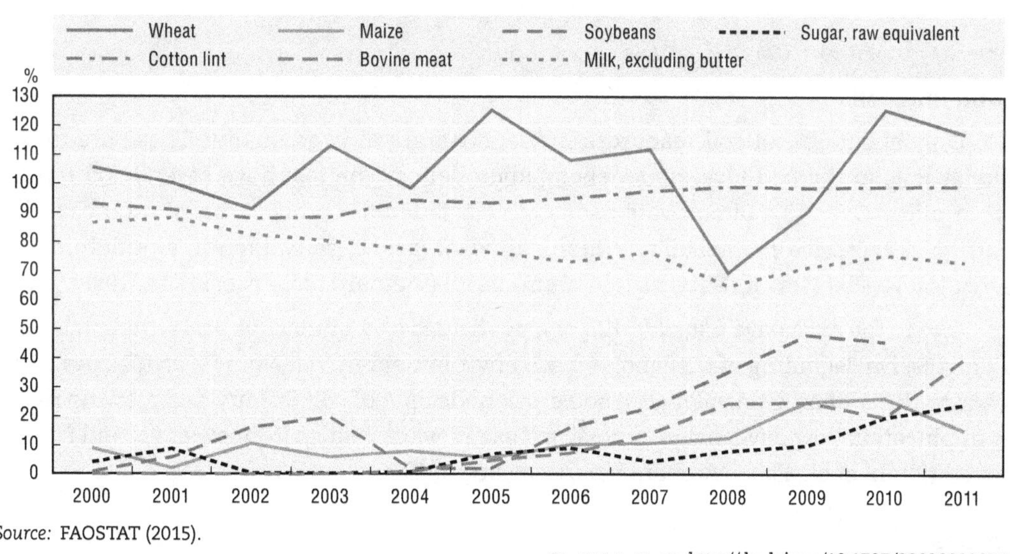

Source: FAOSTAT (2015).

StatLink http://dx.doi.org/10.1787/888933223543

The United States, India, Argentina and Australia are the **major suppliers of agro-food imports** (Figure 1.34). The United States accounts for 13% of the total, supplying a wide variety of products, among them soybeans, brewing or distilling dregs as well as dairy

Figure 1.34. **Main suppliers of agro-food products to Viet Nam, 2011-13 average**

As per cent of total agro-food imports

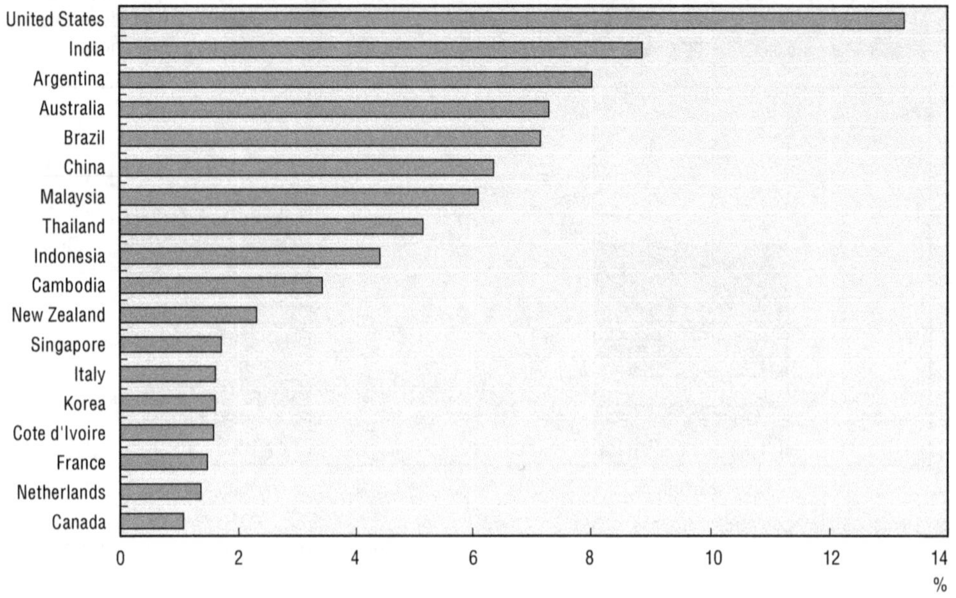

Source: UN (2015), *UN Comtrade database*.

StatLink http://dx.doi.org/10.1787/888933223551

products. India is a major supplier of soybean meal and maize. Argentina's exports to Viet Nam are dominated by soybean meal. Australia is the dominant supplier of wheat with an import share of 87% in 2011. Malt is another noteworthy commodity in Australia's exports to Viet Nam. Among other countries, China and Hong Kong are important suppliers of chicken meat and Malaysia of palm oil.

1.7. Agro-environmental situation

Land and soil

Only about 30% of **soil resources** in Viet Nam are of good quality. These are mostly fluvial soils in the Red River and Mekong River deltas. The rest have several soil fertility constraints. Over 50% of cultivated soils are classified as "problem soils": arenosol (low nutrients, excessive permeability), thionic fluvisol (acidic, salty, high in aluminium) and acrisols (clay-like: low fertility, high in aluminium) (Vietnam Soil Association, 1996).

Due to the **excessive use of fertilisers, pesticides and other chemicals**, there has been a progressive degrading of the land and soil environment in Viet Nam (MONRE, 2014). This leads to the widening prevalence of soil erosion, decline of soil fertility and growing risk of eutrophication (low oxygen and excessive algae in water bodies from excess N and P in the water) (Pham et al., 2006; Vietnam Soil Association, 1996).

Forests

In 2010, 45% of Viet Nam's land surface was covered by forests, 60% more than in 1990 (Table 1.4). However, only 0.5% of these could be considered a primary forest, the most biodiverse form of forest and defined by the FAO as "naturally generated native forest with no clear signs of human activity and undisturbed ecological processes" (FAO, 2010b).

Table 1.4. **Forest characteristics and dynamics**

	1990	2000	2005	2010
	Forest characteristics (1 000 ha)			
Primary forest	384	187	85	80
Planted forest	967	2 050	2 794	3 512
Other naturally regenerated forest	8 012	9 488	10 198	10 205
Total forests	9 363	11 725	13 077	13 797
Total forests as % of country area	28%	38%	42%	45%
	Forest establishment total (ha/year)			
Afforestation	32 260	118 245	138 920	n.a.
Natural expansion of forest	5 720	56 839	543 237	n.a.
Reforestation	116 720	209 540	327 785	n.a.

Source: FAO (2014) *Forestry CountrySTAT Database.*

The country experienced a sustained and intensive **deforestation process** last century (Meyfroidt and Lambin, 2008). Forest cover decreased from 60% of the country's total area at the beginning of the twentieth century to about 25% in the early 1990s. It was in that decade that the government implemented reforestation programmes along with non-government organisations. The most important of these was the Five Million Hectare Reforestation Programme, which aimed to create 3 million ha of production forest, in particular plantations, and 2 million ha of protection forests (watersheds and vulnerable slopes) and special-use forests (national parks, etc.) through plantations, natural regeneration and enrichment planting by 2010. The programme had a strong focus on smallholder reforestation and allocation of forestland to private households, organisations and individuals. These efforts resulted in the recent expansion of the forested area which made Viet Nam one of ten countries with largest annual net gain in forested area in 1990-2010 (FAO, 2010b).

Despite these efforts and successes, over two-thirds of natural forests are considered to be of "poor" or "recovering" quality and low land forests have been almost completely depleted (UN-REDD, 2009). According to the World Wildlife Fund (WWF, 2013), the country still experiences deforestation, forest degradation and fragmentation in the Central Highlands, Central Coast and Southeast regions. Moreover, Viet Nam has one of the highest rates in the world of the deforestation of primary forests.

There are several factors leading to continued **pressures on primary forests**. They include deforestation for infrastructure improvements to support expanding economy, widespread prevalence of illegal logging, weak management of state-owned forestry farms, and expansion of agricultural production as many lower-income farmers still clear forested areas to turn them into agricultural land. Forest conversion to agricultural land is largely due to the expanding area of production of export-oriented commodities, such as coffee and natural rubber. This is particularly the case in the Central Highlights where as much as 79% of new rubber plantations were created on natural forestland (To and Tran, 2014). In turn, in the poorest communities, particularly in the mountainous areas, shifting cultivation continues to be practiced and its population depends on the forests for daily needs, thus also exerting pressures on the remaining forests (REDD, undated note).

Water

Water is relatively **abundant** in Viet Nam. While freshwater availability at around 4 000 m^3/capita/year is twice lower than in Indonesia, it is twice higher than in China and

3.5 times higher than in India. However, due to unevenly distributed monsoon rainfall, the distribution of water resources is highly variable during the year, with about 70-75% of the annual runoff generated in three to four months. These variations, combined with limited storage and flood control infrastructure, lead to damaging floods in the wet season and very low flows in the dry season (FAO AQUASTAT, 2014).

Agriculture exerts significant and growing pressure over the country's available water resources with 95% of used freshwater going to this sector (Table 1.5). The area equipped for irrigation increased by almost 50% between 1996 and 2006. Almost 80% of the irrigated crop area is used to grow rice. The next three crops, in terms of irrigated areas used for their production, are maize, coffee and rubber, but each of them represents barely 3% of the irrigated area in the country (FAO AQUASTAT, 2014).

Table 1.5. **Water availability and utilisation**

	Freshwater availability – volume per year			
	1996	2001	2006	2011
Total [km³]	359.4	359.4	359.4	359.4
Per capita [1000 m³]	4.66	4.4	4.19	4
	Freshwater utilisation (volume per year in km³)			
	1992	1997	2002	2007
Agriculture	47	n.a.	n.a.	77.75
Domestic	n.a.	n.a.	1.05	1.21
Industrial	n.a.	n.a.	3.26	3.07
Total	n.a.	n.a.	n.a.	82.03
	Area equipped for irrigation (1 000 ha)			
	1996	2001	2006	2011
Total	3 150	3 850	4 600	4 600

Source: FAO AQUASTAT (2014).

There are no data on water pollution caused by agricultural practices, but, overall, Viet Nam's water quality is classified as average but deteriorating (ADB, 2010). The biological oxygen demand (the only available indicator) is well above the limit.

Air

Carbon emissions from agricultural production in Viet Nam are growing quickly. Rice cultivation alone accounted for 44% of total CO_2 equivalent emissions from agriculture in 2010. In terms of growth rates, the use of fertilisers is the agricultural activity whose emission impact is growing the fastest (Table 1.6).

Table 1.6. **Emissions of CO_2 equivalent from agricultural activities, gigagrams per year**

	1990	2000	2010	2011	Change 1990-2010 %
Enteric fermentation	6 742.7	7 979.5	9 853.1	9 285.8	46
Manure management	3 063.4	4 446.0	5 859.4	5 759.2	91
Rice cultivation	22 397.6	28 415.0	27 759.4	28 374.8	24
Synthetic fertilisers	2 745.7	8 571.8	7 903.4	5 212.6	188
Manure applied to soils	1 016.5	1 466.8	1 939.4	1 986.6	91
Manure left on pasture	1 908.0	2 401.5	3 103.8	n.a.	63
Crop residues	1 647.3	2 520.6	2 987.6	3 127.2	81
Cultivation of organic soils	947.0	947.0	947.0	947.0	0

Table 1.6. **Emissions of CO_2 equivalent from agricultural activities, gigagrams per year** *(cont.)*

	1990	2000	2010	2011	Change 1990-2010 %
Burning – crop residues	301.1	403.2	425.0	432.4	41
Burning – savannah	136.3	120.9	137.1	85.9	1
Energy use	1 048.6	1 597.3	2 293.5	n.a.	119
Total agriculture, including energy use	41 951.3	58 885.8	63 206.0	n.a.	50.7

Source: FAOSTAT (2015), *Agro-Environmental Indicators: Emissions.*

Biodiversity

The weather, variety of soil types and location of Viet Nam allows for a **great genetic diversity** of crops, flowers and vegetables. Regarding crops, the country is considered as the origin of rice, taro, banana, mango, coconut, tea and specific citrus trees. Pham and Luu (2008) cite survey data reporting over 800 plant species cultivated throughout the country: 41 starchy and 95 non starchy crops, 105 fruits and 55 vegetables, 44 oil crops, 181 medicinal, and 39 spices. Regarding its most important agricultural crop, rice, Viet Nam has over 6 000 varieties of rice, including wild rice species and 700 landraces (i.e. "domesticated" species). Among the most important fruits are banana, pineapple, mango, papaya, mangos teen, cashew and litchi.

There are several initiatives to protect Viet Nam's biodiversity and plant genome, including *in situ* and *ex situ* practices. *Ex-situ* programmes began as early as 1975 with the creation of genome banks, which still preserve most of the country's local plant genome. In the 1990s, various programmes were launched along with various non-governmental organisations and research institutes to introduce diversity management practices among farmers. The main crops included in on-farm management programmes are rice and taro (Nguyen et al., 2005).

Although there is a continued effort by the government to protect crop biodiversity, it still faces several challenges, all due to intensification of agricultural practices in the country. Farmers who are not under the on-site biodiversity management programmes are more likely to replace traditional crop varieties with high-yielding ones. Erosion and soil depletion also represents a threat to the conservation of local varieties.

Climate change

Viet Nam is listed among ten countries potentially the **most affected by climate change**. Climate change scenarios developed by the Vietnamese government predict increases in average temperature, rainfall and sea levels. A recent study by the Ministry of Natural Resources and Environment predicts that in the long term, by 2100, an average temperature will increase by between 1.1 and 1.9 °C in a low-emission scenario, and between 2.1 and 3.6 °C in a high-emission scenario. In turn, a sea level is predicted to rise from 65 cm (low-emission) up to 100 cm (high-emission) and an annual rainfall to increase by between 1 and 5.2% (MONRE, 2009)

The potential impacts are likely to be most serious on agriculture and on water resources, as flood inundation and droughts are predicted to happen more frequently as a result of an increase in rainfall intensity and reduction in number of rainy days. In particular, large cultivation areas in the Mekong and Red River deltas are likely to be affected by salt water intrusion due to sea level rise (ISPONRE, 2009).

A wide range of **adaptation measures** can and should be taken to minimise the negative impacts of climate change, in particular of long-term sea level rise. While Vietnamese farmers have already demonstrated their capacity to adapt, climate change will increase the risks faced by farmers and this "will necessitate improved knowledge and more flexible and diversified farming systems" (WB, 2012). Viet Nam's government is aware of these risks and research has been undertaken to develop and disseminate improved rice varieties which would be more tolerant to saline and flood inundation. In addition, various schemes are being debated to improve flood and water management structures (WB, 2012).

1.8. Agricultural land tenure system

Evolution of the legal framework

In Viet Nam, all land is owned and administered by the state. Therefore, individuals and enterprises that use land are not land owners, instead, they own the **use (*usufruct*)-rights** to the land. Prior to the late 1980s, these use rights were held by collective and state farms. Beginning with the major reform of Resolution No. 10 in 1988, individuals were allowed to hold the use rights. Then, the Land Law of 1993 gave farmers a wide range of rights, including the right to rent, buy, sell, and bequeath land and to use land as collateral with financial institutions for mortgages.

By 2012, most of the country's agricultural land had been allocated to "users". According to the Ministry of Natural Resources and Environment (MONRE), 75% of all land has been mapped and **Land Use Rights Certificates** (LURCs), commonly referred to as a "Red Book", have been issued to cover 85% of agricultural land (CIEM, 2013). However, the progress has been very unequal with the area covered by LURCs ranging from 93.1% in the Long An province in the Mekong River Delta to just 22.7% in the Lai Chau province in the North-West (CIEM, 2013).[12]

The **allocation** process was difficult and contentious. Equity considerations across households were given priority, and other factors such as the number of people in the household and land quality were also considered. Prior land ownership, particularly in the south, was not taken into account. This process led to the fragmentation of the land use pattern, but also to disputes in the allocation of different qualities and locations of land parcels to individual farm households, about equitable treatment in the mix of qualities of land allocated, and because many farms would get lands in different locations (not contiguous). But at least farmers were free to transfer land parcels among themselves, initially only for rental and later (2003) for purchase and sale ("transfers" in the language of the Law), subject to maximum land holdings per household (2-3 ha of annual cropland and 10 ha in the case of perennial cropland).

Revisions of the Land Law made in 1998 and 2003 introduced **restrictions on land use** stating that changes in land use by the farmer were only allowed within the existing physical planning framework adopted by central and local governments. They mostly confine farmers to growing rice on paddy land at the expense of other crops (or fisheries) that could be grown more profitably on the same land. Farmers can apply at the district level for a change in their designated land use, but in practice changes or removals of these restrictions are rarely allowed.[13]

Moreover, the 2003 Land Law revisions allowed the state to appropriate land, including farmland, for economic development purposes. While it was introduced to help encourage industrial and urban development, it resulted in a sharp increase in highly contentious **land disputes**. In the case of expropriations farmers are not only involuntarily losing their base for farming, but also compensations they receive are very low as they are based on the value of land in existing use (agriculture), not its future or alternative use. The new Land Law voted in 2013 made a number of modest improvements, but the essential points of controversy in land disputes remained largely unaddressed (see Section 3.5 for more details).

Farm structures

According to the **2011 agricultural census**, land defined as "agricultural production land" (annual and perennial crops land, 10.1 million ha) is used by individual households (9.1 million ha; 89.4%), Commune Peoples' Committees (0.2 million ha; 2.3%), domestic economic organisations (0.6 million ha; 6.4%), other domestic agencies (0.06 million ha; 0.6%), foreign individuals and organisations (0.009 million ha; 0.1%) and other (0.1 million ha; 1.2%) (GSO, 2012).

Data on non-household type of land-users are scarce. **Enterprise survey** conducted by the GSO indicates that in 2011 there were 955 enterprises in agriculture (excluding those focusing on forestry and fishery) operating on about 0.5 million ha, thus on about 5% of agricultural production land[14] (GSO, 2012). Their efficiency is low (MARD, 2015). Despite ongoing privatisation, large part of them are still state-owned or have a status of state limited companies. Their average size was 543 ha in 2011, with a dominant part of perennial crops (358 ha), followed by forest land (125 ha), annual crops (53 ha) and aquaculture (6.5 ha) (GSO, 2012).

Data on individual households operating in agriculture are provided in two basic ways. One is annual information on large farms provided by the GSO and another is agricultural census data collected every five years and covering all farm households. Data on **large farms** indicate a steady increase in their number from 114 000 in 2005 to 146 000 in 2010 (GSO, 2014). Taking into account a high minimum sales requirement at about USD 2 250 per year to be a large farm, as applied by the MARD in this period, it might be concluded that the large-scale subsector has been growing, even if it remains a tiny fraction of the total number of households engaged in agricultural production. However, the number of such farms declined abruptly to around 20 000 in 2011 (GSO, 2014) as a result of MARD's decision to raise the minimum sales requirement to USD 25 000 in 2011 and after, thus not allowing comparisons over time.[15]

Agricultural census data provide a much more comprehensive picture. According to the 2011 census, out of 15.3 million rural households there were **9.6 million households** engaged in agricultural (excluding fishery and forestry) production, 1.5% less than recorded during the previous census in 2006 (GSO, 2012). To assess changes over a longer period, Figure 1.35 provides the distribution of farm households by the size of their agricultural holding as captured by the 2001 and 2011 censuses. In 2001, 26% of rural households had a total land holding of less than 0.2 hectares. By 2011, the share of the smallest farms had increased by 9 percentage points[16] and the share of the second smallest size of farms had decreased by 7 percentage points, so that the share of farms less than 0.5 ha had been 69%, almost the same as in 2001. The share of the largest farms (at least 2 ha) increased slightly to 6.2% and the share of the second largest land size category (0.5-2.0 ha) fell slightly and accounted for one-fourth of the total in 2011.

Rice production is even more fragmented. In 2011, there were 9.3 million households producing paddy, slightly less than in 2006. Their average size was just 0.44 ha, virtually unchanged from 2006.[17] Half of them held less than 0.2 ha, 35% held between 0.2 and

Figure 1.35. **Distribution of farms by land size, 2001 and 2011**

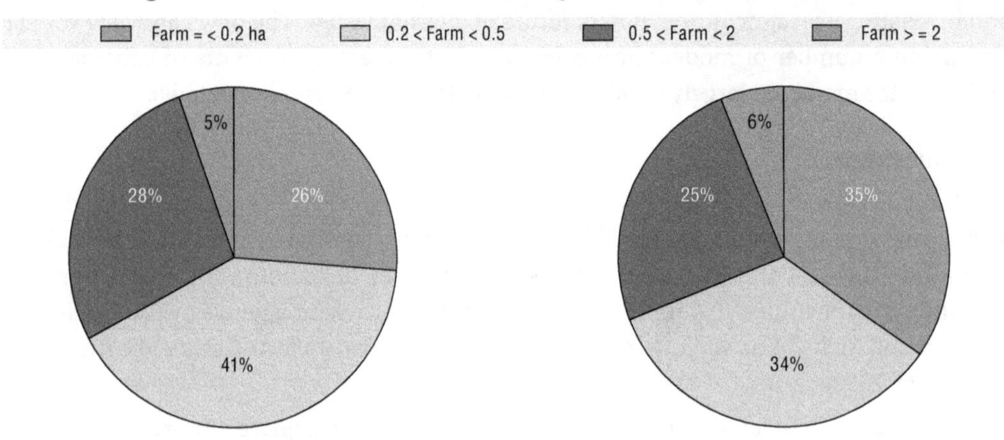

Source: GSO (2012), *Results of the 2011 Rural, Agriculture and Fishery Census*.

StatLink http://dx.doi.org/10.1787/888933223565

0.5 ha, and only 2.3% held a paddy land parcel of 2 ha or more. There are quite important regional differences, including between the Mekong and the Red River deltas, the two key rice-producing regions. In the Mekong Delta an average paddy farm operated on 1.4 ha compared with just 0.2 ha in the Red River Delta. Moreover, while in the Mekong Delta 13% of farms operated on more than 2 ha, in the Red River Delta this percentage was just slightly above zero (GSO, 2012).

Production of **perennial crops** is also dominated by small farms, but their average size is larger than of those producing paddy. These crops were cultivated by 5.1 million households in 2011, i.e. around half of all crop producers. This number fell by 4.2% from 2006. About two-thirds of them reported less than 0.2 ha (a fall of 3.4% compared to 2006) and as much as around three-quarters of them held less than 0.5 ha. For the large scale group of 2.0 hectares or more, the average share was 5%, but it was 20.8% in the Southeast region, and 17% in the Central Highlands (GSO, 2012).

An additional aspect of farm fragmentation in Viet Nam is that each farm consists of several **non-contiguous plots**. According to the 2012 survey conducted in 12 provinces, average farm consisted of 4.4 plots averaging 0.18 ha and just 15% of them had a border with another plot of the same land user. Moreover, in the northern and north-western provinces, where the average size of farm is particularly small, the average number of plots per farm is higher and the average size of plot much smaller than in the south (CIEM, 2013).

The 2011 census data indicate that **livestock production** is also dominated by tiny producers with, for example, half of pig-raising households having just 1-2 pigs, but some consolidation of farm operations has been progressing. For example, while pigmeat production is increasing at high rates (Section 1.3), the number of pig-raising households declined by 35% compared to 2006. Moreover, while the number of households raising less than 10 pigs per year declined by 39%, the number of those raising 50 or more pigs increased by 80% over the same period. Similar tendencies can be observed in poultry production with households producing more than 1 000 chicken increasing 4.32 times between the two censuses (GSO, 2012).

Overall, it might be concluded that while **farm consolidation** in livestock production has started, consolidation of crop production is in its very early stages. Census data show that farming continues to be dominated by tiny units and that very few farms have grown to the "large scale" category of just 2 ha. This is true both for rice as well as for perennial crops, the two commodity areas where the country has demonstrated very rapid production and export growth. It also shows how large is the task of increasing farm size within Viet Nam.

1.9. Competition and structural change beyond the farm gate

In Viet Nam, like in all ex-centrally planned economies, the problem of adequate levels of competition in the upstream and downstream sectors is substantially exacerbated by the **prevalence of SOEs** that were set up to be large to achieve economies of scale, but which in a market economy end up having no competitors or at least enjoy substantial market power.

The privileges of SOEs are well known (Coxhead et al., 2010), and include access to cheap capital, close connections to government regulators and policy-makers and limited competition. While the influence of the SOE sector across the economy has declined (Section 1.2) including in the agro-food sector, SOEs still play an important role in input supplies, processing, storage, and marketing, including exports (see also Table 2.1, Section 2.2). Moreover, many SOEs have been only **partly privatised** through the equitisation process (Section 1.2). The main caveat is that such an equitised firm may appear to be private but is likely to be only marginally so. In addition to the fact that the newly-created equity shares may be held by the state, the firm may continue to hold advantages from ex-SOE status, such as continued market power and easier access to credit.

There is a widely held view that Viet Nam suffers from **low quality** and low prices in the export market, that it gains low value-added in its exports, and that the processing and exporting sector firms, primarily state-owned enterprises, are not performing well enough to deal with this problem (WB, 2012; Dao and Nguyen, 2013; Coxhead et al., 2010). Most of these studies argue for a reduction in the dominant role of SOEs, with added competition in processing and exporting from foreign firms and truly private domestic firms.

This section examines this issue through a short review of the key components of the agro-food value chain: input markets (fertilisers, seeds and machinery), marketing channels for three key export commodities (rice, coffee and rubber), retailing and agricultural trading enterprises.

Inputs

Fertilisers

The fertiliser market in Viet Nam is large and important, both economically and politically. While there are more than 500 fertiliser producers, only 3 or 4 of them account for the vast majority of the total market. For example, in 2012, Petrol Fertilizer and Chemical Company accounted for 88% of nitrogen production and the remaining 12% was supplied by Ha Bac Nitrogen and Chemical Company (Agroinfo, 2013).

Government policy has been to encourage domestic fertiliser production through **subsidised prices of natural gas, electricity and coal**, all fixed by the government and made available to the large state-owned chemical companies (PetroVietnam Group (PVN) and the Vietnam National Chemical Group (Vinachem) and their subsidiaries. For example, the subsidised price of natural gas was just 50% of the market price in 2012 (Nguyen Hang T., 2013). Smaller SOEs have mostly been transformed into joint stock companies (equitised).

These firms have also increased their exports, which by 2013 were approximately 1 million tonnes, compared to 2.5 million tonnes of fertiliser imports. Domestic production is primarily urea fertilisers (45% of total consumption), phosphate fertilisers (26%), and NPK (19%). Potash accounts for 9% of total consumption and is imported (Nguyen Hang T., 2013; Ken Research, 2014a).

Domestic producers face stiff competition from cheaper fertiliser **imports from China**. At the end of 2013, the Vietnamese government increased the import duty on selected nitrogen fertilisers from zero to 3%, but China cut its export tax to 2%, thus imports remained competitive. The Viet Nam Fertiliser Association is aware that the industry is based on obsolete technology, produces at high costs and lacks competitiveness, which may undermine its position in the domestic market in the future (Viet Nam News, 2014).

Seeds

The seed market within Viet Nam is **highly fragmented** in terms of firms, with state, private and foreign firms (Nguyen Manh Hai, 2013). As of 2011, there were 240 companies in seed marketing and distribution, 76 crop variety centres (government), and 99 other business units for a total of 415. There were eight multinational companies involved in the seed business in Viet Nam.[18] Among major domestic firms, Vinaseed (NSC) and Southern Seed (SSC) are two of most dominant firms, both SOEs/joint stock companies and the only ones listed on the domestic stock exchange. NSC is the industry leader in rice seed (85% of its sales), with a 12% market share in the North, while SSC has a 10% market share in the South, and a nearly equal split in its sales between corn and rice (Nguyen Tien Duc, 2011).

Seeds are typically sourced from equitised national seed companies, government seed stations, farmer groups, co-operatives, and imports. The seed companies and seed centres then distribute seeds to farmers through private agents (80% of total seed sales), co-operatives and the agricultural extension system, shops in seed stations, and other companies' branches (Nguyen Trung Kien, 2012).

Within the current seed market structure, where domestic SOEs are relatively dominant but where private domestic investment in seed production occurs and imports are common, **seed prices are market determined**. There appears to be little in the way of quantitative import restrictions, although there are lists of approved varieties. Foreign firms can operate by importing and selling seeds, subject to certain required conditions (Nguyen Manh Hai, 2013). Judging by the degree of import penetration in hybrid seed varieties, the border restrictions are minor, or are easily bypassed.

Through its extension services, MARD takes an active role in promoting the use of **improved seeds**. Their activities include demonstration models, dissemination of information via mass media and exhibitions, and training through extension staff. The agricultural research system is also active in seed selection and creation (traditional plant breeding, seed imports, and some biotech applications) through 18 seed research institutes and 6 universities. Although hundreds of new varieties have been tested, registered and introduced, the system still suffers from issues such as small budgets, limited numbers of highly skilled staff and an excessive focus on rice and maize. The result has been low levels of quality control, insufficient capacity for seed testing and certification, poor information sharing, and an inability of the domestic seed industry to meet domestic seed demand. All these components of the seed system could be improved substantially with potentially high social payoffs (Nguyen Mau Dung, 2013).

Machinery

Production methods in Viet Nam's agriculture rely heavily on human and animal labour, with **little mechanisation**. Machinery is used primarily for improvement of yield and quality for rice, sugar, corn and legumes (through soil preparation, threshing, water pumping, and transport), in addition to the development of greenhouse systems (buildings and climate control). Secondly, it is used to reduce losses and ensure quality in post-harvest operations, especially drying.

While the level of mechanisation is still very low, it is growing quickly, partly in response to rising wage rates in agriculture but also due to the increasing demand for food safety and quality. Quantitative evidence is difficult to obtain over the whole period since 1990, but some data exist to show this growth. First, tractor numbers were collected until 2000 (World Bank, *WDI Database*). This is merely a count of tractors, independent of size or horsepower, but they show numbers growing at the rate of 20.6% per year over the 1990-2000 period and the total number had reached 163 000 by 2000. Nguyen Quoc Viet (2011) claims "nearly 500 000 tractors" by 2009, which suggests a 12-13% annual rate of growth in the 2000s.

Since 2000 there are three observations (2001, 2006, 2011) on machinery horsepower used in agricultural field cultivation plus soil cultivated for forestry plants (Nguyen Duc Long, 2013), with data sourced from the General Statistics Office (2012). This shows that machinery horsepower growth continued in the last decade, growing at a slower rate (4.6% per year) from 2001 to 2006, but more than twice as quickly from 2006 to 2011, at 11.2% per year.

This quite rapid growth rate, sustained over at least 20 years, has been further enhanced by various **government programmes**. These include projects to reduce post-harvest losses of agricultural and aquatic products, to encourage horticultural development, and to "make breakthroughs in agricultural modernisation and rural industrialisation". Subsidised credit was potentially a more influential form of support at the farm level (Nguyen Duc Long, 2013). However, as discussed in Chapter 2, actual budgetary expenditures on these programmes were small, thus their impact should not be expected to be large.

Another indirect form of government support for agricultural mechanisation is through the advantages that accrue to the agricultural equipment manufacturing industry. All major farm equipment makers appear to be equitised SOEs, so the access to credit that such enterprises enjoy give them some advantage.[19]

The **demand for agricultural machinery** will continue to grow and its adoption, whether it is water pumps, tractors, or rice threshers, will increase agricultural productivity in Viet Nam. Government efforts to encourage this adoption should be crafted with care. In general, they should be limited to providing support through extension and training support, better road infrastructure, plus *possibly* agricultural credit. At the manufacturing level, agricultural engineering research is also typically very productive. But existing manufacturing firms have shown such disappointing performance that any attempt to protect or subsidise them is likely to fail. Import protection for this sector will only deny farmers cheaper and more productive options, thereby reducing machinery adoption rates. In addition, it will likely slow the modernisation and evolution of an efficient farm equipment manufacturing industry, especially given the current dominance of SOEs. Possible solution may come from unsubsidised joint ventures with foreign equipment makers who can provide technology alongside the Viet Nam SOEs that could possibly provide low cost assembly labour, an activity that has already been proposed by VEAM (VEN, 2014).

Marketing channels for major commodities

Rice

Marketing channels for rice differ somewhat between the Mekong River Delta and the Red River Delta. While in the Mekong River Delta two-thirds of paddy produced at the farm level is marketed, in the Red River Delta it is approximately half, with the rest used on-farm. More than 95% of marketed paddy is sold to **private assemblers**. They operate at relatively small-scale, usually within a perimeter of about 10 km, and sell to medium to large scale millers (Mekong River Delta) or first to millers, then sell the processed rice to wholesalers and retailers (Red River Delta, rest of country). All evidence points to a substantial number of traders within the major growing regions, giving farmers good choice in selecting traders to whom to sell, or millers, or even small retailers and consumers (Mehta et al., 2011; Dao The Anh et al., 2013).

Rice milling involves both private and state-owned mills, but most are small scale operations that serve small rice farmers' own consumption needs on a contract basis. So-called large mills employ 5-10 workers, and are located largely in the Mekong River Delta and the southeast. Export rice is polished at such mills and is found only in the Mekong River Delta. Most commonly, medium and large mills buy rice from assemblers and after milling they sell it to wholesalers and SOEs. Wholesalers act as intermediaries between various traders, millers and SOEs, and can be very large, such as in the Southeast (Mehta et al., 2011; Dao The Anh et al., 2013).

SOEs, including regional and provincial food companies, play a major role in **wholesale and international trade** (Mehta et al., 2011). Even after regulatory restrictions on new entrants to rice exporting were removed in 1997 (removing the SOEs' legal monopoly), their dominance and market power in that activity remains substantial. In 2009, SOEs accounted for more than 80% of rice exports; they also accounted for about 35% of rice processing nationwide (WB 2012). The World Bank assumes that private companies could secure a 10% higher export price for rice than that obtained by the SOEs, due to their "higher efficiency and higher responsiveness to consumer requirements" (WB, 2012).

The privileges of SOEs include access to subsidised loans for paddy procurement, tax advantages and, through the **Viet Nam Food Association**, a major voice in the allocation to specific firms of export quantities, quotas, and export contract allocations, in particular within government to government contracts. In addition, SOEs have the power to refuse export permission to other firms at any time, both making it difficult for private firms to sign overseas contracts and adding uncertainty to all participants. Would-be private rice exporters are mostly so small that they do not qualify for rice exports as specified in Decree 109 because their storage and milling capacities are too small. The Vietnam Food Association has more than 200 exporters registered with it, but the two largest SOEs, Vinafood 1 and Vinafood 2, account for 53% (2007-09) of the value of the country's rice trade. The top ten exporters account for 70% of the total trade. Fully state-owned enterprises in 2008 accounted for 79% of trade, with the private sector and companies with minority government ownership stakes accounting for 19% (Tran Thang et al., 2013).

Coffee

Even though the coffee sector now operates through apparently private firms and channels, with **minimal public support**, the role of the state has a strong legacy. In 2004, SOEs were still important in the coffee processing and irrigation sectors, and handled "a significant percentage of the exports" (WB, 2004).

At present, 95% of the total growing area is in private hands and there are now about 150 registered coffee processors and exporters (Ipsos, 2013). The number of **foreign investors** that specialise in coffee exporting has increased quickly since 2010. "Enterprises with some level of foreign investment now account for 60-65% of the total coffee exported each year" (Ipsos, 2013). The role of foreign firms may diminish somewhat in the future as a result of a June 2012 Ministry of Industry and Trade regulation which banned them from direct purchasing of coffee supplies from farmers (Ipsos, 2013). However, this would only protect domestic firms, and would appear not to stop the gradually increased role of private firms in the coffee processing and export business.

Rubber

The structure of the rubber sector within Viet Nam is heavily oriented toward **state enterprises.** Private sector activity within the sector is mostly limited to smallholder (farm) production and private holdings of shares in joint stock (SOE) companies. The role of state-owned firms is shown by comparing Viet Nam with Indonesia and Malaysia. In Indonesia 85% of the rubber-producing area is held by smallholders, while in Malaysia it is 93%. In Viet Nam as of 2013, the smallholder production area was 49% of total rubber land and 44% was operated by the government sector, notably the Vietnam Rubber Group. The private sector share was 6%. Further, taking into account new plantings, the government share will soon surpass the smallholder share (Ngo Kinh Luan, 2013).

Many of the SOE rubber plantations also operate processing operations, and smallholders typically process their rubber at (or sell it to) the plantation processing plants. The **Vietnam Rubber Group** is the largest SOE in the sector and it controls 40% of the country's total rubber area, 41% of production, and 85% of its export production. It also controls most of the large rubber companies, with more than a 60% stake in those rubber companies that are equitised. The role of foreign ownership appears limited to only investing in the existing companies. In four large companies of the Vietnam Rubber Group, the largest share of foreign ownership is 33% (Viet Capital Securities, 2011).

Supermarkets

There has been a supermarket "revolution" in developing countries around the world since the mid-1990s, as an increasing number of international retailers have moved their operations into these countries, including Asia. It is relevant for the agricultural sector for several reasons. When supermarkets enter the market, they tend to rely on imports as it is initially quite difficult for small farmers to meet standards for product specification, delivery times and food safety. As a result, small farms must either meet these **standards** or receive inferior prices in traditional wet markets. For consumers, supermarkets offer a level of **convenience and food safety** that middle-income consumers find attractive, despite higher prices. This creates a sometimes challenging environment that prepares the primary food sector (farmers, traders, and processors) to upgrade product quality and thus compete more successfully against imports in the domestic market and improve export competitiveness via higher value-added farm and food products (Reardon et al., 2010).

Viet Nam is still in a relatively **early stage** of the process of retail modernisation and the vast majority of food continues to reach consumers through traditional channels such as small shops and wet markets. However, the share of modern retailing has been growing and according to various estimates varied between 4% and 20% in 2013 (GAIN, VM3062, 2013). Geographically, large stores started in Ho Chi Minh City and Hanoi, but as competition among supermarkets grew in those two locations, retailers have moved to smaller cities and towns across the country. Sample analysis indicates that initially 95% of foods were imported, but by 2008 the proportion of domestic food products had grown from that initial 5% to 70%. Dry items were stocked first, followed by packaged and processed food products and non-food products. Fresh foods have been sold more recently but with only a limited range of products and they account for a small percent of supermarket sales (Maruyama and Trung, 2011).

Foreign-owned firms played a major role in this supermarket revolution. Their institutional capacities, including supply chain contacts and logistics, managerial expertise and customer service, are high and gave them a strong advantage over domestic retailers. This led the government to protect the retail market by limiting the number of licenses given to foreign retail firms. Only after Viet Nam's 2009 accession to the WTO did the retail sector become fully open to foreign retailers. However, there are some regulatory limits on foreign retailers (e.g. number of stores) to protect domestic firms. The industry is still very diverse on a national scale with the five largest retailing groups accounting for only a 5% market share (Maruyama and Trung, 2011).

This expansion of supermarkets can be expected to contribute to the **upgrading** of farm production and the whole food supply chain across Viet Nam as it has already done in many neighbouring countries. To respond to this process, MARD has made it a priority to strengthen capacities and incentives for quality and food safety management (MARD, 2012). It has also encouraged co-operatives and other forms of aggregating farm product supplies to allow large scale purchases by the chains, and to impose higher food safety standards upon member farmers. It necessitates an important investment in know-how at the farm level and in processing plants and requires increasing government's involvement in surveillance, inspection, and education, at the farm, processing, and consumer levels.

Agricultural trading enterprises

There are many cases where SOEs, often referred in the trade context as STEs (state trading enterprises), **dominate exports and imports** within the agricultural sector, possibly distorting prices where they have some degree of monopoly or monopsony power. As discussed above, this is the case of rice, rubber and fertilisers where private firms play a minor role in trading. In coffee, private firms play a somewhat greater role in exports, but SOEs remain important. In the cases of cashews, pepper, and cassava, there are little data on the ownership or market shares of individual exporters, but within lists of exporters one can find firms that are joint stock companies (equitised SOEs). Those firms also tend to be very large, often selling a range of agricultural products such as tea, vegetables, and cinnamon (cassia) in addition to the cashews, pepper and cassava.

Private trading firms tend to be small and medium-sized enterprises (SMEs), often engaged also in agro-processing. In cashews, small private firms are entering the export trade but they are niche-oriented, in contrast to the large joint stock companies that are the main buyers and exporters (EDE Consulting for Coffee, 2006). The SMEs' growth is constrained by lack of access to land and loans, necessary permits or licenses, ability to enforce supply contracts, quite apart from the marketing challenges in world markets (FAO, 2012b).

Value-added in Vietnamese agriculture

Viet Nam's agro-food exports are commonly derived from **low-value commodity sales**. This "commodity" approach to exports is long on quantity growth, but short on quality and value added. It is recognised in Viet Nam that moving up the value scale in food markets allows exporters, and usually farmers, to capture higher prices without having to increase production or find more inputs such as scarce land. For this reason, it is one of the main pillars of the Government's Agricultural Restructuring Plan (ARP), which includes the improved quality of basic farm commodities, and food processing into innovative products (Chapter 2).

The problem is particularly striking when for example rice export prices are compared across key exporters (Figure 1.36). Viet Nam's prices are often at the bottom, and this phenomenon has persisted since the country started exporting rice in the late 1980s. If quality can be improved at relatively low cost, then it would help improve farm incomes, given the size of the rice sector. However, efforts to raise rice quality must include a balancing of the costs with the benefits. There may be other commodities, even among only exports, where improved quality can be obtained with a better benefit cost calculation. Jaffee (2011) compared ratios of value added to output in various food processing industries and found that for rice this ratio is the lowest among five commodities listed. It is the highest for natural rubber and for fruit and vegetables. Any investment strategy to increase value-added in Vietnamese agriculture would want to look very closely at these commodities, and others, to see where the **gains in value-added** can be achieved most cost-effectively. High attention is being paid to fruits and vegetables, where production and revenue growth has recently been high.[20]

Figure 1.36. **Rice export prices, 2012-14**

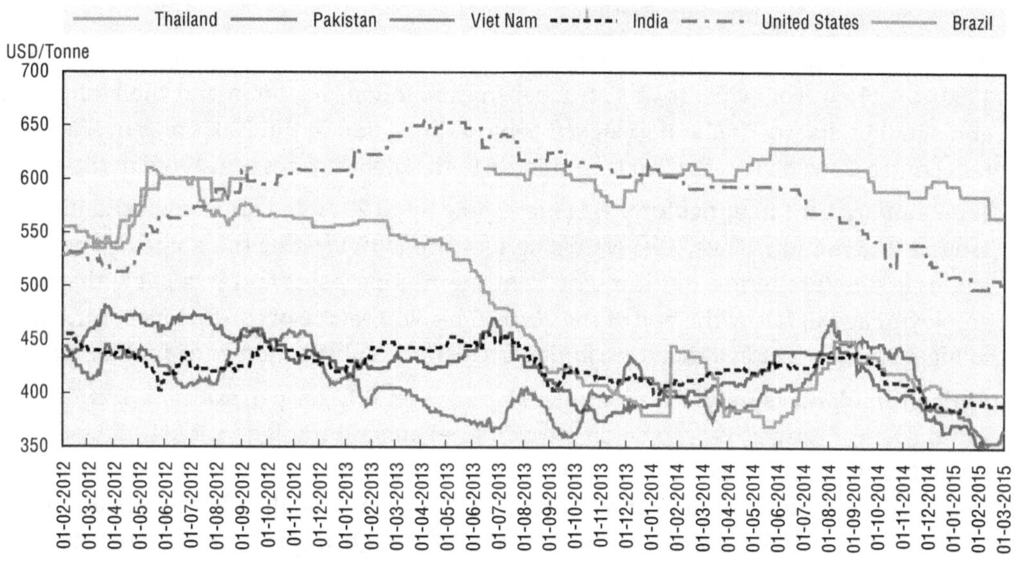

Note: To make the quality comparable, export prices of 5% broken rice was taken into account for Viet Nam, Thailand, India and Brazil and for the United States No 2.4% broken.
Source: International Grain Council, IGC (2015).

StatLink http://dx.doi.org/10.1787/888933223578

1.10. Summary

- Viet Nam's **impressive economic growth** over the last two decades stimulated demand for food and eased absorption of farm labour by other sectors of the economy. Vietnamese economy is moving from being heavily agricultural to a diverse mix of agriculture, services and industry. Still, agriculture accounted for 20% of GDP and 47% of total employment in 2012.

- To sustain high rates of economic growth and to improve agriculture's enabling environment **structural reforms** are needed. In particular, Viet Nam's competitiveness has been undermined by weak institutions (including corruption and weak property rights), low quality infrastructure, underperforming higher education and training, weak financial market development, and slow adoption of the latest technologies by enterprises.
- Economy-wide reforms undertaken in the mid-1980s and sector-based reforms, including the de-collectivisation of farms mandated in 1988 and the land rights issuance in 1993, created conditions for **strong supply response** to growing domestic demand and to commodity price boom in international markets in the 2000s. As a result, agricultural production almost doubled in volume terms between 1990 and 2012, outperforming all its major competitors in Asia.
- By 2011-13, Viet Nam had become the **world's largest exporter** of cashews and black pepper, the second largest exporter of coffee and cassava, the third largest exporter of rice and fisheries and the fifth largest exporter of rubber. The total value of agro-food exports increased six-fold between 2000 and 2012. This contributed to the positive balance of agro-food trade at almost USD 12 billion in 2012.
- There has been a shift away from staple foods to perennial crops such as coffee and rubber and to livestock production, in particular pigmeat. This reflects **export orientation** of perennial crops and changing preferences of consumers to higher value products. However, rice remains the most important commodity accounting for 36% of total value of Viet Nam's agricultural production.
- **Agricultural total factor productivity** (TFP) growth has been strong and sustained in 1990-2010. However, while it was stronger than in Indonesia, India and the Philippines and equal to that in Thailand, it lagged behind China and in the 2000s also in Malaysia, reflecting a slowdown in the 2000s compared to the highest rates registered in the 1990s.
- Total **employment in agriculture** was increasing up to 2009 and since then stabilised at around 24.5 million. Thus, Viet Nam's agriculture is not yet at the stage of shedding labour in absolute terms, but it might be at the turning point and it is likely that farm employment will fall by the end of the 2010s. This will be one of the factors contributing to higher labour cost in agriculture and to accelerated substitution of capital for labour.
- **Agricultural land is very scarce** in Viet Nam at just 0.12 ha per capita, a sixth of the world average. Largely due to deforestation, total agricultural land increased by 61% in 1990-2012. Arable land expanded in the 1990s, but since then remained relatively stable. This might indicate that almost all accessible arable land is already in cultivation and further production growth can be achieved through higher yields, already high compared to Viet Nam's Asian peers.
- While capital inputs remain relatively small **fertiliser use per hectare increased** by 80% in 2010-12 compared to 1990-92 and is currently almost twice as high as the average in Southeast Asia. Annual data suggest that application rates stabilised in recent years which might indicate that arable land is becoming saturated with fertiliser applications and that, on average, higher rates are unlikely to generate substantially higher yields.
- **Real rural incomes have been increasing**, but are just half of those of urban residents. This is also the main reason for almost four times higher poverty rates in rural than in

urban areas. However, both rates (defined nationally) have been falling and are currently (2012) at 15% for rural and 4% for urban residents.

- Viet Nam has made astounding **progress in fighting undernourishment**. The proportion of undernourished in total population has fallen from 45.6% in 1990-92 to 12.9% in 2012-14. This is one of most rapid falls for all countries, just after Thailand and faster than in China. However, still 11.9 million people in Viet Nam were undernourished in 2012-14. Most of them live in rural areas.
- Rapid economic growth, combined with rising population and expanding agricultural production, is exerting **massive pressures on the environment**. Expansion of agricultural land has led to considerable deforestation which is only partly remedied by reforestation efforts undertaken over the last 15 years. While the overall forested area increased, primary forests continue to disappear. In the long term, **climate change** is likely to have strong negative impact on the Vietnamese agriculture.
- Farming is dominated by **tiny farms** of about 0.8 ha on average, each consisting of about 4 plots. Very few farms have grown to the "large scale" category of 2 ha or above. Farm consolidation in livestock production has been progressing, in particular in pigmeat and poultry production.
- All **land is owned and administered by the state**. The Land Law of 1993 gave farmers a wide range of rights, including the right to rent, buy, sell, and bequeath land and to use land as collateral with financial institutions for mortgages. Holders of these rights are entitled to Land Use Rights Certificates. By 2012, rights to 85% of agricultural land had been certified, but land disputes are frequent due to expropriations and low compensations for lost access to land.
- While Viet Nam has been successful in boosting exports of agricultural commodities, it suffers from **low quality, low prices and low value-added component** in its exports. This is partly due to the strong involvement of SOEs in the processing and exporting.
- Prices of many commodities exported by Viet Nam declined over the last 2-3 years and are projected to fall further over the next decade. Most of easy sources for expansion (arable land, cheap labour, higher rates of fertiliser use) are exploited and negative environmental impacts are increasingly seen. These will become major **challenges** for the next decade, but will also open **opportunities** to adopt new technologies, to give incentives for larger farms and to focus attention on quality and higher value added products.

Notes

1. According to the Government Decree No. 72/2001 dated 5 October 2001 on the classification of urban centres, an urban commune must meet three criteria at the same time: the non-farm labour force accounts for more than 65% of the total labour force; the commune's total population is more than 4 000; and the density is higher than 2 000 people/km^2. Rural communes are those which do not meet those criteria (Cling J.P. et al., 2010). Accordingly, urban residents are those who live in areas defined by the state as urban and rural residents are all remaining ones (GSO, 2014).
2. In Viet Nam the definition of agriculture includes not only crop and livestock production, but also forestry and fisheries. This national definition is reproduced by most international databases, including by the World Bank's World Development Indicators quoted in this section.
3. For example, using long term estimates from Darvas (2012), the real effective exchange rate appreciated from 1995 to 1998 by about 25%, then declined to early 2004 by another 25%, and has appreciated since then to 2012 by one-third.

4. Fertiliser imports fluctuated from 2.9-4.5 million tonnes in 2005-12 (MARD, 2013), but then fell to about 2 million in 2013, partly due to urea supplies coming from two new fertiliser plants put in operation in 2012. Their joint capacity at 1.36 million tonnes doubled the total capacity of the industry increasing prospects for urea exports from Viet Nam (Nguyen Hang T., 2013).
5. The reason that corn seed demand is almost totally met by the formal sector is that hybrid corn dominates farm plantings. In rice, about half the seeds used (46%) are certified (from both the formal system and farm-saved sources) and half are non-certified, although this varies considerably by region. In the Mekong River Delta, two-thirds of seeds used are certified, in the Red River Delta three-fourths are certified, whereas in the Central Highlands and mountainous regions only 20% are certified (Nguyen Trung Kien, 2012).
6. During the 2008-09 economic slowdown, many workers, in particular from ethnic minority groups and women, returned to agriculture, which is a common phenomenon in times of crisis and loss of jobs (OECD, 2014).
7. Very high level of farm labour productivity in Malaysia is driven by two factors: the domination of high value products (oil palm, rubber, livestock) in the total value of agricultural production and a low level of farm employment accounting for just 12% of total employment in 2012 (Figure 1.3).
8. In Viet Nam, there are two main national approaches to measuring poverty. A consumption-based approach, developed in co-operation with the World Bank, has been used by the GSO to examine poverty changes over the longer term. A separate income-based approach is used by the Ministry of Labour, Invalids and Social Affairs (MOLISA) to create a classification applied for determining anti-poverty programme eligibility and for poverty monitoring over the short term. In this section a modified MOLISA line is used as developed and a published systematically by the GSO. It applies a cost-of-basic-needs methodology, making it similar to the first one. More specifically, the poverty line is based on expenditure for a reference food basket and basic non-food allowance, using a caloric norm of 2 100 kcals per person per day. This line is calculated based on income obtained from regular Household Living Standards Survey and adjusted for inflation year-to-year, thus allowing for comparisons over time (Demombynes and Linh Hoang Vu, 2015).
9. Indeed, food and foodstuff prices in January 2012 were by 140.2% higher than in January 2007 compared to the overall increase in the Consumer Price Index by 89.8% over the same period (GSO Consumer Price Indexes, 2015).
10. The World Bank estimates that rice consumption in Viet Nam could have been at 135 kg/capita in 2010 if out of house consumption of rice is included. However, the declining trend in per capita rice consumption is evident and faster than Viet Nam's population growth rate, thus the absolute consumption of rice in Viet Nam has begun to fall, but very slowly (WB, 2012 and personal communication with Steven Jaffee, March 2015).
11. Even before the signing of the ASEAN-China preferential trade agreement in 2010, Viet Nam's trade with China had been stimulated by liberalised border-trade arrangements. In particular, Guanxi, a Chinese province along the Vietnamese border, benefited from free trade zone regulations that allowed border citizens to trade goods under a limited amount per day (Arita and Dyck, 2014).
12. One explanation for the low level of registration in some mountainous provinces is that the plots are situated in areas with challenging topography and steep slopes, which makes the measurement of plots problematic (CIEM, 2013).
13. Effects of land use restrictions are not small in effect. It is estimated (Vasavakul 2006) that the land restricted to produce rice, as a percentage of agricultural cropland, is 35% for Viet Nam as a whole, but 75% for the Mekong River Delta region and 68% for the Red River Delta. For other regions, the percentage of crop land affected is 40% in the North Central Coast, 23% in the South Central Coast, 18% in the Northeast and Northwest (mountainous regions), 10% in the southeast, and 5% in the Central Highlands. Furthermore, the income losses to farmers from this restriction are very large on a per hectare basis. As estimated by the World Bank, farmers' income could have been by 123% higher on average if they were not subject to the rice land designation policy, including by 120% in Mekong River Delta and by 181% in Red River Delta, the largest two rice producing regions (World Bank, 2012). Giesecke et al. (2013) estimate using a dynamic CGE model that the removal of this policy would not only increase real private consumption by an average of 0.35% per year over 2011-30, but would also reduce poverty, improve food security and contribute to more nutritionally balanced diets among households.
14. This compares with 2 536 enterprises operating in a broadly defined agriculture, it means including forestry and fisheries. In total they operated on 2.9 million ha in 2011, 76% of it was forest land (GSO, 2012).

15. Both farm size limits are very high, especially for a developing country. For example in the USA, the threshold for defining a farm is at USD 1 000 of annual sales, making the Vietnamese farm size limits prior to 2011 high, and after 2011, very high.

16. This is due to growing rural population in absolute terms and continued subdivision of already small farms (GSO, 2012).

17. The actual size could be somewhat larger due to quite active subleasing market among rice producers, in particular in the Mekong River Delta.

18. They include: Syngenta (crop seeds, and the largest foreign maize seed supplier), Bioseed Research (hybrid maize), CP Seed Company (hybrid maize), Ease West Seed (VN) Company, and representative offices for Monsanto Thailand (maize), Siminis Vegetable Seeds, Nong Huu Seed Company and Bayer.

19. However, the companies themselves complain that they do not have sufficient government financial support to compete against the large number of farm machinery imports that come largely from China. The Vietnam Engine and Agricultural Machinery Corporation (VEAM) estimates that there are 550 000 agricultural machines currently in use, mostly rice-harvesters, of which "more than 60%" are from China (Vietnam News, 2012). The company claims that it needs government loans that have not been forthcoming, in order to continue developing machines that are in demand in Viet Nam. Farm equipment sellers say that farmers prefer imported machines because domestic machinery is priced 15-30% higher than imports and is of lower technology, quality or reliability.

20. Much progress has been made in Viet Nam in some sectors. Drawing on Census data from 2001 and 2006, there has been a shift in the Mekong Delta into fruit production when land use regulations allow it, and from fruit into shrimp production. Similarly in the southern parts of the Central Highlands, there has been a shift from coffee into fruit, likely to serve growing demand in Ho Chi Minh City (IFAD-IFPRI, 2011).

ANNEX 1.A1

Viet Nam: Production and trade performance for major agricultural commodities

This annex examines 13 farm commodities, chosen to cover all products where the value of production is at least 1% of the country's total agricultural output. These are rice, coffee, rubber, cashews, cassava, black pepper, tea, pigmeat, cattle, eggs, poultry meat, maize, and sugar cane. With the exception of cassava, they are also covered by the OECD Producer Support Estimates discussed in detail in Chapter 2. The first eight are all net export commodities (arranged from highest to lowest value of exports), and the latter five are net import items (arranged from highest to lowest value of imports). For each, production evolution (area cultivated, production, yield, and geographical location), price changes and basic trade flows are reviewed. Trade flows are represented by export and import values, but it should be borne in mind that some of this value growth is due simply to rising world prices, especially in the period from 2007 when world prices at least doubled over the following two-three year period. As at the moment of writing data for 2013 are still preliminary or not yet included in international databases, analysis is provided up to 2011 or 2012 with only occasional references to the most recent developments.

Rice

Rice in Viet Nam is an important food staple but is also a strikingly successful export commodity. Viet Nam's rice production (dried paddy equivalent) increased from 19.2 million tonnes in 1990 to 43.7 million in 2012, which is a compound annual growth rate of 3.8% over 22 years. This growth has been remarkably steady, only slowing slightly during the second of these two decades. It has been achieved with modest changes in harvested area. In 1990 Viet Nam's area under rice cultivation was 6.0 million ha while in 2012 this number had grown to 7.7 million ha, but the area increase occurred until 1999. Since then rice area harvested has been basically flat, and the sustained production growth has come from yield increases (FAOSTAT, 2015).

Historically, the two most important rice-producing regions were the Mekong Delta in the south and the Red River Delta in the north. This is still true for rice production, but the land area planted to rice has changed. First, the Mekong Delta has become even more dominant by expanding its planted area as well increasing its yield so that it now accounts for 54% of the total land area planted to rice in Viet Nam, and 56% of its rice production in 2012. Second, although the Red River Delta is still the second largest region for production (16% of total), it is almost equalled by the North Central and Central Coastal region (15%). These three regions account for 87% of all rice production (85% of rice land) in the country (GSO, 2014).

Rice (paddy) yield was growing at 2.5% per year in 1990-2013, and by 2013 it had reached 5.6 tonnes/ha. Such a yield is among the highest in Asia, although still less than China's average yield of 6.7 tonnes/ha. For example, Indonesia's average yield was 5.2 tonnes/ha and Bangladesh's was 4.4 tonnes/ha in 2013 (FAOSTAT, 2015). This steadily increasing trend has to a large extent been facilitated by the introduction of improved, high yielding rice varieties (Section 1.9).

Growth in rice prices has been one of the elements that has contributed to Viet Nam's dramatic expansion of rice production. Producer prices increased from USD 123/tonne in 2000 to USD 316/tonne in 2011 (Figure 1.A1.1). The average compound growth in prices was therefore 8.9% per year in nominal terms. The real annual growth rate was lower by the average rate of (US) inflation over that period (2.1% per year), and so equal to 6.7% per year in real terms. Even if the Vietnamese dong has been appreciating somewhat (slightly more than 1% per year) over the period, this is still a strong inducement to expand output. The serious growth in rice prices has come during the commodity price boom in 2007-11. However, in 2012-13 producer prices fell by about 20% to reach USD 261/tonne in 2013 (Figure 1.A1.1), reflecting a fall of rice prices on international markets. Looking forward, rice prices are projected to slide further down over the next decade. This reflects the large supplies accumulated in the early 2010s, in particular by major rice exporting countries in Asia. It is projected that it will take a few years to offload the market and will likely weight on international markets until 2015. After this fall, the nominal world price for rice is projected to recover but to continue to fall in real terms (OECD-FAO, 2014). Even if the transmission of world prices to the domestic market is sometimes delayed, it can be expected that producer prices in Viet Nam will by and large follow the world market trend.

Figure 1.A1.1. **Rice: Producer prices and trade flows, 2000-13**

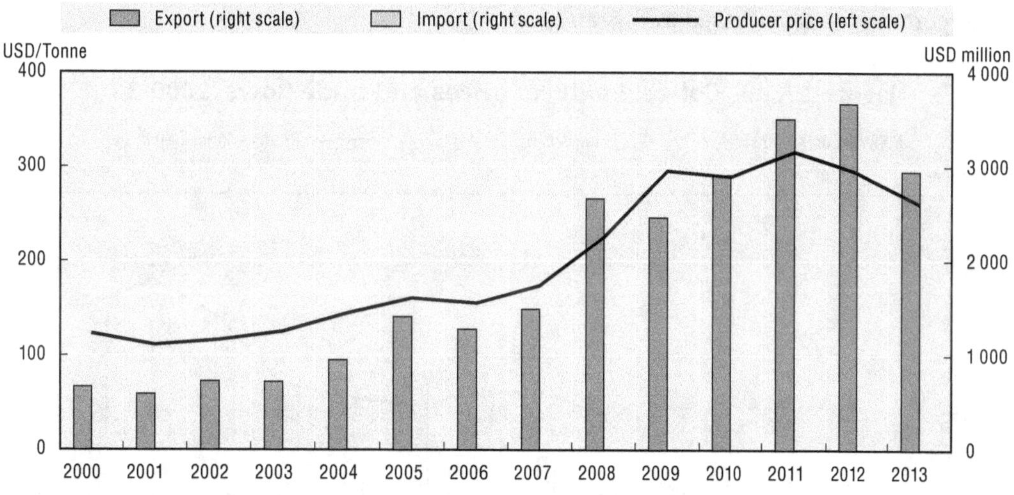

Source: OECD (2014), "Producer and Consumer Support Estimates", *OECD Agriculture Statistics Database.*

StatLink http://dx.doi.org/10.1787/888933223581

Rice is the largest agricultural export and has grown very quickly. Export quantities rose from 1.6 million tonnes in 1990 to 8.0 million in 2012, representing an annual growth rate of 7.5%, to be compared to the growth in total production of 3.8% per year. The value of rice exports rose from USD 0.3 billion in 1990 to USD 3.7 billion in 2012, which is a 12% growth rate in nominal USD terms (about 10% in real terms). This rate of growth may appear less impressive than the rates for some other commodities discussed in the subsequent sections, but what makes rice exports so impressive is that the starting point was already above one million tonnes, not some tiny initial volumes as for most other commodities.

The top buyers of Vietnamese rice exports vary from year to year, depending on production volatility, and which country has had a bad or good production year. In 2011, for example, Indonesia was by far the largest buyer. But over the past ten years, the top five buyers were the Philippines, Indonesia, Malaysia, China, and Cote D'Ivoire.

Coffee

Viet Nam is the biggest producer of Robusta coffee in the world. Its production growth since 1990 has been impressive: from just 92 000 tonnes in 1990 to 1.29 million tonnes in 2012, an annual growth rate of 12.8% (FAOSTAT, 2015). This increased production has come mostly from growth in the area planted to coffee which has increased by 10.7% per year, while yields have only grown at 2% per year. Most of the area growth took place up to 2000, with much slower (1.6%/year) growth since then. Most coffee in Viet Nam is grown in the Central Highlands. In fact, four provinces of the five Central Highlands provinces account for three-fourth of the country's total land planted to coffee. With such increases in production and with more than 95% of production exported, Viet Nam is now the world's second largest coffee exporter, after Brazil.

As in the case of rice, part of the striking growth in coffee production is due to growing producer prices. However, unlike rice, producer coffee prices reflect a coffee price cycle, for example they halved from 2000 to 2002, fell again in the late 2000s, and again in 2013 (Figure 1.A1.2). Despite these fluctuations and even from the relative high point of 2000, an average annual growth rate of 12.0% can be observed in 2000-12. In real terms this is 9.8% per year, 46% more than the real growth in rice prices and very high by comparison with other commodities discussed in this annex.

Figure 1.A1.2. **Coffee: Producer prices and trade flows, 2000-13**

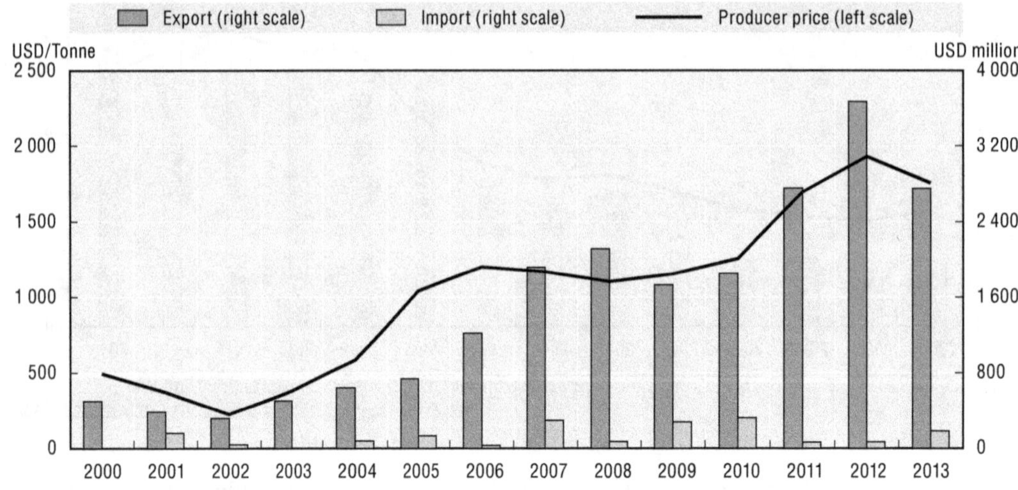

Source: OECD (2014), "Producer and Consumer Support Estimates", *OECD Agriculture Statistics Database.*

StatLink http://dx.doi.org/10.1787/888933223594

Coffee exports are one of five commodities where the growth in export values and quantities was in the double-digit range. This has made Viet Nam the second largest exporter of coffee, with a one-seventh world market share, after Brazil. In 1990-2011, the quantity of exports rose by 12% per year, almost the same as the 13% growth rate in production.

Unlike rice, Viet Nam's coffee exports go mostly to high-income OECD countries. The largest two buyers for the past decade have consistently been the United States and Germany, which in 2011 imported USD 506 million and USD 457 million, respectively, of Vietnamese coffee. The third and fourth largest buyers were also in Europe: Italy (USD 215 million) and Spain (USD 214 million). The fifth largest buyer was Japan (USD 134 million) (UN, *UN Comtrade Database*, 2014).

Rubber

Viet Nam has become the world's fifth biggest natural rubber producer, after Thailand, Indonesia, Nigeria and Malaysia. Vietnamese rubber production has increased from 58 000 tonnes in 1990 to 864 000 tonnes in 2012, a compound annual growth rate of 13%. This growth can be decomposed into 8.7% annual growth in area planted to rubber, and a 4.0% growth rate in rubber yields (FAOSTAT, 2015).

Rubber production is concentrated in four southern provinces of Binh Phuoc, Binh Duong, Tay Ninh and Dong Nai, together accounting for almost 60% of Viet Nam's rubber planted area. These four provinces lie sandwiched between the Central Highland coffee growing provinces to the northeast, Cambodia to the north and west, and Ho Chi Minh City to the south. Viet Nam's biggest rubber exporting enterprise, Vietnam Rubber Group, is expanding plantings as well in Laos and Cambodia (one-fourth of its 2013 plantings).

As with rice and coffee, producer price growth has played an important role in production expansion. Over the period from 2000 to 2011, producer rubber prices have grown at 12.8% per year in nominal terms, 10.4% in real terms. However, unlike the cases of rice and coffee, producer prices grew at a relatively moderate rate up to 2010 (7% real), followed by a dramatic price spike in 2011, then falling rapidly to 2009 levels in 2013 (Figure 1.A1.3).

Figure 1.A1.3. **Rubber: Producer prices and trade flows, 2000-13**

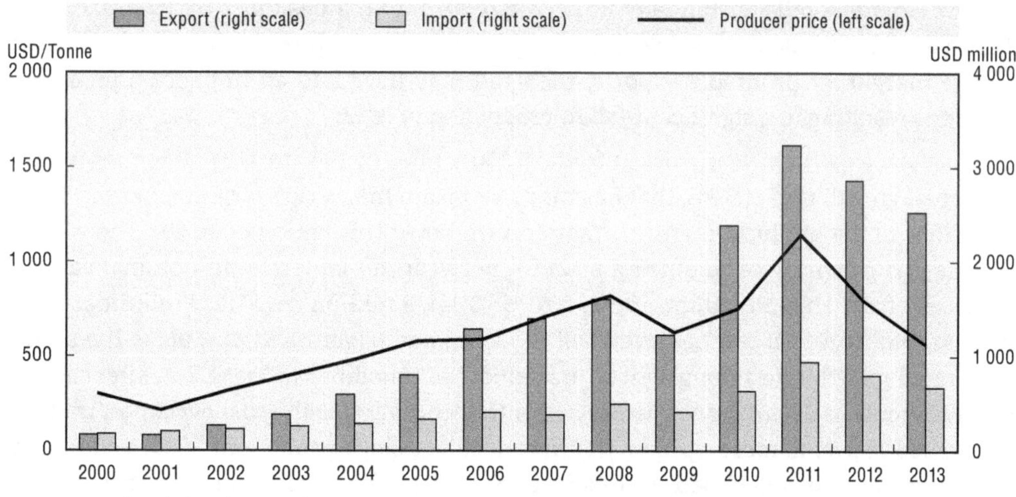

Source: OECD (2014), "Producer and Consumer Support Estimates", *OECD Agriculture Statistics Database.*

StatLink http://dx.doi.org/10.1787/888933223607

Rubber shows a similar pattern of very rapid export growth as for coffee. In terms of export quantities, rubber exports grew at 12% per year, which compares closely to the annual growth rate of overall rubber production (13%) in 1990-2011. The path of export values is shown in Figure 1.A1.3, starting at USD 166 million in 2000, and rising to USD 3.23 billion in 2011, partly driven by a price spike on world markets in 2010 and 2011. The fall in export values in 2012-13 largely reflects a fall in prices. Among the buyers of Viet Nam's rubber, China is easily the largest customer accounting for about 40% of all Viet Nam rubber exports. In 2011, China's imports amounted to USD 1 053 million, followed by Malaysia (USD 261 million), Chinese Taipei (USD 134 million) and Germany (USD 117 million) (UN, UN Comtrade database, 2014).

Cashew nuts

Cashew production has grown from 140 000 tonnes in 1990 to 1.2 million tonnes in 2012, with the most dramatic growth occurring between 1999 and 2007. The total area under cultivation of cashews in Viet Nam increased from 140 000 ha in 1990 to 306 000 ha in 2011. After a significant dip in area harvested in 1992 (by two-thirds), land area in cashews has grown steadily to 2009. It has since declined by about 10% to 2011. The average growth rate for cashew production over the total period is 10.2% per year, composed of a) land harvested growing at 3.6% per year, and b) yield growth almost double that at 6.4% per year, although subject to some quite large swings from year to year. Viet Nam is easily the largest cashew nut producer in the world, accounting for about 30% of world production. By comparison, Nigeria's share of world production is 20% and India's 16%.

Within Viet Nam, production of cashews occurs mainly in the south, including the southern part of the Central Highlands. The six provinces that are the largest sources of Vietnamese production are Binh Phuoc, Dak Nong, Dak Lak, Dong Nai, Binh Thuan, and Binh Duong.

Producer prices for cashews have moved up fairly steadily over the 2000-11 period but with a significant decline from 2000 to 2002 and a sharp rise in 2011. The nominal price growth is 4.0% per year, but in real terms the growth rate drops to 1.8% per year, much slower than for rubber, coffee and rice. However, world market prices faced by Vietnamese cashew exporters fell substantially in 2012 and 2013 from a peak in 2011 (OECD, PSE/CSE database, 2014). While they were still well above the domestic producer prices, thus leaving a large margin for profitable exports, their fall may have had an impact on producers' decisions resulting in a significant fall in production in 2013.

Cashew exports have increased from 25 000 tonnes in 1990 to 179 000 tonnes in 2011, an annual growth rate of 9.9%, that has made Viet Nam the world's largest exporter. This is very close to the production growth that occurred over this period of 10.2%. The value of exports has grown twice as quickly, at 21.9% per year in real terms. In nominal values it increased from USD 15 million in 1990 to USD 1 473 million in 2011. Preliminary data suggest that in 2013 it reached even USD 1 630 million (Figure 1.A1.4). This is the second most rapid growth rate among agricultural export commodities in Viet Nam, after cassava. However, much is occurring within the sector that complicates the cashew export trade. As mentioned above, domestic production has declined since peaking in 2009, and planted area to cashews has declined by 10% in 2009-13. This has even prompted government attention to try to reserve some of the land base for cashews. What has helped the cashew exporters to maintain their volume in recent years is the increasing import of raw cashews from abroad. In other words, the cashew export sector is increasingly becoming a processing specialty sourcing raw nuts from domestic and foreign suppliers.

Figure 1.A1.4. **Cashew nuts: Producer prices and trade flows, 2000-13**

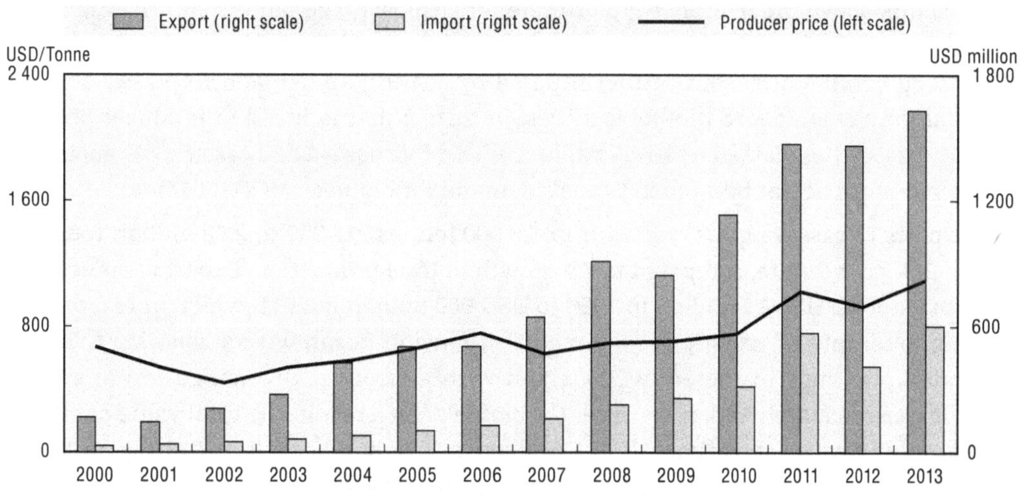

Source: OECD (2014), "Producer and Consumer Support Estimates", *OECD Agriculture Statistics Database.*

StatLink http://dx.doi.org/10.1787/888933223613

Vietnamese cashews exports are directed mainly to the United States. In 2011, about 30% (USD 406 million) of the country's cashew export revenue came from sales to the United States. Half this amount was exported to the Netherlands (USD 198 million) followed by exports to China (USD 111 million) and to Australia (USD 97 million).

Cassava

Cassava is a major crop that is now grown on 551 000 ha of land, almost as much as for coffee and 10% more than for rubber. Production of cassava has shown strong but not spectacular growth, rising from 2.3 million tonnes in 1990 to 9.7 million tonnes in 2012. This constitutes annual growth of 6.8% per year, 3 points higher than for rice but about half the growth rate for coffee and rubber production. When decomposed into increases in land area harvested and yield growth, these two sources are roughly equal, 3.5% annual growth in land area and 3.2% increases in yield. Total production stayed roughly constant from 1990 to 2000, then it grew almost five-fold to 2008, after which production has increased by only 5% to 2012. Land area devoted to cassava has followed this pattern closely, as has yield growth that was roughly flat to 2000, followed by rapid then attenuating growth to 2012. About 90% of this production is consumed domestically in the form of food, animal feed, cassava powder, and more recently and increasingly, bio-ethanol. There are now 6 biofuel facilities processing cassava, absorbing almost 40% of cassava production in 2011. The balance is exported (AgroInfo, 2012). However, taking into account that vast majority of bio-ethanol is exported, almost exclusively to China, the dependence of the industry on exports is much stronger and could be close to half of production in certain years.

Production of cassava is widespread throughout the country, often in remote areas with poor transport conditions. But the six major producing provinces, in order of their 2012 production, are Binh Duong, Gia Lai, Kon Tum, Binh Thuan, Tay Ninh, and Dak Lak, mostly in the Central Highlights and in the South East. Together they account for 45% of total country-wide cassava production.

Prices to producers for cassava have increased from USD 49/tonne in 2000 to USD 125/tonne in 2011. This represents an annual growth rate of 8.9% in nominal terms. Removing inflation the real rate of price growth was 6.6% per year, the same as was observed for rice, and 4 points above the real price growth rate for cashews. The pattern of prices has been steadily upward up to 2011, with the exception of the spike in 2008, as was observed for many food commodities, but further inflated by soaring up demand for cassava as raw material for newly opened biofuel facilities. In turn, a dramatic fall in producer prices in 2012-13 is largely explained by a 20% fall in demand for cassava and cassava products from China, but in particular by a fall in China's demand for biofuels (MOIT, 2014).

Exports of cassava grew tenfold from 28 000 tonnes in 1990 to 2.68 million tonnes in 2011, a 24% growth rate, compared to 7% growth in total production. Export revenues grew even faster, from USD 15 million in 1990 to USD 960 million in 2011, which in real terms is a 30% annual rate of growth. A real export expansion began with a brief lag following expanded plantings in the early 2000s and with a stronger diversification of exports towards cassava starch and, more recently, biofuels. As a result, the total value of cassava exports could have reached almost USD 1.4 billion in 2012, but then fell in 2013 (Figure 1.A1.5). This made Viet Nam the second largest exporter of cassava on the world market by 2012, after Thailand. About 80-90% of all cassava exports go to China. Only two other countries, Korea and Australia, consistently buy Vietnamese cassava.

Figure 1.A1.5. **Cassava: Producer prices and trade flows, 2000-13**

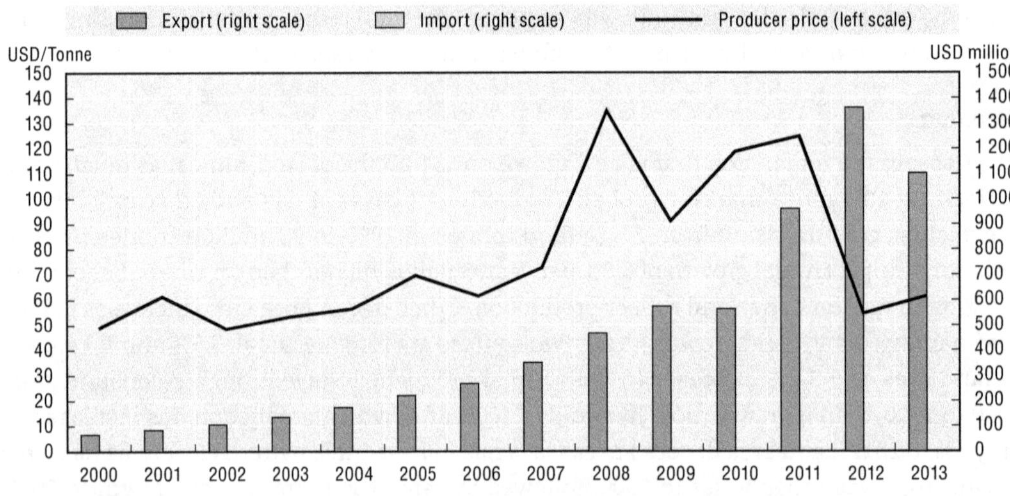

Source: OECD (2014), "Producer and Consumer Support Estimates", *OECD Agriculture Statistics Database.*

StatLink http://dx.doi.org/10.1787/888933223627

In addition to strong price fluctuations, there are a variety of other challenges facing the industry. Yields have started levelling off for reasons of too many farms using older technologies and seeds, and extensive production methods. The wider use of better seeds, greater application of fertiliser, better informed cassava farmers, and vertical integration from farm to processor would improve farm productivity, and the quality of cassava produced.

Black pepper

Pepper is another production and export success story for Viet Nam where production increased almost 14 times over the two decades from 1990 to 2012. This represents an annual growth rate of 12.6%, which is the third fastest production growth rate for agricultural commodities in Viet Nam over this period, after rubber (13.1%) and coffee (12.8%).

Vietnamese pepper is mostly black pepper, not the more valuable white pepper. Production did not start growing significantly until 1997, but since then it has increased in almost all years with a steep trajectory. The harvested area in pepper has followed a similar pattern, and with the exception of the year 2000, it has increased every year since 1994. The average rate of growth per year is 7.7%. Even so, the current land area in pepper production is only 47 000 ha. Yields have grown more slowly than land area planted (4.5% per year), which explains just more than one-third of total production growth, except for spikes in yields that occurred in 1997 and the 2000-03 period. Viet Nam's two biggest pepper-producing provinces are Dong Nai and Ba Ria-Vung Tau which are in the southeast region of the country, just east of Ho Chi Minh City.

Producer pepper prices followed very closely those on international markets. They have grown over the 2001-11 decade, but the first half of the decade featured falling nominal prices. Only after 2006 and the commodity price boom did pepper prices start rising in a quite sustained fashion (Figure 1.A1.6). Over this decade, prices rose on average at 5.1% nominal, or 2.9% real terms. Mid-2014 domestic prices have been pushed to new highs, USD 5 700/tonne, and are forecast, at least in the two main producing provinces to soon exceed USD 6 000. This would appear to be approaching a high point in the pepper price cycle, but the trade forecasts a continued strong international market for pepper.

Figure 1.A1.6. **Black pepper: Producer prices and trade flows, 2000-13**

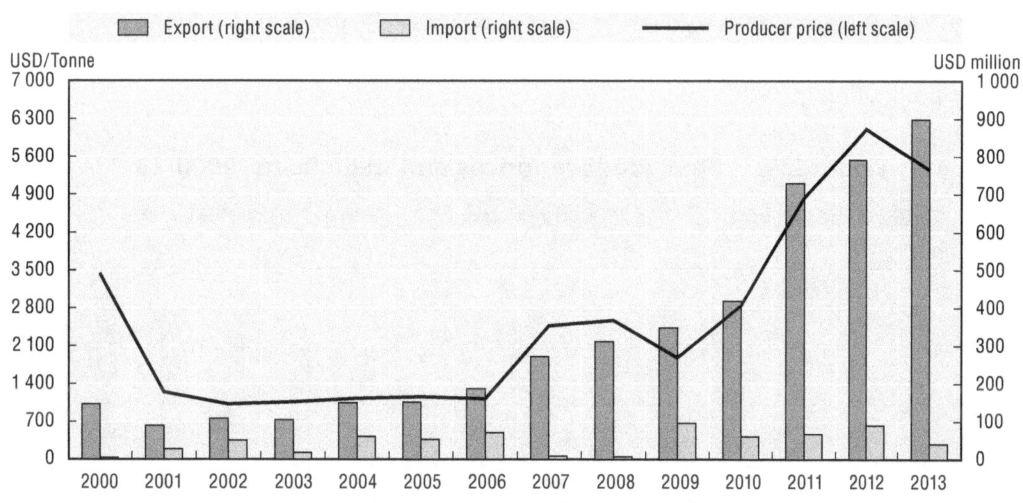

Source: OECD (2014), "Producer and Consumer Support Estimates", *OECD Agriculture Statistics Database.*
StatLink http://dx.doi.org/10.1787/888933223637

Pepper exports rose from 9 000 tonnes in 1990 to 124 000 tonnes in 2011, a 13% rate of growth (for both export quantities and production). Those exports in value terms increased from USD 13.9 million to USD 732 million over this period, which in real terms is an 18% growth rate. This is the third fastest export revenue growth of all Vietnamese agricultural commodities and has placed Viet Nam as the top exporter of pepper in the world (above Indonesia and Brazil). The export value path steepens in the last five years, particularly from 2009 to 2011. Recent data suggests that the value growth for these exports had continued to climb quickly, and that pepper exports are likely to achieve USD 1 billion/year by the end of 2014. While Viet Nam's exports concentrate on lower-priced black pepper, it increasingly produces and exports white pepper and powder pepper.

For 2011, the main importers of Vietnamese pepper were, first, the United States (USD 127 million), Germany second (USD 74 million), and then the Netherlands (USD 49 million).

Tea

Viet Nam is the fifth largest producer of tea in the world. Total production of tea has grown rapidly, from 32 000 tonnes in 1990 to 217 000 tonnes in 2012, which represents an annual growth rate of 9.1%.

The land area harvested has shown the same kind of steady increases, at an annual rate of 4.5% per year. Tea plantings occupy currently 116 000 ha. Yields are somewhat more variable from year to year, as is normal with varying environmental conditions annually, but increasing in almost all years over the past two decades at an average growth rate of 4.4%/year and growing somewhat more quickly since 1998 in a near-linear fashion. Lam Dong province in the south accounts for about one-fourth of the tea-cultivation area and the country's output, but other important tea-growing provinces are Nghe An (north-central region), and the six northernmost provinces, especially Ha Giang and Yen Bai. In sum, this industry is broadly based in a regional sense, and shows steady increases in average productivity.

Producer tea prices fell by one-third in 2000-03, but then were gradually and mostly steadily increasing up to 2012, largely in line with the trend in the world tea prices (Figure 1.A1.7). The nominal growth in this period was 4.5% per year, while the real growth rate was 2% per year. This relatively modest, but sustained, financial incentive motived steady production growth.

Figure 1.A1.7. **Tea: Producer prices and trade flows, 2000-13**

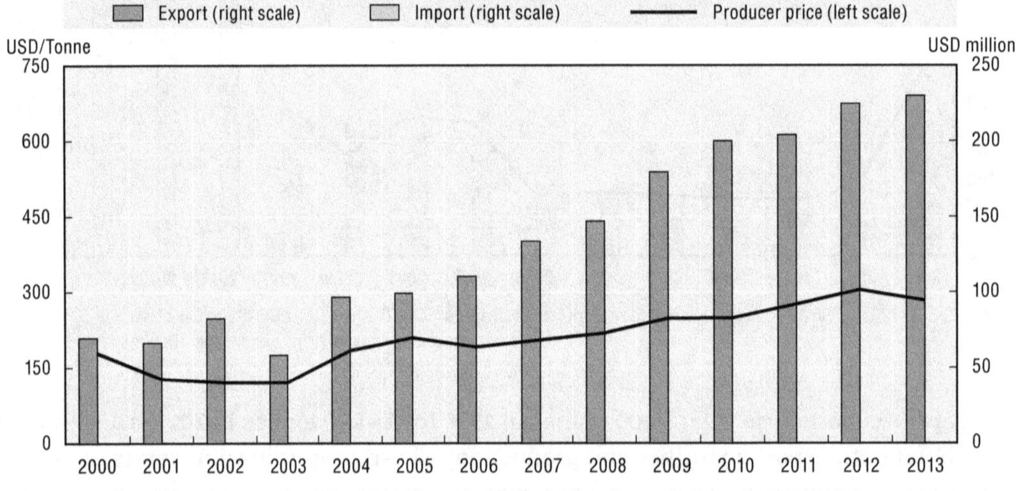

Source: OECD (2014), "Producer and Consumer Support Estimates", *OECD Agriculture Statistics Database*.
StatLink http://dx.doi.org/10.1787/888933223647

Tea is the seventh largest agricultural export in value terms, rising from USD 25 million in 1990 to USD 204 million in 2011, a real increase of 8% per year. Export quantities grew from 16 000 tonnes to 134 000, an annual growth rate of almost 11%. The pattern of increase is quite steady for both quantities and export value. Major destinations of these exports for 2011 were, in order, Pakistan, Chinese Taipei and Russia.

Pigmeat

Pigmeat production is the second largest sector in Vietnamese agriculture accounting for about 15% of the total, but it is almost all produced for the domestic market. Its dominant size is indicated by the fact that domestic pork represents about three-fourth of total livestock production in Viet Nam. Production has grown steadily since 1990 to 3.16 million tonnes in 2012 (FAOSTAT, 2015). The annual rate of growth in total meat production is 6.9% over the whole period, but is slowing over the last three years from 2009.

The national pig herd is distributed widely, but is found more heavily in the northern regions with around 70% of the national hog inventory found in the Northern Midlands and Mountain areas, the Red River Delta, and the North Central and Central Coast areas.

Producer prices have increased from USD 828/tonne in 2000 to USD 2616/tonne in 2012, i.e. by 10.1% annually in nominal terms (Figure 1.A1.8), around 7.8% in real terms if adjusted for the US inflation rate. This is one of the highest growth rates in real producer prices of all commodities reviewed above, which partly explains the rapid growth in pigmeat production within Viet Nam. Available data suggest that domestic producer prices remain much below those registered in China (*OECD PSE Database*). This would indicate a scope for competitive exports, but low quality and, in particular, persistent diseases such as porcine reproductive and respiratory syndrome, foot and mouth disease, and piglet fever, undermine export prospects. Despite these issues and despite a ban on live pig imports from Viet Nam imposed in 2003 by the Chinese Ministry of Agriculture, the media reports large-scale smuggling of live pigs to China at 600 000 annually to the Guangxi province alone (Global Times, 2014). On the import side, restrictive barriers are applied (Chapter 2) that isolate the domestic pork market from international competition. Moving forward, an issue for the industry is the effect that a Trans Pacific Partnership trade agreement would have if trade barriers to imported pork are reduced or removed. Potentially this could not only lower prices but also reinforce the need to raise pork quality to be competitive with imported pork.

Figure 1.A1.8. **Pigmeat: Producer prices and trade flows, 2000-13**

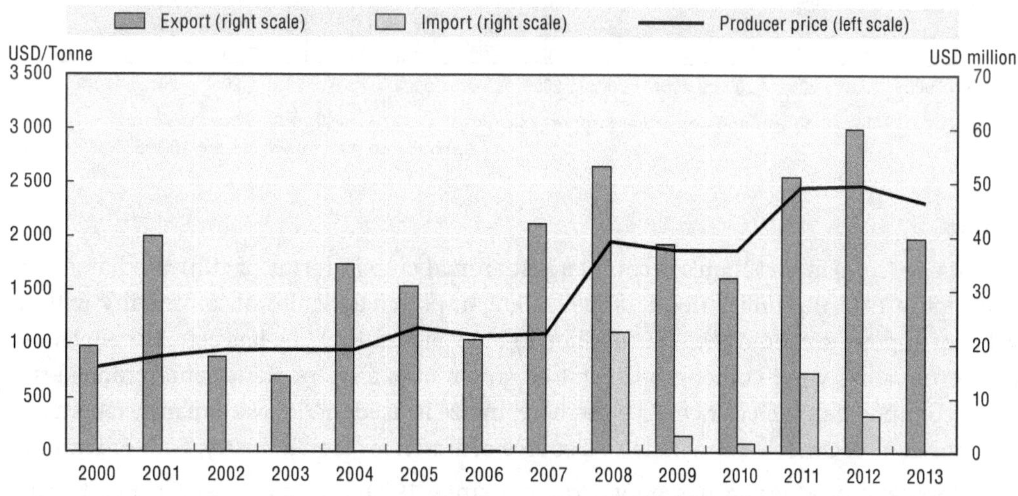

Source: OECD (2014), "Producer and Consumer Support Estimates", *OECD Agriculture Statistics Database*.

StatLink http://dx.doi.org/10.1787/888933223650

Eggs

The egg industry in Viet Nam is relatively small but has shown similarly favourable growth characteristics as all the other commodities so far reviewed. Shell egg production has grown from 1.9 billion eggs to 7 billion eggs over the 1990-2012 period, a 6.1% growth in production per year. Measured in tonnes, the industry has grown from 96 700 in 1990 to 350 000 in 2012 (FAOSTAT, 2015). The industry has a dual structure, with many back-yard, free-range, small scale egg producers all over the country, along with a growing number of large scale, industrial-type poultry egg operations often financed by international investors.

Producer prices are shown in Figure 1.A1.9 for the period from 2000-13. They had grown steadily by 2008, but then fluctuated and in 2013 were just slightly higher than in 2008. The nominal price growth for the whole period was 4.6% annually, slightly more than 2% per cent in real terms, thus less than for most commodities examined above. The egg market in Viet Nam is almost completely isolated from international competition with no imports and marginal exports. This is partly due to border protection, including an import quota (Chapter 2), but probably also to its still local character. Vietnamese producer prices for eggs are significantly higher than those in China (*OECD PSE database*). This would indicate that with the current production and cost structure, Vietnamese egg production would not be competitive in China, potentially its largest export market.

Figure 1.A1.9. **Eggs: Producer prices and trade flows, 2000-13**

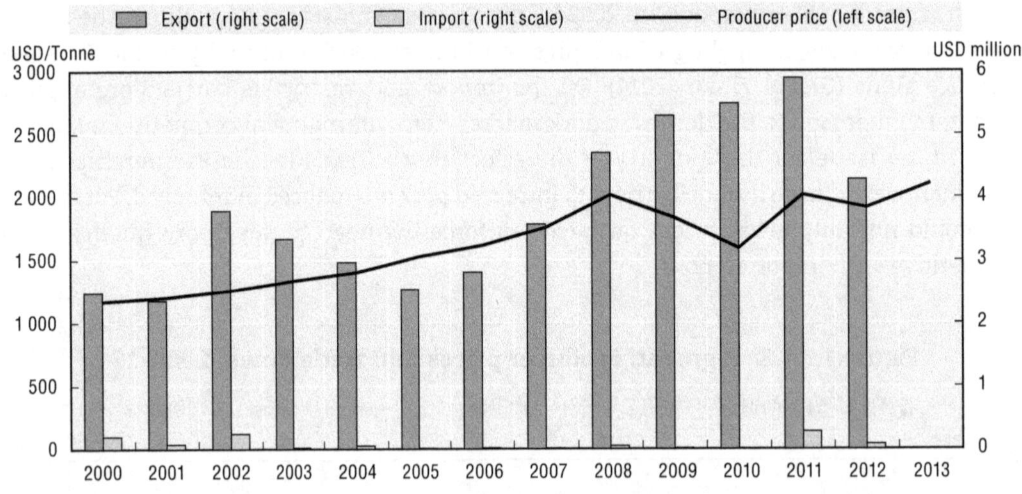

Source: OECD (2014), "Producer and Consumer Support Estimates", *OECD Agriculture Statistics Database.*

StatLink http://dx.doi.org/10.1787/888933223669

Maize

Maize (corn) is Viet Nam's second largest annual crop in terms of cultivated area, after rice (Figure 1.10). But unlike rice, it is the leading imported agricultural commodity, with the value of imports exceeding USD 600 million in 2013 (Figure 1.A1.10). It is usually cultivated in less fertile areas where other crops can't be grown. It is also frequently intercropped with other commodities such as rice. As elsewhere, maize is used in Viet Nam mainly (about 80%) for animal feed. Other uses include as a source of starch for food industry, and for textiles.

Domestic production has grown rapidly since 1990, rising almost 7 times to 2012, i.e. an annual growth rate of 9.4%. This rise has been quite steady, although flattening since 2008. This is due to maize profit margins being lower than those obtained with competing

Figure 1.A1.10. **Maize: Producer prices and trade flows, 2000-13**

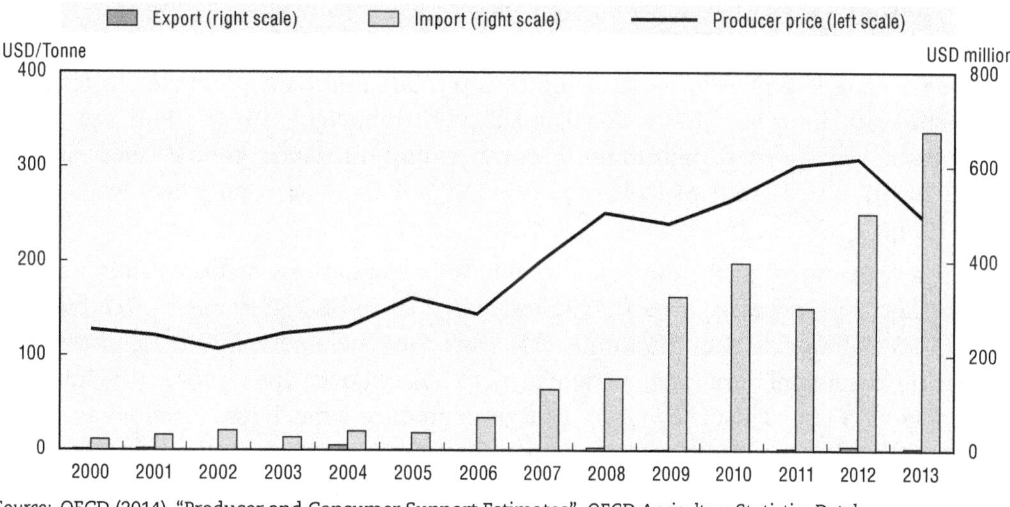

Source: OECD (2014), "Producer and Consumer Support Estimates", OECD *Agriculture Statistics Database.*

StatLink http://dx.doi.org/10.1787/888933223670

field crops, causing relatively modest land shifts into maize. The maize area harvested has grown at slightly less than half the overall production growth, but it too has flattened since 2005. Yield growth, on the other hand, has risen steadily by 4.7% per year, only showing some signs of diminishing recently (FAOSTAT, 2015).

The Northern Midlands and Mountain areas are the largest growing region, followed by the North Central and Central Coastal areas and Central Highlands, each with half as much land in corn. These three regions account for 82% of the country's total planted area in maize (GSO, 2014). Government programmes are now encouraging a shift from rice into maize in the interests of diversification.

Maize prices have grown since 1991 to 2012, but only significantly since about 2004, but then declined by 20% in 2013, partly due to a significant increase in imports. Over the period 1991-2011, nominal producer prices grew at 6.0% per year (3.9% in real terms), but from 2004 to 2011 the nominal growth rate was 12.7% (Figure 1.A1.10). Imported corn prices constrain or determine domestic prices, and typically are somewhat lower (*OECD PSE Database*).

The demand for maize has been growing at high rates, especially from animal feed processors, and is not met by the domestic production. As a result, Viet Nam increasingly imports maize (Figure 1.A1.10), mainly from Brazil, Argentina, US, India, and Thailand.

Sugar cane

Sugar cane was grown on about 0.3 million ha in 2012, making it the sixth largest crop by land area in Viet Nam (after rice, coffee, cassava, rubber, and cashews). Total production of sugar cane has grown more slowly than that of other crops, at 5.9% per year, increasing from 5.4 million tonnes in 1990 to 19 million tonnes in 2012 (FAOSTAT, 2015). The pattern of growth is the reverse of many of the commodities reviewed above: most of the growth occurred in the 1990s, and since 1999, production has been trendless (Figure 1.6). It would appear that as the comparative advantages of the more rapidly growing crops were being exploited, sugar cane production levelled off. The land area devoted to sugar cane is consistent with this. It also rose steeply in the 1990s, particularly from 1993 to 1999, after which sugar cane harvested area declined by at least one-seventh in the subsequent 13 years.

Averaged over the whole 22 years, land area grew by 3.8%. But yields per ha grew slowly but consistently over that whole period, averaging 2.0% per year, to offset some of the acreage declines in the 2000s (FAOSTAT, 2015).

Sugar cane is grown in most of the country, but the main provinces in terms of production are, in order, Thanh Hoa, Gia Lai, Tay Ninh, Nghe An, and Phu Yen. These provinces are in the north central/north coast, central highlands, central coast, and the south. Together this group of five account for 44% of all sugar cane grown in Viet Nam (GSO, 2014).

Sugar cane prices at the farm gate level have been relatively stable, even in nominal terms. They have increased from USD 31/tonne in 2000 to USD 52/tonne in 2011, but then fell to USD 43/tonne in 2013 (Figure 1.A1.11). Over the whole period 2000-13, prices were increasing in nominal terms at an annual rate of 2.5%, but once one removes (US) inflation, the real growth rate is just above zero. This gives producers much less incentive to stay in sugar cane when there are considerably more profitable prospects in the crops described above. As a result, farmers are moving to crops where they have more of a competitive advantage, where both productivity and prices are increasing. Another explanation might be weak transmission of refined sugar prices from the wholesale level back to the farm gate. Available data indicate that wholesale prices of refined sugar increased at much higher rates than those at the farm gate and that they remained quite strongly above those on international markets determined in the region by Thai sugar exports (*OECD PSE database*). High wholesale prices compared to those on international markets are mostly due to sugar import quota and other border measures applied by Viet Nam to protect domestic sugar industry (Chapter 2).

Figure 1.A1.11. **Sugar cane: Producer prices and trade flows, 2000-13**

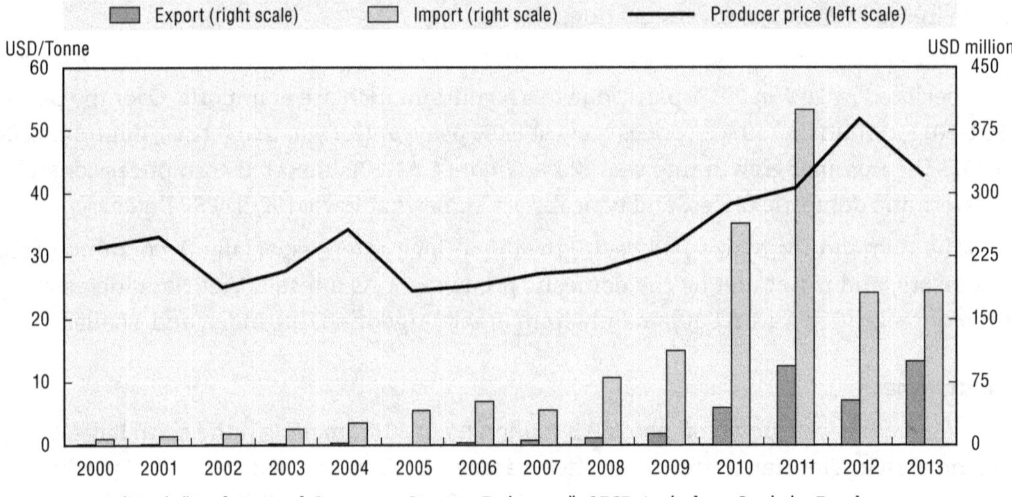

Source: OECD (2014), "Producer and Consumer Support Estimates", *OECD Agriculture Statistics Database*.

StatLink http://dx.doi.org/10.1787/888933223684

Viet Nam is a net importer of sugar with a value of imports peaking at around USD 400 million in 2011, but then falling to USD 170-180 million in 2012-13. Sugar exports tend to grow as well and amounted to USD 100 million in 2013 (Figure 1.A1.11). While sugar imports are almost exclusively from Thailand, exports are destined almost uniquely to China.

Poultry meat

The poultry meat industry in Viet Nam includes both chicken and ducks. It has more than tripled in terms of output tonnage from 1992 to 2012, to 618 000 tonnes. This represents an annual growth rate of 5.8%, very strong output expansion. The pattern of output growth has been quite steady: aside from declines in 2004 and 2010, poultry production has increased in all years since 1994 (FAOSTAT, 2015).

The distribution of poultry within the country has been relatively uniform, dominated by the Red River Delta at 81 million birds, but followed by the North Central and Central Coastal areas, the Northern midlands and mountain areas, and the Mekong River Delta, all with above 60 million birds in 2012 (GSO, 2014).

Producer price data show strong overall growth in prices for poultry meat (in carcass weight equivalent) up to 2011 followed by a strong fall in 2012-13 (Figure 1.A1.12). The nominal price more than doubled from 2000 to 2011 giving annual growth of 7.8%, and real growth of 5.5%. However, the inflation adjustment is only capturing US inflation, and does not capture the increased feed costs that followed the commodity price boom of 2007/08. Thus, the increased poultry prices that mostly occurred in 2006-11 were partly eaten up by the substantially increased cost of feed that occurred at the same time, given that poultry production on large modern enterprises is very feed-intensive. Price changes in 2012-13 would indicate reversed tendencies with falling poultry prices partly attenuated by lower feed costs. When compared with international prices, for example China's FOB prices for poultry meat, it can be seen that Vietnamese poultry prices are significantly higher which indicates lack of international competitiveness of this industry and explains Viet Nam's net import position (*OECD PSE Database*). Poultry meat imports, mostly of frozen chicken and offal, were negligible until 2006, peaked at USD 127 million in 2008 and then fluctuated between USD 70-100 million in 2009-13 (Figure 1.A1.12). China and the United States are the main suppliers.

Figure 1.A1.12. **Poultry meat: Producer prices and trade flows, 2000-13**

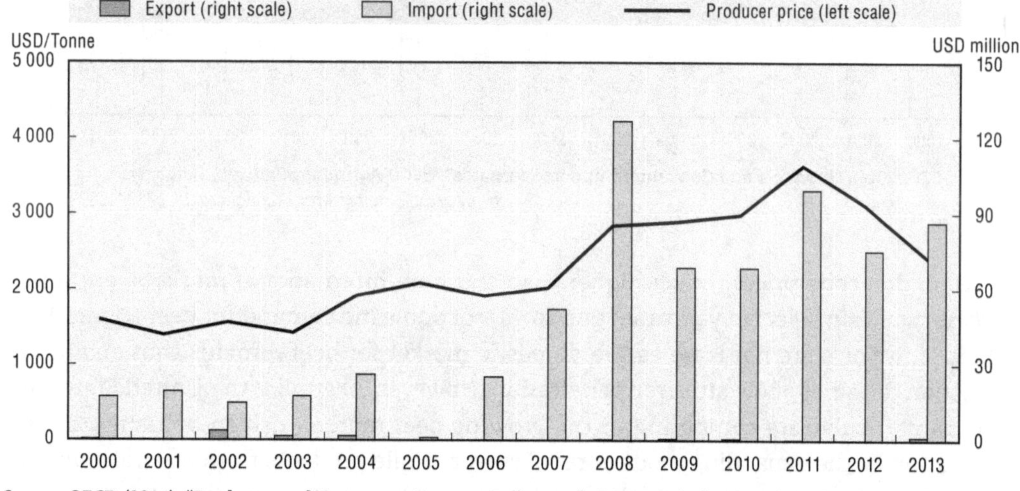

Source: OECD (2014), "Producer and Consumer Support Estimates", *OECD Agriculture Statistics Database.*

StatLink http://dx.doi.org/10.1787/888933223699

Cattle

Cattle production in terms of meat produced has grown from 75 000 tonnes in 1990 to 294 000 tonnes in 2012. This is only 9% of the pigmeat produced in 2012. But the cattle industry has grown at almost the same rate as the hog industry, at 6.4% per year in total production (slaughtered meat output). Production grew slowly to 2002 when it first exceeded 100 000 tonnes, but has since almost tripled in the last 10 years (FAOSTAT, 2015).

Estimates for 2012 indicate 5.2 million head of cattle throughout the country, compared to 1.7 million cattle slaughtered that year. In addition to cattle there were also 2.6 million water buffalo in 2012, used mainly for tilling rice fields but also for providing milk rich in fat and protein (GSO, 2014). Cattle are heavily concentrated in the North Central and Central Coastal area, and the Northern midlands and mountain areas. Those two areas account for 58% of all cattle, and 87% of all buffaloes (GSO, 2014).

Producer price data for cattle are scarce and not clearly defined. Available data would indicate that prices tended to decline in the first half of the 2000s, growing at high rates since then, peaking in 2009 and again in 2011 (Figure 1.A1.13). An increase from USD 2 600/tonne in 2005 to almost USD 7 000/tonne (both in carcass weight) in 2011 would suggest an annual growth rate at 18% in nominal USD terms. This shows a quite dramatic price increase and would appear that prices are profitable to cattle farmers, enough to boost production growth but not enough to meet growing demand.

Figure 1.A1.13. **Cattle: Producer prices and trade flows, 2000-13**

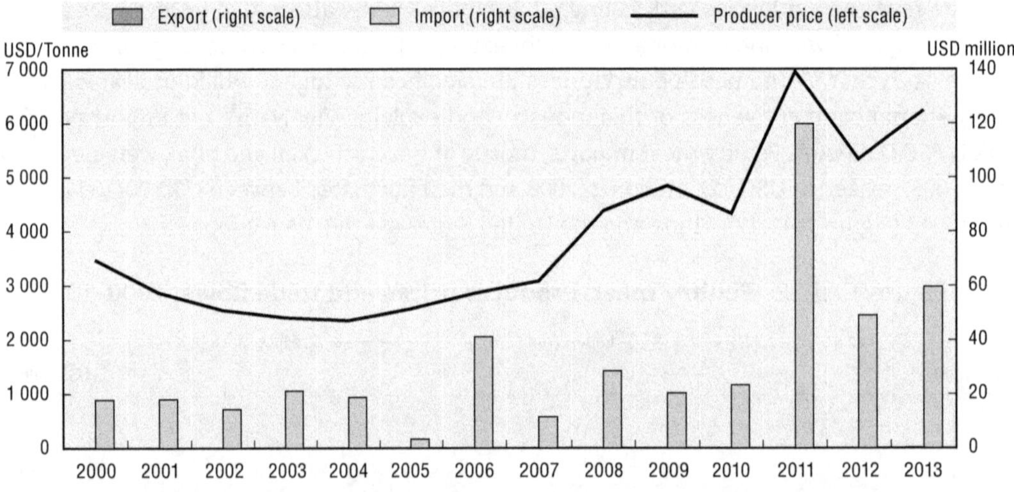

Source: OECD (2014), "Producer and Consumer Support Estimates", *OECD Agriculture Statistics Database.*

StatLink http://dx.doi.org/10.1787/888933223704

High domestic prices, much higher than those on international markets, encourage imports, which in selected years meet up to 40% of domestic demand for beef (Figure 1.33). Values of imports are not large as the domestic market for beef remains thin and border protection does not allow stronger penetration. However, Australia, the United States and India in particular are competing in the growing beef market, driven by rising tourism inflows and by the expanding modern retail sector. While products from Australia and the United States compete at the higher end, India is the largest supplier of frozen beef. Australia is also a large supplier of live cattle for slaughtering into fresh beef. Australia benefits from lower duties through the ANZ-ASEAN Free Trade Agreement, while the US does not, although the Trans-Pacific Partnership negotiations are likely to address this issue. It is worth noting that some beef imports into Viet Nam are "re-exported" across the border into China.

ANNEX 1.A2

Viet Nam: Projected production, consumption and trade for major commodities by 2023

This annex presents the main projections for the major agricultural commodities produced, consumed and traded by Viet Nam during the next ten years as embedded in the OECD-FAO *Agricultural Outlook* 2014-2023 report. The main purpose of the report is to build consensus on global prospects for the period 2014-23, for the agriculture, fisheries and food sectors, and on emerging issues which affect them. A jointly developed modelling system, based on the OECD's Aglink and FAO's Cosimo models facilitates consistency and analysis of the projections. The fully documented outlook database, including historical data and projections, is available through the OECD-FAO joint Internet site *www.agri-outlook.org*.

Viet Nam: The main macroeconomic and policy assumptions underlying the baseline projections

The *Outlook* is presented as a baseline scenario that is considered plausible given a range of conditioning assumptions. These assumptions portray a specific macroeconomic and demographic environment which shapes the evolution of demand and supply for agricultural and fish products. These general factors for Viet Nam are as follows:

- population is assumed to increase from 92 million in 2013 to 98 million in 2023
- inflation is expected to average around 7% in the next ten years
- Vietnamese Dong is expected to depreciate in nominal terms relative to USD from VND/USD 21 668 in 2013 to VND/USD 28 841 in 2023
- GDP is expected to grow at 5.5% per year
- the policy framework, including the level of tariffs, is assumed to remain as defined within Viet Nam's WTO commitments until 2023.

Main findings

Before providing specific results for Viet Nam, it is worth indicating that global supply and demand projections up to 2023 point to slowly declining real prices on international markets. Figure 1.A2.1 shows that in the decade 2014-23 average prices for most livestock commodities will be higher than average prices in the decade of 2004-13, but for all crops average prices are projected to be lower. Moreover, prices for all commodities, both crops and livestock, are projected to be lower than in 2011-13. Thus, for Viet Nam external conditions will become much more challenging compared to the 2000s when growing prices on international markets provided strong incentives for agricultural production and export growth.

Figure 1.A2.1. **Price trends in real terms for agricultural commodities to 2023**

Per cent change in average real prices relative to different base periods 2011-13 and 2004-13

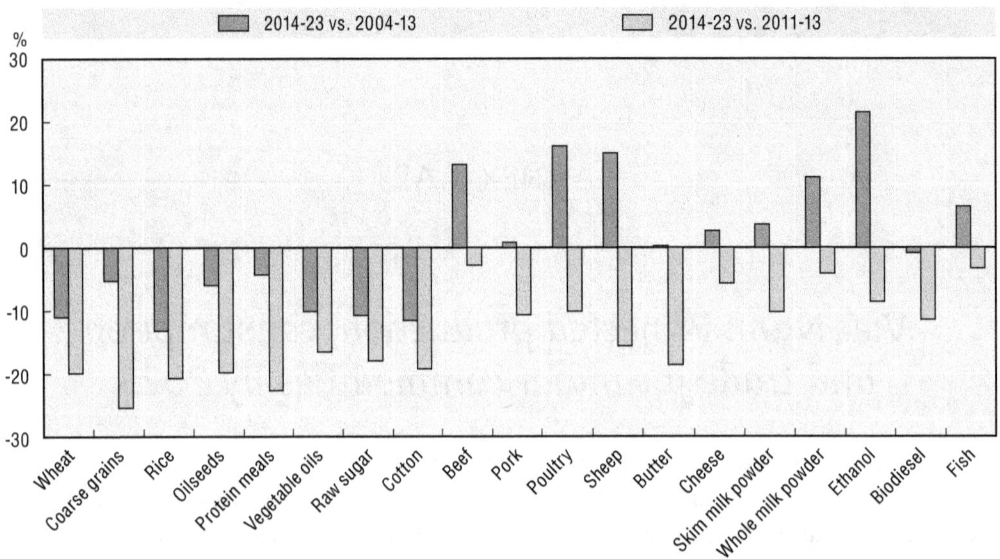

Source: OECD-FAO (2014), *Agricultural Outlook 2014-2023.*

StatLink http://dx.doi.org/10.1787/888933223710

Viet Nam's position on selected world markets will increase

The evolution of production in Viet Nam in the coming decade is largely above the rates projected for world production of such commodities as protein meals, pigmeat, sugar, vegetable oils, poultry meat and coffee (Figure 1.A2.2). In particular, production growth rates of coffee, pigmeat and sugar are expected to be more than twice stronger in Viet Nam than in the world.

Figure 1.A2.2. **Production: Per cent change 2023 compared to 2011-13 average**

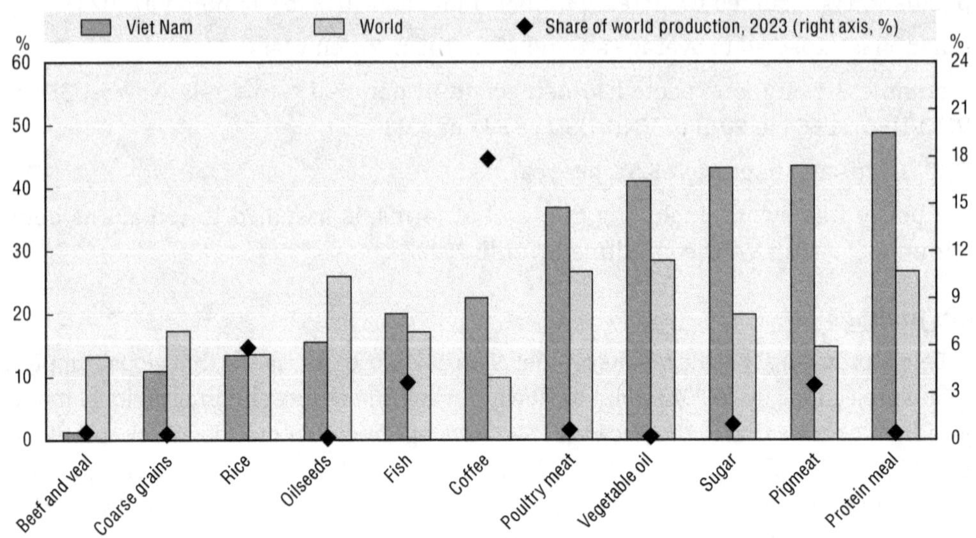

Note: 2019 for coffee, FAO data.

Source: OECD-FAO (2014), *Agricultural Outlook 2014-2023.*

StatLink http://dx.doi.org/10.1787/888933223728

Main trends for selected commodities

Viet Nam is projected to consolidate and improve its position of net exporter for a number of commodities, but in particular for rice and coffee (Figure 1.A2.3).

Figure 1.A2.3. **Projected net trade for selected products**

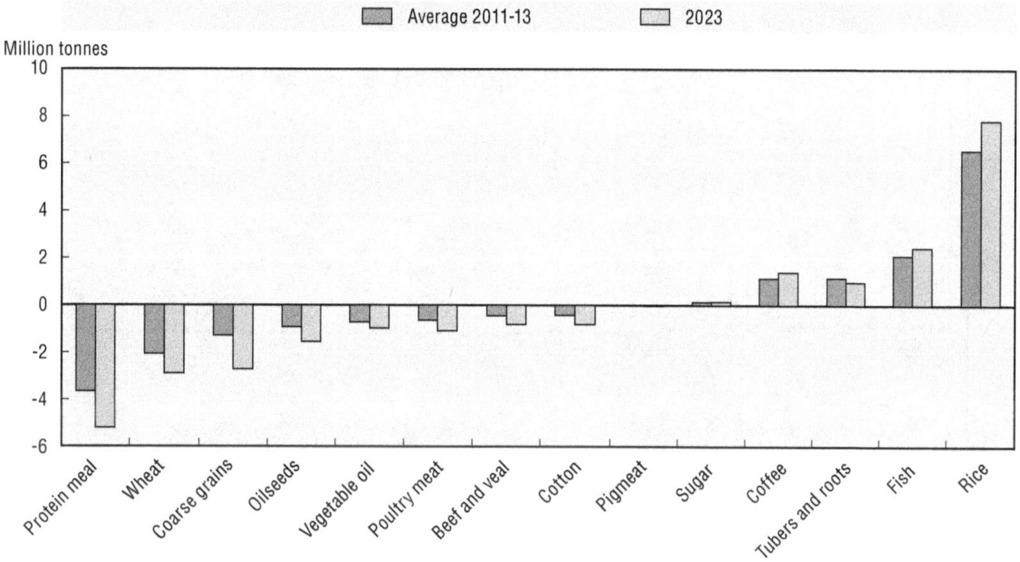

Note: 2019 for coffee, FAO data.
Source: OECD-FAO (2014), *Agricultural Outlook 2014-2023.*

StatLink http://dx.doi.org/10.1787/888933223738

Viet Nam shipped 6.5 Mt of rice in 2013 and this amount is predicted to increase to almost 8 Mt in 2023. Production is also projected to increase, exclusively due to higher yields. While Viet Nam's average yields are already one of the highest in South East Asia, further improvements are to result from the use of modern high-yielding rice varieties and improved irrigation facilities.

Viet Nam's production of coffee is projected to increase to almost 1.6 Mt in 2019, which will bring Viet Nam's share of the world's production from 16% in 2013 to 19% in 2019 and will enable the country to maintain its position as one of the key exporters in the world. The government is developing policies that encourage sustainable production of coffee. Thus, it can be assumed that the increase in production will result mainly from improvements in input use and better production practices (replacing old tree areas) and not from the expansion of area planted. Around 1.4 Mt of coffee is projected to be exported, around 90% of production. A small part, stable in the forecast period, is to be consumed domestically.

For some other commodities Viet Nam is projected to remain a net importer, including for protein meals, wheat and coarse grains (Figure 1.A2.3). With respect to protein meals Viet Nam's imports are projected to expand to 5 Mt by 2023 – more than doubling from the base period. For wheat and coarse grains, imports are projected to reach 2.8 Mt and 2.7 Mt, respectively, in 2023. This increase is linked to the development of the livestock industry.

In response to increasing demand, total meat production in Viet Nam is projected to reach 6 Mt by 2023, an increase of 38% from the base period, mainly of pigmeat and poultry meat. Total meat consumption on a retail weight basis will grow by 2.6% p.a. It is projected that over the *Outlook* period, Viet Nam's total meat consumption per capita will get close to the OECD average and its fish consumption per capita has been higher than the OECD average since the mid-2000s (Figure 1.A2.4).

Figure 1.A2.4. **Meat and fish consumption in Viet Nam, kg/capita, 2001-23**

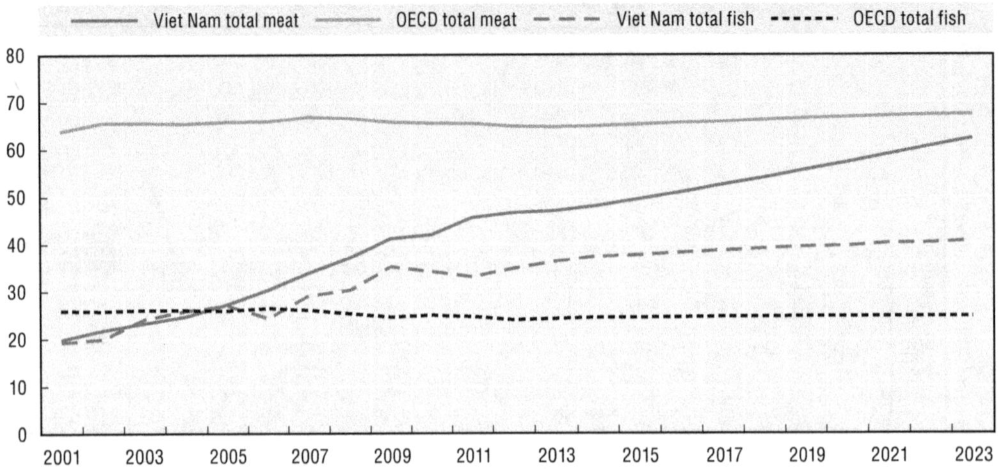

Viet Nam total meat | OECD total meat | Viet Nam total fish | OECD total fish

80 70 60 50 40 30 20 10 0

2001 2003 2005 2007 2009 2011 2013 2015 2017 2019 2021 2023

Source: OECD-FAO (2014), *Agricultural Outlook 2014-2023.*

StatLink http://dx.doi.org/10.1787/888933223747

References

ADB (Asian Development Bank) (2014), *Key Indicators for Asia and the Pacific 2014, www.adb.org/statistics.*

ADB (2010), "National Performance Assessment and Subregional Strategic Environment Framework in the Greater Mekong Subregion", *Consultant's Report*, January 2010.

ADB (2005), *Key Indicators 2005, www.adb.org/statistics.*

Agroinfo (2013), *Fertilizer 2012 and Outlook: Annual Report.*

Arita Shawn, S. and J. Dyck (2014), "Vietnam's Agri-Food Sector and the Trans-Pacific Partnership", *Economic Information Bulletin*, No. 30, ERS/USDA, October.

Bloomberg (2013), *Vietnam Tightens Land Seizure Law After Farmers Protest, www.bloomberg.com/news/2013-12-08/vietnam-tightens-land-seizure-law-after-protests-southeast-asia.html.*

CIEM (Central Institute for Economic Management) (2013), *Characteristics of the Vietnamese Rural Economy: Evidence from a 2012 Rural Household Survey in 12 Provinces of Viet Nam*, Hanoi.

Cling, J.P., Nguyen Huru Chi, M. Razafindrakoto and F. Roubaud (2010), "Urbanization and Access to Labour Market in Vietnam: Weight and Characteristics of the Informal Sector", Ton Nu Quynh Tran, Quertamp Fanny, Miras Claude de, Nguyen Quang Vinh, Le Van Nam, Truong Hoang Truong (eds.), 2012, *Trends of Urbanization and Suburbanization in Southeast Asia*, Centre for Urban and Development Studies (CEFURDS), Regional Conference, Ho Chi Minh City, 9-11 December 2008, Ho Chi Minh City General Publishing House, 328 pp.

Coxhead, I., Nguyen Viet Cuong and Vu Hoang Linh (2015), "Internal Migration in Vietnam: New Evidence from Recent Surveys", Paper prepared for Vietnam Rural-Urban Migration Survey (VRUMS) Conference, Hanoi, 13-14 January 2015.

Coxhead, I., Kim Ninh, Vu Thi Thao and Nguyen Thi Phuong Hoa (2010), *A Robust Harvest: Strategic Choices for Agricultural and Rural Development in Vietnam*, The Asia Foundation, *http://asiafoundation.org/resources/pdfs/VNARobustHarvest.pdf.*

Dao The Anh and Nguyen Van Son (2013), "Vietnam Agricultural Value Chain in the FTA of Asian Region", in *FFTC.NACF International Seminar on Threats and Opportunities of the Free Trade Agreements in the Asian Region*, Sept. 29-Oct. 3, 2013, Seoul, Korea, *http://ap.fftc.agnet.org/ap_db.php?id=106 &print=1.*

Dao The Anh, Thomas Reardon, Kevin Chen, Thai Van Tinh, Vu Nguyen, Nguyen Ngoc Vang, Nguyen Van Thang and Le Nguyen Doan Khoi (2013), *Rice Value Chain Study in Mekong River Delta, Viet Nam*, CASR@D, IFPRI and ADB, Hanoi, May.

Darvas, Z. (2012), "Real Effective Exchange Rates for 178 Countries: A New Database", *Working Paper*, Department of Mathematical Economics and Economic Analysis, Corvinus University of Budapest, Budapest, *http://web.uni-corvinus.hu/matkg/*.

Demombynes, G. and Linh Hoang Vu (2015), "Demystifying Poverty Measurement in Vietnam", *Vietnam Development Economics Discussion Paper*, World Bank Group, Country Office Vietnam, February.

Diaz-Bonilla, E., D. Orden and A. Kwieciński (2014), "Enabling Environment for Agricultural Growth and Competitiveness: Evaluation, Indicators and Indices", *OECD Food, Agriculture and Fisheries Papers*, No. 67, OECD Publishing, Paris, *http://dx.doi.org/10.1787/5jz48305h4vd-en*.

EDE Consulting for Coffee (2006), "Support for the Development of the Cashew Sector in Dak Lak", Fact Finding Mission Report Prepared for GTZ, Hanoi, February, *www.sme-gtz.org.vn/Portals/0/AnPham/Fact%20findings%20on%20Support%20for%20Cashew%20sector%20Dak%20Lak.pdf*.

EIU (Economist Intelligence Unit) (2015), *Country Report: Vietnam*, London, 23 March.

FAO (Food and Agriculture Organization) (2014), *Forestry CountrySTAT Database*.

FAO (2012a), *FAO Yearbook of Fisheries and Aquaculture Statistics*.

FAO (2012b), "Agriculture Investment Trends-The Role of Public and Private Sector in Vietnam", *www.fao.org/fileadmin/templates/tci/pdf/CorporatePrivateSector/Vietnam_-_Private_Sector_Investments_in_Agriculture__Final_Report.pdf*.

FAO (2010a), *Selected Indicators of Food and Agricultural Development in the Asia-Pacific Region 1999-2009, www.fao.org/docrep/013/i1779e/i1779e00.htm Part V; www.fao.org/docrep/013/i1779e/i1779e04.pdf*, Tables 149-153.

FAO (2010b), "Global Forest Resources Assessment 2010: Main Report", *FAO Forestry Paper* 163, Rome.

FAO AQUASTAT (2014), *FAO's Global Water Information System, www.fao.org/nr/water/aquastat/main/index.stm*.

FAO AQUASTAT (2013), *Viet Nam: UN-Water Country Brief*, 13 June, *www.unwater.org/fileadmin/user_upload/unwater_new/docs/Publications/VNM_pagebypage.pdf*.

FAO-IFAD (International Fund for Agricultural Development) – WFP (World Food Programme) (2014), *The State of Food Insecurity in the World 2014. Strengthening the enabling environment for food security and nutrition*, Rome, FAO.

FAOSTAT (2015), *FAOSTAT Database, http://faostat.fao.org/*.

Fuglie, K. and N. Rada (2013), *International Agricultural Productivity Dataset*, USDA Economic Research Service: *www.ers.usda.gov/data-products/international-agricultural-productivity*.

GAIN-VM 3062 (2013), "Vietnam: Retail Foods, Sector Report 2013", FAS, 12 November.

Giesecke, J.A, Nhi Hoang Tran, E.L. Corong and S. Jaffee (2013), "Rice Land Designation Policy in Vietnam and the Implications of Policy Reform for Food Security and Economic Welfare", *The Journal of Development Studies, http://dx.doi.org/10.1080/00220388.2013.777705*.

GSO (General Statistical Office) (2014), *Statistical Yearbook of Vietnam 2013*, Statistical Publishing House, Hanoi, *www.gso.gov.vn*.

GSO (2013), *Household Living Standards Survey 2012*, Statistical Publishing House, Hanoi.

GSO (2012), *Results of the 2011 Rural, Agriculture and Fishery Census*, Statistical Publishing House, Hanoi.

Henson, S. (2013), *Meeting Standards – Winning Markets: How to Address the Compliance Challenges of Seafood Value Chains*, Presentation at the Workshop, Hanoi, Viet Nam, 22 March 2013.

IFAD-IFPRI (2011), "IFAD-IFPRI Partnership Newsletter", *Market Access*, November.

ILO (International Labour Organisation) (2011), *Vietnam Employment Trends 2010*, National Centre for Labour Market Forecast and Information, Bureau of Employment, ILO Office in Vietnam.

IMF (International Monetary Fund) (2013), "IMF Executive Board Concludes 2013 Article IV Consultation with Vietnam", *Press Release* 13/304, August 9.

IMF (2012), "Vietnam", *IMF Country Report* No. 12/165.

Ipsos Business Consulting (2013), *Vietnam's Coffee Industry*, July.

ISPONRE (Institute of Strategy and Policy on Natural Resources and Environment) (2009), *Vietnam Assessment Report on Climate Change*, With support of the United Nations Environment Programme (UNEP).

Jaffee, S. (2011), *Vietnam Rice Economy, Trade and Food Security. Research Consortium: Topics, Approaches, and Major Findings*, World Bank Policy Consultation Workshop, 13 June.

Ken Research (2014a), *Vietnam Fertilizer Industry Research Report*, January 2014.

Ken Research (2014b), "Vietnam Seed Industry Outlook to 2018-Rice Seed and Hybridization to Drive Future Growth", 5 August, *www.kenresearch.com/agriculture-food-beverages/agriculture-industry/vietnam-seed-market-research-report/553-104.html*.

Kompas, T., Tuong Nhu Che, Ha Quang Nguyen and Hoa Thi Minh Nguyen (2012), "Productivity, Net Returns and Efficiency: Land and Market Reform in Vietnamese rice Production", *Land Economics* 88(3): 478-495.

Le Canh Dung, Nguyen Van Sanh, Vo Van Tuan, Pham Thi Tam (2011), *A Study on Vietnam's Mekong Delta Region: Malnutrition Amongst Plenty*, Can Tho University.

MARD (Ministry of Agriculture and Rural Development) (2015), Comments received from MARD on the Review, May 2015.

MARD (2013), *Statistical Yearbook of Agriculture and Rural Development 2012*, MARD, Agricultural Publishing House, Hanoi.

Maruyama, M. and L.V. Trung (2011), "Modern Retailers in Transition Economics: The Case of Vietnam", *Kobe University Discussion Paper* DP2011-08, March.

Mehta, Meeta Punjabi, Nguyen Minh Hien and Do Troung Lam (2011), "Rice Value Chain in Vietnam", unpublished study commissioned by the Asian Development Bank and implemented by IFPRI.

Meyfroidt, P. and E.F. Lambin (2008), "Forest Transition in Vietnam and its Environmental Impacts", *Global Change Biology* 14, 1319-1336.

MOIT (Ministry of Industry and Trade) (2014), *Production and Export of Cassava in 2013*, *www.moit.gov.vn/en/News/508/production-and-export-of-cassava-in-2013.aspx* (accessed 9 October).

MONRE (Ministry of Natural Resources and Environment) (2014), *http://vea.gov.vn/en/EnvirStatus/C-CEnvironment/Pages/The-current-stituation-of-environment-pollution-in-Viet-Nam-.aspx* (accessed 18 August).

MONRE (2009), *Climate Change, Sea Level Rise Scenarios for Viet Nam*.

Ngo Kinh Luan (2013), "Natural Rubber Industry Report", Fpt Securities.

Nguyen Do Anh Tuan (undated presentation), *Rice Market and Policy in Vietnam*, IPSARD.

Nguyen Duc Long (2013), "Vietnam's Agricultural Mechanization Strategies", presentation to UNESCAP-CSAM, *Regional Forum on Sustainable Agricultural Mechanization in Asia and the Pacific 2013*, *www.unapcaem.org*.

Nguyen Hang, T. (2013), "Fertilizers Manufacturing in Vietnam", *VietinBankSc Industry Report*.

Nguyen Manh Hai (2013), "The Role of Inputs Policy in Transforming Agriculture in Vietnam", presentation to ReSAKSS-Asia Program, International Conference, Siem Reap, Cambodia, 25-27 September.

Nguyen Mau Dung (2013), "Study of Seed Industry in Vietnam", Initiative on Role of Fertilizer and Seeds in Transforming of Agriculture in Asia, ReSAKSS-Asia Program, IFPRI, Siem Riep, 2013, *www.slideshare.net/resakssasia/nguyen-mau-dung-seed-sector-vietnam-26949258*.

Nguyen Ngoc Hue and F. Goletti (2001), "Explaining Agricultural Growth", Technical background report submitted by Agrifood Consulting International to ANZDEC Ltd. as part of ADB TA3223-VIE, Hanoi, June.

Nguyen Quoc Viet (2011), "Current Status of Agricultural Mechanization in Vietnam", *www.unapcaem.org*, *Sustainable Agricultural Mechanization Roundtable: Moving Forward on the Sustainable Intensification of Agriculture*, 8-9 December.

Nguyen Thi Ngoc Hue, Truong Van Tuyen, Nguyen Tat Canh, Pham Van Hien, Pham Van Chuong, Bhuwon Ratna, Sthapit and Devra Javis (2005), "In Situ Conservation of Agricultural Biodiversity on Farm: Lessons Learned and Policy Implications", Proceedings of Vietnamese National Workshop, 30 May-1 April 2004, Hanoi, Viet Nam, International Plant Genetic Resources Institute, Rome, Italy.

Nguyen Tien Duc (2011), *Vietnam Seed Industry – NSC & SSC*.

Nguyen Trung Kien (2012), "Seed Development Industry in Vietnam," ADBI, *www.adbi.org/files/2012.12.11.cpp.day.1.sess3.4.kien.seed.devt.industry.viet.nam.pdf.*

OECD (2014), *Social Cohesion Policy Review of Viet Nam*, Development Centre Studies, OECD Publishing, Paris, *http://dx.doi.org/10.1787/9789264196155-en.*

OECD (2013), *Southeast Asian Economic Outlook 2013: With Perspectives on China and India*, OECD Publishing, Paris, *http://dx.doi.org/10.1787/saeo-2013-en.*

OECD-FAO (2014), *OECD-FAO Agricultural Outlook 2014*, OECD Publishing, Paris, *http://dx.doi.org/10.1787/agr_outlook-2014-en.*

OECD-The World Bank (2014), *Science, Technology and Innovation in Viet Nam*, OECD Publishing, Paris, *http://dx.doi.org/10.1787/9789264213500-en.*

Pham Quang Ha, M. McLaughline and I. Oborn (2006), "Nutrient Recycling for Sustainable Agriculture in Viet Nam", *Improving Plant Nutrient Management for Better Farmer Livelihoods, Food Security and Environmental Sustainability*, Proceedings of a Regional Workshop (Beijing, China), FAO, Regional Office for Asia and the Pacific, Bangkok.

Pham Thi Sen and Luu Ngoc Trinh (2008), *Viet Nam: Second Country Report on the State of the Nation's Plant Genetic Resources for Food and Agriculture*, Food and Agriculture Organization, Plant Resources Center.

Reardon, T., P.C. Timmer and B. Minten (2010), "Supermarket Revolution in Asia and Emerging Development Strategies to Include Small Farmers", *Proceedings of the National Academy of Sciences of the USA*, Vol. 1109 (31), *www.pnas.org/content/109/31/12332.full.*

REDD (Reduced Emissions from Deforestation and Forest Degradation) (undated note), *The Situation of Land Use, Forest Policy and Governance in Viet Nam.*

Technoserve (2013), Technoserve for the Sustainable Coffee Program, *Vietnam: A Business Case for Sustainable Coffee Production*, December.

To Xuan Phuc and Tran Huu Nghi (2014), *Rubber Expansion and Forest Protection in Vietnam*, Hue, Viet Nam.

Tran Thang, Dinh Linh, and Vu Phuc (2013), *Vietnam Rice Sector Policy Analysis*, SNV Netherlands, May.

Tran Van Tho (2013), "Vietnamese Economy at the Crossroads: New Doi Moi for Sustained Growth", *Asian Economic Policy Review* No. 8, 122-143.

UN-REDD (2009), *Revised Standard Joint Programme Document*, UN-REDD Viet Nam Programme.

USDA (2014), *Farm Household Well-Being: Glossary*, *www.ers.usda.gov/topics/farm-economy/farm-household-well-being/glossary.aspx#.VCvdNPnoQjM.*

Vasavakul, T (2006), "Agricultural Land Management Under Doi Moi: Policy Makers' Views", Chap. 11 in Marsh, S.P., G. MacAuley and P.V. Hung (eds.), *Agricultural Development and Land Policy in Vietnam*, Australian Centre for International Agricultural Research, Sydney.

VEN (Vietnam Economic News) (2014), "VEAM Expects to Achieve Annual Targets", 14 August.

Viet Capital Securities (2011), "Vietnam Rubber Group", 28 March, *www.vcsc.com.vn.*

Viet Nam News (2014), "Fertiliser Market to Face Stiff Competition", 28 August.

Viet Nam News (2012), "Chinese Farm Machinery Dominates", 5 April.

Vietnam Soil Association (1996), *Vietnam Soils*, Agriculture Publishing House, Hanoi.

WB (World Bank) (2015), *World Development Indicators*, *http://data.worldbank.org/data-catalog/world-development-indicators.*

WB (2012), *Vietnam Rice, Farmers and Rural Development: From Successful Growth to Sustainable Prosperity.*

WB (2004), "The Socialist Republic of Vietnam: Coffee Sector Report", *World Bank Report* No. 29358-VN, June.

WEF (World Economic Forum) (2014), *The Global Competitiveness Report 2013–2014*, *www.weforum.org/reports/global-competitiveness-report-2013-2014.*

WHO (World Health Organisation) (2015), UNdata, *http://data.un.org/Data.aspx?d=SOWC&f=inID:106.*

WWF (World Wildlife Fund) (2013), *REDD+ Country Profile: Vietnam. Forest and Climate Initiative*, WWF-Germany, May 2013.

Chapter 2

Trends and evaluation of Viet Nam's agricultural policy

The focus of this chapter is on major developments in agricultural policy in Viet Nam since 2000. It describes the framework of agricultural policy with regard to key policy objectives, the major phases of policy development, and the legal and institutional arrangements for administering agricultural policy. Domestic agriculture-related policies are then described, with polices grouped in accordance with the indicators of agricultural support developed by the OECD. It is followed by a detailed examination of trade policies relating to the agro-food sector. Support provided to agriculture and the cost that these policies impose on Vietnamese consumers and taxpayers are then estimated. The final section summarises the main conclusions.

2.1. Introduction

This chapter examines agricultural policy and the support provided to agricultural producers in Viet Nam since 2000. The current priorities of agricultural policy are to achieve high quality output and competitiveness, raise rural incomes and maintain food self-sufficiency. The Ministry of Agriculture and Rural Development (MARD) has the primary role in developing and implementing policies to achieve these objectives. Other central government ministries and agencies along with local government also have significant roles. A range of input subsidies on irrigation, seeds, credit, etc., is used to support producers. The budgetary cost of these measures has grown rapidly since the mid-2000s. A number of initiatives have been introduced to deal with disease outbreaks and natural disasters. In 2012 a direct payment per ha was introduced to encourage farmers to maintain land in rice production. Expenditure on irrigation, both capital works and operations and maintenance, accounts for a relatively large proportion of government spending on agriculture.

Tariffs have fallen significantly over the period. The average tariff on agricultural product imports has fallen from 24% in 2000 to 16% in 2013. Import monopolies, licensing requirements and export restrictions on agricultural products that were still remaining following the reforms of the 1990s were removed in the early 2000s. Import requirements imposed for food safety and quarantine purposes are becoming more stringent. They are often implemented in a non-transparent manner and add to the cost of importing. The current system for controlling rice exports reduces competition in the market.

The level of support to producers as measured by the ratio of policy-related transfers from taxpayers and consumers to gross farm revenues (percentage Producer Support Estimate, %PSE) averaged 7% in 2011-13, varying between -21% in 2008 and 16% in 2009. This variation reflects the government's efforts to stabilise domestic prices and to balance the interests of producers and consumers in the context of price volatility on international markets. The total value of transfers arising from support to agriculture was equivalent to 2.2% of GDP in 2011-13.

2.2. Agricultural policy framework

This section provides an overview of the agricultural policy framework in Viet Nam. A summary of the current key policy objectives for the sector is followed by a description of agricultural policy developments since reunification in 1976. These can be divided into five broad phases that follow the broader economic policy transition in Viet Nam from a centrally planned to a socialist-oriented market economy. The roles of various government agencies are then discussed. While MARD has the main responsibility for agricultural policy formation, a number of other ministries formulate policies directly affecting the sector. Local governments are primarily responsible for service delivery. State-owned enterprises (SOEs) and co-operatives have been used in the past to implement policies but their influence has fallen in recent times. In Viet Nam, the agricultural sector is broadly defined to include forestry and fishery (both aquaculture and capture) production.[1] This study uses a narrower definition of the agricultural sector encompassing crop (annual and perennial) and livestock production only. Similarly, the study does not examine broader rural development initiatives.

Agricultural policy objectives: current

The overall goals for agricultural policy as set out in the MARD plan for the five years 2011-15 are to achieve sustainable development with high quality output; improve the living standard of people living in rural areas, especially the poor; and protect and effectively utilise natural resources and the environment. These high-level goals are refined into the following **six key objectives**, with specific targets and various actions and programmes for each objective:

- achieve sustainable, high quality growth of the sector with improvement in productivity, quality and competitiveness of products
- improve living standards and conditions of population living in rural areas, especially the poor
- develop infrastructure to meet requirement of the agricultural production and serve people living in rural area
- strengthen competitive capacity and international integration of the sector
- use and protect natural resources and the environment in a sustainable and efficient manner
- improve the government's managerial capacity of the sector in an efficient and effective manner.

Phases of agricultural policy development

Reunification (1976-1986)

In the initial years following reunification, the Communist Party of Viet Nam (CPV) attempted to extend its **socialist centrally planned system** to the whole country. The government's priority goal was to develop heavy industry. Agriculture's role was to support this by providing goods at low prices while at the same time achieving food self-sufficiency in rice and other staple foods (Nguyen and Grote, 2004; MARD, 2014a). Private farming was essentially abolished. Agricultural production was organised into co-operatives focusing on annual crops and state farms focusing, in general, on perennial crops. The upstream and downstream sectors were reorganised as SOEs. Co-operatives managed production and distribution decisions in accordance with targets developed by central governing authorities. They also controlled marketing functions, collecting and selling surplus product either to the state at "negotiated prices" or in the unorganised free market. Within co-operatives, production targets were assigned to brigades who allocated labour supply to activities (Kirk and Nguyen, 2009).

It was obvious by the early 1980s that these arrangements for organising agriculture were not working. Production levels were well below targets. Surpluses were sold on the informal private market rather than through the state procurement system that offered much lower prices. The government was forced to increase food imports at a time when Western and Chinese aid was declining.[2] Drawing lessons from the success of illegal "underground contracts" that were spreading throughout the country, the CPV Central

Committee issued Directive No. 100 CT/TW dated 13 January 1981 in an **attempt to improve the efficiency** of the collective system.[3] The Directive allowed households and co-operatives to enter into contracts that permitted households to farm land owned by the co-operative in exchange for delivering an annual production quota to the co-operative. Quotas were based on the productivity of the land during the previous three years. Households were responsible for planting, tending and harvesting. Co-operatives provided services such as preparing land and providing water, seeds, fertilisers and pesticides, paid for by the selling of contracted output (Dao et al., 2005). Any surplus produced by the household above the quota could be sold to the state via SOEs or in the private market. State procurement prices of agricultural goods were gradually increased to the same level as market prices (Nguyen, 2010).

At first this partial reform seemed promising: agricultural production grew 11% in 1982. Success, however, was short-lived. The reforms were not deep enough to give farmers real incentives to produce more. The government maintained strict control over prices and trade for both inputs and outputs through SOEs and internal trade restrictions. Household quotas were sometimes raised by co-operatives following their observation of increased production. Due to supply shortages, the allocation of inputs to farmers from co-operatives fell far short of their requirements (Kirk and Nguyen, 2009). Many farmers did not obtain adequate output to pay their duty to co-operatives. In addition, the government issued the Agricultural Tax Ordinance in 1993 to unify and rationalise the tax base across the country. This introduced another output-based tax on farmers: 6-14% for paddy; 10-30% for fruit trees and 12% for industrial and other crops (Barker et al., 2004). Farmers started to give back land. The **incentive to produce was lost**. By the mid-1980s, large areas of the country were experiencing near-famine conditions, and food shortages were resulting in widespread suffering. Inflation became a serious problem and the failure of the so-called "price-salary-money reform" initiated in 1985 led the economy into crisis with hyperinflation (Vo, 2008).

Renovation (1986-1993)

The VIth National Congress of the CPV in 1986 recognised that the centralised management mechanism was failing and began a process to **renovate (Doi Moi)** economic management institutions and policies (Pham, 2006).[4] The reforms had sweeping goals: they sought to stabilise the economy, develop the private sector, increase and stabilise agricultural output, shift the focus of investment from heavy to light industry, focus on export-led growth, and attract foreign investment. Importantly, to stabilise the socio-economic situation, the development of agriculture, forestry and fisheries, and the rural economy in general, was elevated to the task of primary importance.

Resolution No. 10/1988/NQ-TW on renovation of agricultural management dated 5 April 1988 shifted the focus of agriculture and rural development **from co-operatives to farm households**. Resolution No. 10 obliged agricultural co-operatives to contract all but 5% of farmland to households for 15 years for annual crops and 40 years for perennial crops.[5] Although the terms of the land allocation varied across Viet Nam, in most instances land was allocated on the basis of family size. This was done to ensure that each household had enough land to meet its subsistence requirements. A further egalitarian feature was that land of different qualities was allocated to each household, meaning that, depending on the geographical features of the area, households could be farming as many as 15 different plots of land scattered throughout the village (Dang et al., 2006). In addition, households were allowed to buy and sell animals, equipment, and machinery. They still had to meet production quotas, but the production amounts and prices were fixed for five years. Farmers were given the ability to make their own decisions concerning production in response to market demand and the private sector was allowed to engage in food marketing (Nguyen, 2010). Co-operatives were limited to the roles of trading (mainly inputs) and providing services (irrigation, plant protection, extension) to farmers (Dang et al., 2006). Many co-operatives simply disappeared in the wake of Resolution No. 10.

A large number of reforms quickly followed. These reduced government control over prices and **opened markets** to both greater domestic and international competition. The reforms increased the effectiveness of Resolution No. 10 by raising prices for agricultural outputs and lowering them for farm inputs. Import tariffs were introduced in 1988; the border trade with China was reopened in 1989; the ability of private enterprises to engage in foreign trade was authorised in 1991; and prices for most goods and services were opened to market determination in 1992. However, prices remained regulated for a limited number of products that were deemed to be economically and/or socially essential for the country, including fertiliser, sugar and rice. The government reduced its control on export and import activities to quotas applicable to 12 main commodities. Viet Nam switched from a fixed exchange rate regime to one in which the rate is permitted to float within a band determined by the State Bank of Viet Nam (SBV). This resulted in a sharp devaluation of the currency, making Vietnamese exports much more competitive on international markets. Agricultural production jumped. Viet Nam, which had imported more than 460 000 tonnes of food in both 1987 and 1988 to meet shortfalls in national production, became the world's third-largest exporter of rice in 1989 (Nguyen, 2010). Success in agriculture became a key driver of overall economic growth and lead to a much stronger emphasis being placed on the role of agriculture (Pham, 2006).

Expansion (1993-2000)

Having ensured food supplies at the national level, efforts were made to expand food production for export to generate foreign exchange earnings. A number of limitations were recognised. Farmers did not have long-term rights to their land, making it difficult to grow commercial crops such as coffee, rubber, cashew nut, and pepper. Many rural households, especially poor smallholders, had difficulty obtaining access to production technologies, inputs, and capital for production (Nguyen, 2010). Financial institutions refused to accept existing land-use rights as collateral, preventing households from acquiring loan funds for agricultural investment. The government promulgated a range of decrees aimed at **institutional reform and improving investment** and technological innovation including:

- Decree No. 13/1993/ND-CP on agricultural extension dated 2 March 1993 stipulated the establishment and development of agricultural extension to transfer technology to farmers
- Decree No. 14/1993/ND-CP on credit policy for family farms dated 2 March 1993 allowed rural households to borrow loans from commercial institutions
- Land Law 1993 extended land use rights to 20 years for annual crops and 50 years for perennial crops; granted households land use rights certificate (red book); and gave households the rights to exchange, transfer, lease, inherit and use land use rights certificate as a mortgage for loans
- Law on Agricultural Land Use Tax 1993 replaced both the compulsory quota system and the agricultural output tax with a land use tax

- Decision No. 151/1993/QD-TTg established the Price Stabilisation Fund (PSF) to regulate and stabilise prices of essential commodities, including urea, paddy and rice, coffee and sugarcane
- Law on Co-operatives 1996 clarified the co-operatives' role as service providers and established a legal framework for them within a multi-sectoral commercial economy.

The remaining market restrictions on key agricultural products were gradually liberalised. Most important was the relaxation of **restrictions on rice exports**. The export quota was increased from less than one million tonnes in 1992 to 4.5 million by 1998. However, the right to export was limited, allocated to two central government established SOEs – Vinafood I (also known as the Northern Food Corporation) and Vinafood II (Southern Food Corporation) – and a number of provincial SOEs (Kirk and Nguyen, 2009).[6] Internal barriers to trade in rice that had restricted the flow of rice from the south to the north were relaxed. Especially important in this regard was Decree No. 140/1997/ND-TTg, implemented in March 1997, which lifted internal trade restrictions on rice, and eliminated some licenses and controls on transport. Viet Nam signalled its commitment to trade liberalisation by entering into a large number of bilateral and regional trade agreements and partnerships.

The improved policy environment was supported by a rapid **increase in budgetary expenditure** for agriculture, which quadrupled in real terms during the 1990s (Baker et al., 2004). Increased funding was provided to the Viet Nam Bank for Agriculture and Rural Development (VBARD) to support the opening of commercial credit to farmers. Several large-scale state agricultural projects and programmes were implemented during this period, such as the *VND 50 Million/Hectare program*, the *One Million Tons/Year Sugar program*, and the *Building Canals for All Rice Fields* program (Phan, 2014; Ellis et al., 2010).

Consolidation (2000 to 2008)

In 2000 the government set the goal of becoming a modern industrialised country by 2020. The IXth National Congress of the CPV held in April 2001 proposed to strengthen market price transmission and mobilise essential resources to step up agricultural and rural modernisation and industrialisation. In response the government issued Decree No. 5/2001/ND-CP on stimulating agricultural and rural **modernisation and industrialisation** in 2001-10 (Pham, 2006). In this period, agricultural production transformed from an expansion phase toward objectives of higher yield, better quality and higher value in order to create jobs and raise income for people in the rural areas. To achieve the national agricultural objectives, four broad policies were implemented: 1) encourage domestic production of primary and processed commodities, 2) encourage quality improvement, 3) encourage domestic and international trade, and 4) increase investments from various sources in physical and social infrastructure (Phan, 2014). Further international integration – bilateral (e.g. the United States-Vietnam Bilateral Trade Agreement signed in 2000 and in effect since late 2001), regional (e.g. adopting AFTA commitments) and multilateral (negotiations to become a WTO member) – both locked in previous reforms and obliged further actions (Vo, 2008).[7]

After 2000, through active support from the government, the **livestock sector** developed rapidly. The government intensified its investments in the sector, and at the same time some major direct foreign investments were made in feed milling and livestock operations. The effort to satisfy high demand growth since 2000 triggered some selected government support to remedy the shortages. For dairy this support was directed at artificial insemination and the importation of dairy breeds to upgrade the traditional yellow Vietnamese cattle and increase its dairy potential (JICA, 2012). To overcome the challenges imposed by land fragmentation, the government issued Decision No. 150/2005/QD-TTg on 25 June 2005 urging land accumulation to be finished early so that large and modern commercial production areas can be developed to replace small-scale farm household production. However, difficulties in the land transfer procedure prevented this occurring (Tran et al., 2013).

Reorientation (2008 onwards)

Despite these successes, the government remains concerned about the unsustainable direction in which agriculture is headed. The competitiveness of the sector is low and relies on low labour cost and natural advantages; value added is limited. There is a high dependence on some traditional export markets; excessive uses of chemical inputs are polluting the environment; and the large agricultural labour force remains unskilled and unstable (Tran and Dinh, 2014a). At the VIIth Conference of Central Party Committee No. 10, the Central Committee issued Resolution No. 26/2008/NQ-TW on agriculture, farmers and rural areas dated 5 August 2008, commonly referred to as the ***Tam Nong*** resolution. This is the CPV's current orienting document for agriculture, rural development and farmer livelihoods. It states that development in all three areas will be based on the market economy with socialist orientation. Both general and specific objectives to be attained by 2020 are laid out, including the following principal goals:

- to build up a comprehensively developed agriculture sector in a modern and stable manner with large-scale commodity production, high yield, good quality, better efficiency, high competitiveness, along with the development of industry and services in rural areas, to ensure food security
- to build up new rural areas with modern socio-economic infrastructure; rational economic structure and production organisations, linking agriculture with the rapid industrial, service and urban development based on planning; stable rural society rich in traditional culture; enhancing the intellectual level and protecting the ecological environment
- to improve spiritual and physical life of rural residents; farmers are trained and act as the leaders in the rural community.

Alongside *Tam Nong*, and prompted by the sharp rise in international food prices during 2007-09, Resolution No. 63/2009/NQ-CP to **ensure national food security** was issued on 23 December 2009. The objectives of the Resolution include: ensuring adequate food supply sources for immediate- and long-term national food security, meeting nutrition needs and putting an end to food shortage and hunger; improving food consumption structure and quality and stepping up intensive rice farming; and ensuring that rice producers earn higher profits. To meet these objectives, specific production targets for 2020 are set for a variety of products, such as protecting 3.8 million ha of rice land to yield 41-43 million tonnes of rice, covering all domestic demand along with exporting about 4 million tonnes of rice per year.[8] The achievement of these targets will be done through food production planning and rice land planning; infrastructure, scientific and technological development including construction of irrigation works and new dyke systems, construction of warehouses for food reserve and preservation, selection, creation and production of adequate plant varieties and animal breeds of high yield and quality, etc.; human resource training; consolidation of food circulation and export system;

renovation of the organisation of food production forms; etc. In addition, it includes a commitment to ensure farmers receive a profit from rice production of at least 30% above the cost of production (Tran and Dinh, 2014a).

Framework for policy implementation

The broad guidelines and direction for all policy, including that for agriculture, are established by resolutions of the CPV made at their five-year national congresses and annual meetings of the Central Committee. The government develops legislation and regulation, and ten-year **Socio-Economic Development Strategy** (SEDS) and the five-year Socio-Economic Development Plan (SEDP) to implement these directions. The latest ten-year SEDS for the period 2011-20 approved in January 2011 reinforces the overall objective of making Viet Nam a modern industrialised country by 2020. It has been developed to give effect to Resolution No. 26/2008/NQ-TW and Resolution No. 63/2009/NQ-CP. Specific actions relating to agriculture include:

- increase investment in agricultural production and rural economy
- continue to improve mechanisms and policies to renovate the operation of collective economy, farm-based economy, and craft villages towards achieving sustainable development
- maintain rice cultivation area at 3.8 million ha while issuing specialised mechanisms and policies to support localities and rice growers to ensure national food security, especially in the context of climate change and sea level rise impacting the delta provinces
- continue to invest in agricultural products and locally advantageous products and products that can substitute imported ones
- attach importance to vocational training aimed at training one million rural labourers a year.

As a step towards implementing the SEDS and SEDP, Decision No. 124/2012/QD-TTg approving the **master plan for agricultural** production development through to 2020 with a vision toward 2030 was issued on 2 February 2012 and entered into force on the date of its signing. The master plan has four general objectives:

- to develop the agricultural sector towards modern, sustainable, large-scale commodity production on the basis of comparative advantage
- to apply science and technology to increase productivity, quality, effectiveness and competitiveness to ensure national food security in both the short and long term while adapting to the diverse needs of domestic and exports
- to improve the effectiveness of land use, water, labour and capital
- to raise incomes and living conditions of farmers, fishermen, salt producers and foresters.

As a further move towards implementing Resolution No. 26 and the SEDS, Decision No. 899/2013/QD-TTg approving the **plan of restructuring the agricultural sector** (often referred to as the Agricultural Restructuring Plan, ARP) towards improving value-added and sustainable development was issued on 10 June 2013. The long-term objectives for agriculture will be reflected in three sustainable pillars:

- economic: maintain robust agricultural growth and improve sectorial competitiveness, primarily via advances in productivity, efficiency, and value addition, and better meet the needs and preferences of consumers

- social: Continue to raise farmer incomes and rural living standards, reduce the incidence and severity of rural poverty, and ensure household and national food and nutrition security
- environmental: Improve natural resources management, reduce impacts, contribute to get environmental benefits and improve capacities to manage weather-related and other natural hazards in the context of Viet Nam.

The restructuring plan is viewed as a **major turning point** in agricultural policy. It signals an important change in emphasis: from extensive development based on quantity to one focused on quality and efficiency improvement. It also identifies a changing role for the government: from service provider to facilitator. Based on the perspective that sector restructuring should be in line with the overall national process of adopting the market mechanism and guaranteeing fundamental benefits for farmers and consumers, the state will play a supportive role in order to enable a favourable environment for the activities of social and economic sectors from central to local levels, promote public-private partnerships (PPP) and co-management mechanisms, and enhance the role of community organisations (FAO, 2013).

Institutional arrangements for administering agricultural policy

Central government ministries and agencies

MARD has the **main responsibility** for formulating, implementing and administering agricultural policy.[9] It is also responsible for performing the state management functions in relation to the forestry, fisheries and salt production sectors, irrigation/water services and rural development nationwide. It was formed in 1995 with the merger of three ministries: Forestry, Water Resources, and Agriculture and Food Industries, with the aim of reducing policy overlap across the primary sectors. MARD has undergone considerable changes since its establishment as new responsibilities have been added (e.g. the Ministry of Fisheries was merged into MARD in 2008) and the process of decentralisation has been implemented. The main tasks of MARD are to:

- Submit to the government legal projects, draft resolutions of the National Assembly (NA), ordinances, draft resolutions of NA Standing Board; draft resolutions, decrees based on approved programmes, annual law plans of MARD and projects assigned by the government and the Prime Minister.
- Submit to the Prime Minister the development master plans and strategies; annual, five-year and long term plans as well as key programmes and projects within MARD's mandated areas.
- Issue decisions, directives, and circulars within MARD's mandated areas; guide how to implement these documents and inspect the implementation process.
- Guide, supervise and organise the implementation of legal documents, strategies, master plans, programmes, projects, standards, techno-economic norms relating to agriculture, forestry, salt industry, fishery, irrigation/water services and rural development. Disseminate information and raise public awareness about regulations within areas covered by MARD's mandated areas.

The **organisational structure** of MARD is arranged to carry out these responsibilities. MARD is comprised of 26 units, consisting of 11 professional departments/state management offices, 9 functional departments and 6 "non-productive" units (Figure 2.1). The main

Figure 2.1. **Level flow chart of the Ministry of Agriculture and Rural Development, 2013**

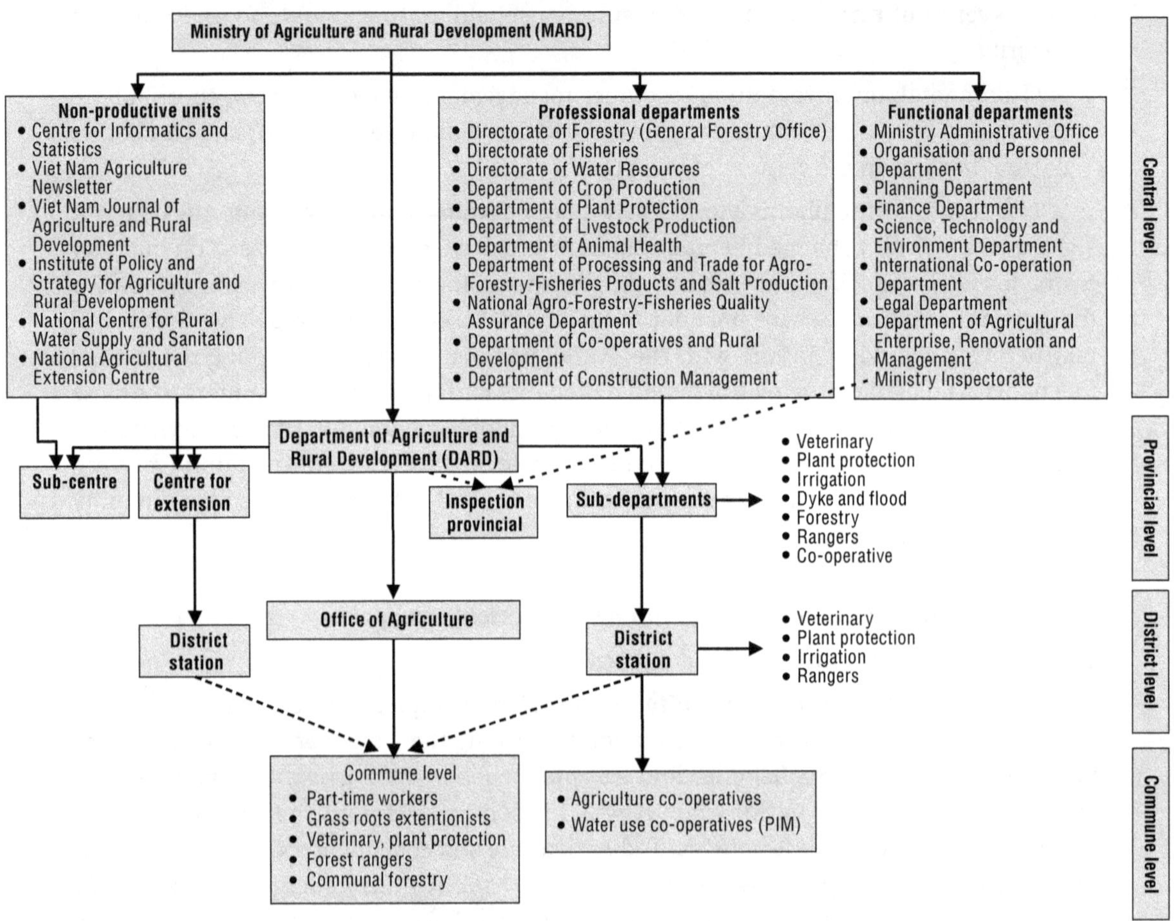

Source: MARD International Cooperation Department, *http://icd.mard.gov.vn/tabid/270/Default.aspx*.

functions of the 26 units are set out in Table 2.A1.2. Also directly under the umbrella of MARD are institutional agencies such as research institutions, universities, colleges, secondary schools, media organisations, etc.

There are 63 **Departments of Agriculture and Rural Development** (DARD), one for each of the 58 provinces and 5 municipalities. Established in 1997, these operate as local branches of MARD, in conjunction with the respective Provincial People's Committee (PPC). DARD provides advice, administration and instruction at the provincial and district levels to plan and implement central government policies including agricultural land registration, land allocation, extension, irrigation and rural development planning. At the same time, DARD's also give effect to the socio-economic development plans of the PPC.

A large number of **other central government line ministries** or public institutions have responsibilities for policies that directly impact on the agricultural sector. For example, the annex to the *Tam Nong* resolution on agriculture and rural development lists almost 20 distinct government agencies with roles to play. The main ones include:

- **Ministry of Planning and Investment (MPI)** is responsible for the state management of planning, development investment and statistics. The Department of Agricultural Economy is in charge of allocating state investment in agriculture and the Central

Institute for Economic Management (CIEM) has a direct reporting role about economic policies, including food prices and agricultural policies.

- **Ministry of Industry and Trade (MOIT)** is responsible for the state management of industry and trade in both the domestic market and internationally. The Agency of Foreign Trade (Department of Export and Import) is responsible for issuing import, export, duty exemption and quota certificates, and managing tariff-rate quotas. The Domestic Market Department manages the rice trade in the domestic market. The Science and Technology Department is responsible for the food safety of products under MOIT jurisdiction.
- **Ministry of Finance (MOF)** is responsible for the finances of the state, including managing the national budget, tax revenue, state assets and the finances of state corporations. The Department of Price Management is responsible for monitoring and implementing appropriate policies to stabilise the domestic prices of selected commodities. The Department of Tax Policy is responsible for the agricultural land use tax and import/export taxes. The General Department of State Reserves is responsible for managing the state reserves of specific products, including rice, seeds, veterinary medicines and crop protection chemicals.
- **Ministry of Science and Technology (MOST)** is responsible for the state management of science and technology including research, intellectual property, standards and meteorology. The Department of Intellectual Property manages the registration of trademarks, origin branding such as geographic indication, and the collective branding of agricultural products.
- **Ministry of Health (MOH)** is responsible for the governance and guidance of the health, healthcare and health industry including nutrition. The Viet Nam Food Administration is responsible for the food safety of products under MOH jurisdiction.

Local government

A complex structure of representation and responsibility, mirroring the national structure, exists at the subnational level. At each of the **three levels** (provincial, districts and communes),[10] inhabitants elect a local People's Council who in turn elect a local People's Committee. The local People's Committee is responsible for implementing the Constitution, laws, and documents issued by higher state organs and resolutions of the local People's Council, and for issuing and implementing decisions and directives within their areas of competence (WTO, 2013). A system of dual subordination operates at all subnational levels. The local People's Council is accountable to its electorate and to the upper level legislative body. Similarly, the local People's Committee is accountable to their respective legislative body at the same level and to the upper level executive body (Bjornstad, 2009).

Prior to the implementation in 2004 of Law No. 01/2002/QH11 on the State Budget dated 16 December 2002, the fiscal management of local budgets was highly centralised. Although local government in 2000 executed almost 40% of investment from the state budget, it was primarily a budget line distribution to the provincial level government to carry out central government policy (World Bank, 2012). The State Budget Law 2002 advanced **fiscal decentralisation** by assigning more authoritative responsibilities to provincial government and by guaranteeing revenue sources to commune governments. In particular, responsibility for service delivery was shifted from central government to the provincial level, leaving the organisation of expenditures at the subnational level up to the provincial government. This provides a large amount of discretion to the provincial government to adapt to their specific conditions, but less so for sub-provincial level governments (Bjornstad, 2009).

As a result, some provinces issue additional policies to encourage agricultural production depending on their natural conditions or their own objectives for development. For example, in 2010 Nghe An province increased investments in transportation networks and irrigation system for tea production. The provinces of Lam Dong, Nghe An, Lao Cai, Phu Tho, Lang Son, Son La, Tuyen Quang, and Thai Nguyen encouraged farmers to plant new high-yielding tea varieties by subsidising about 20%-25% of the total new variety cost (Phan, 2014). However, **large differences in budget revenue** between provinces have begun to emerge as a result of the revenue sharing arrangements (Tran, 2014b).[11] These arrangements benefit locations where there are better social-economic conditions. Moreover, Viet Nam law currently applies the origin principle whereby the taxes, fees, etc. which enterprises pay are allocated to the provinces where the head office of the enterprise is located. This creates further inequality between provinces.

Parastatal institutions

Prior to the *Doi Moi* reforms the economy was largely dominated by **SOEs**. SOEs were awarded effective monopolies in many key industries and service sector activities including agricultural input supply, the storage and marketing of outputs, and to the extent it existed, further processing. While markets have been progressively liberalised and reforms carried out to increase private ownership of SOEs, they remain an important and influential part of Viet Nam's agricultural sector. Although SOEs do not generally play a major role in agricultural production, there are some notable exceptions such as rubber, tea, coffee and to a less extent sugar, and several others are involved in processing and trade, as well as supplying inputs to farmers (Table 2.1). In addition, SOEs enjoy near-monopoly status in the production of several goods and services providing agricultural inputs, specifically fertiliser (99% of total output), electricity and gas (94%) and water supply (90%) (World Bank, 2012).

Farmer organisations

Prior to Resolution No. 10, **co-operatives** were the primary entity around which agricultural policy was centred. With this reform, they subsequently lost their *raison d'être*. Their number fell from a peak of over 126 000 in the early 1980s to just under 14 000 in 1997. Recognising farmers, particularly small-scale producers, would still benefit from co-operative institutions to provide inputs and assist with marketing, the first Law on Co-operatives dated 20 March 1996 and effective 1 January 1997 was enacted to provide a new direction. The Law required existing co-operatives to be transformed into membership-oriented service co-operatives promoting the income of their members rather than as a delivery mechanism for government; otherwise they had to be dissolved. It also provided the option for farmers to establish new agricultural service co-operatives from scratch and broadened the scope of activities that could be undertaken (Wolz and Pham, 2010). Despite these changes, the number of agricultural co-operatives has continued to fall to around 10 400 at the end of 2013. The vast are located in the northern half of the country: 36% in the Red River Delta (RRD), 19% in the Northern midlands and mountainous regions and 27% in the Northern Central and Central coastal areas (Table 2.2).

Table 2.1. **Major state-owned enterprises involved in agriculture**

Name	State owner-ship (%)	Activities
Viet Nam Northern Food Corporation (Vinafood I)	100	Involved in the purchase, processing, import and export of a range of food, wood and salt products. Imports fertilisers, animal feed products and other agricultural inputs. Reformed into a limited liability company under Decision No. 1544/2009/QD-TTg dated 25 September 2009.
Viet Nam Southern Food Corporation (Vinafood II)	100	The largest exporter of rice, with capacity to process 3 million tonnes of rice per year, storage facilities for over one million tonnes, and exclusive supplier status for government-to-government contracts. It also processes and exports a range of other agricultural commodities including cassava, maize, beans, cashew nuts and coffee, as well as seafood and fish. It imports and processes wheat into consumer products, and operates a chain of retail stores and a hotel/resort system. Reformed into a limited liability company under Decision No. 979/2010/QD-TTg dated 25 June 2010.
Viet Nam Rubber Group (VRG)	100	VRG is the largest natural rubber company in Viet Nam with 40 subsidiaries, 39 farms, and 30 processing plants. Through its subsidiaries VRG controls about 270 000 ha of rubber plantations, corresponding to 40% of the national total and 85% of total export production. Established in its current ownership structure by Decision No. 249/2006/QD-TTg dated 30 October 2006 following the restructuring of the Viet Nam Rubber Corporation as a multi-ownership Group. VRG is also involved in a livestock production, plantation forests, wood processing, electricity, engineering, managing seaports, etc.
Viet Nam National Coffee Corporation (Vinacafe)	100	Established by Decision No. 251/QD-TTg dated 29 April 1995, Vinacafe is the biggest state-owned corporation specialising in coffee production, processing, exporting in Viet Nam, and carrying out general business operations. It is made up of 56 companies, enterprises and agricultural fields. It produces 50 000 tonnes of coffee beans and exports 250 000-300 000 tonnes of coffee beans per year – accounting for 20-25% of Vietnamese coffee bean exports. Vinacafe also exports peppers and cashew nuts, imports fertilisers and facilities for the coffee industry, undertakes research into coffee production, and assists producers by providing seeds and advice.
Viet Nam National Tea Corporation (Vinatea)	100	The biggest state-owned producer, manufacturer and marketer of tea, exporting around 70 000 tonnes each year to more than 50 countries. It owns more than 100 000 ha of tea plantations, manages over 60 tea enterprises and 6 joint ventures with foreign partners. Involved in tea research and providing extension services.
Sugarcane and Sugar Corporation No. 1 – Joint-stock Company (Vinasugar I)	64	Growing sugar cane and production of sugar products, confectionary, spices and beverages. Trading in fertiliser, agrochemicals, food products, and machinery and spare parts for the sugar industry. Provision of construction and investment services for sugar mills and business warehousing. Capital invested into a number of other sugar processing companies. Reformed into a joint-stock company under Decision No. 1913/2012/QD-TTg dated 21 December 2012.
Sugarcane and Sugar Corporation No. 2 – Joint-stock Company (Vinasugar II)	64	Growing sugar cane and production of molasses, refined sugar, confectionary, alcohol, soft drink, wine, beer, micro-organic fertiliser, plywood MDF, cattle feed; sugar technical services, providing goods for material areas. Producing food containers/wrapping, textile industry. Manufacturing the mechanical products and tools for sugar industry. Investment consulting and construction for sugar industry. Building, repairing, developing and expanding sugar factory. Wholesale and retail of the products of food manufacturing, the specialised machines and spare-parts, materials, and consumer products. Hostel business, office leasing, warehouse leasing and house trading. Capital invested into a number of other sugar processing companies. Reformed into a joint-stock company under Decision No. 1914/2012/QD-TTg dated 21 December 2012.

Source: Data collected from a variety of websites including the home pages of the listed SOEs and financial investment company reports.

In addition to formal agricultural co-operatives, there also exist more informal **collaborative groups** among farmers. There were almost 62 500 such groups in 2013, up from about 50 000 in 1996. In comparison to formal co-operatives, many of these are located in the Mekong River Delta (MRD) region (29%). Collaborative groups are registered at the commune level only. In general, they focus on organising soil preparation and irrigation, as they are not allowed to conduct business activities (e.g. marketing) on their own.

Table 2.2. **Agricultural co-operatives by sector and region, 2013**

Sector	Region						Total
	Red River Delta	Northern midlands and mountainous areas	North Central and Central coastal areas	Central Highlands	South East	Mekong River Delta	
Number							
Agricultural services	3 633	1 842	2 691	412	282	917	9 777
Forestry	4	31	8	0	0	10	53
Aquaculture	101	105	48	6	20	211	491
Fisheries capture	1	1	23	0	2	6	33
Salt industry	25	0	24	0	1	2	52
Total	3 764	1 979	2 794	418	305	1 146	10 406
Share of total agricultural co-operatives (%)							
Agricultural services	34.9	17.7	25.9	4.0	2.7	8.8	94.0
Forestry	0.0	0.3	0.1	0.0	0.0	0.1	0.5
Aquaculture	1.0	1.0	0.5	0.1	0.2	2.0	4.7
Fisheries capture	0.0	0.0	0.2	0.0	0.0	0.1	0.3
Salt industry	0.2	0.0	0.2	0.0	0.0	0.0	0.5
Total	36.2	19.0	26.8	4.0	2.9	11.0	100.0

Source: Ministry of Agriculture and Rural Development (2014), *Statistical Yearbook of Agriculture and Rural Development 2013*.

While this might be a disadvantage, these entities can work fairly flexibly on an ad hoc basis, need only a simple management structure and do not have to pay taxes (Wolz and Pham, 2010).

Despite government support, the role of co-operatives **remains insignificant**. A revised Law on Co-operatives was introduced in 2013.[12] The new Law is intended to support the innovation and development of co-operatives through the training of personnel management, provision of technical assistance and technology transfer, and assistance with market development and trade promotion. However, many agricultural co-operatives are still passive in reforming and adapting to the market economy and economic integration (Tran, 2014b). Many farmers are reluctant to participate in co-operatives because of their past experience when co-operatives were forced on them, preferring autonomy over dependence on others. Consequently, many rural households do not act collaboratively in terms of commercial matters, e.g. negotiate contracts, lodge complaints or settle disputes, despite potentially being better off by doing so.

The Viet Nam Farmers Union (VFU) is a **socio-political organisation** of the CPV established in 1930. However it is very weak at the grassroots level, and only operates in an administrative manner at the central level. Its main roles are to dissemination and explain new policies and mobilise the farmers in general. According to a survey conducted in 2005, the impact of the VFU on rural life ranks third out of four, which clearly shows that it has not really become an organisation supported by farmers, effectively acting in their interest. Although rural residents account for almost 70% of the total population they are yet to play an important role in socio-economic life and in the policy making process (Nguyen, 2010).

2.3. Domestic policies

This section discusses in detail the domestic policy measures that **provide support** to agriculture in Viet Nam. It begins by examining the policies through which transfers are directly received by producers, i.e. included in the measurement of the Producer Support

Estimate (PSE), from price support measures and input subsidies through to disaster relief. Trade policies can also provide support to producers and these are discussed in Section 2.4. Three important policies providing support to the agricultural sector as a whole are then discussed: extension, research and development and infrastructure. These are included in the General Services Support Estimate (GSSE). The final sub-section discusses policies that are provided to consumers specifically for the purposes of reducing the price of the goods they consume. These are included in the Consumer Support Estimate (CSE).

Price support measures

The government has operated a **State Reserve** system for certain products since reunification.[13] Agricultural-related goods held in the reserve are rice and various production inputs. The strategy for the State Reserve out to 2020 has set the following annual reserve targets to be held by 2015: 500 000 tonnes of paddy and rice (paddy equivalent); 10 000 tonnes of rice seed; 1 500 tonnes of maize seeds; 130 tonnes of vegetable seeds; 600 tonnes of pesticides; 10 million doses of vaccines, and 1 million litres of antiseptics for the prevention and suppression of cattle diseases.[14] The current reserve is maintained for the purposes of preventing and overcoming the consequences of natural disasters and epidemics affecting people, plants or animals, and ensuring economic, social and national security. Stocks are held throughout the country under the responsibility of 22 Department of State Reserves, each covering 2-3 provinces. The annual quantity of paddy/rice to be brought into the reserve is set by Prime Ministerial decision. This can be brought from farmers or traders, with a maximum price set by MOF.

In 1992, the government **deregulated prices** for most goods and services in the economy.[15] Some exceptions, however, remained. These goods and services were divided into two lists: one for which the state determined fixed prices and one subject to framed (floor or ceiling) prices. Fixed prices were set for electricity, postal fees, domestic telephone, water, natural resource; land rent and residential premises owned by the government. Framed prices for agricultural-related commodities included: maximum prices for selling rice in major domestic markets, for transporting food from the south to the north and to mountainous areas, for transporting fertiliser from the north to the south, for importing urea (fertiliser) in foreign currencies; and minimum prices for buying paddy from farmers and for exporting rice in foreign currencies.

A **Price Stabilisation Fund** (PSF) was established in 1993 with the objective of regulating and stabilising domestic prices.[16] There was no set list of goods and services subject to price stabilisation, but it was generally applicable to essential goods such as paddy and rice, coffee, rubber, sugarcane, cashew nuts, petroleum-related products, iron and steel, and fertiliser (WTO, 2006). Customs surcharges were used as an instrument to both finance the PSF and stabilise domestic prices. The government determined when commodities would be subject to a surcharge, the application period and the surcharge rate (which was based on the difference between external and domestic prices). Surcharges applied to both exports (e.g. unprocessed cashew nuts, rubber latex and coffee) and imports (e.g. petroleum, iron and steel for construction purposes, DAP fertiliser and sheet steel). In addition to surcharges, stocks held by SOEs were used to stabilise prices. The PSF, together with Export Reward Fund, was transformed into the Export Promotion Fund (EPF) in 1999.[17]

In 2004 the policy of setting framed prices was abandoned and replaced with a more general commitment to **stabilise market prices** for essential goods and services when their prices "abnormally fluctuate".[18] Policy measures that could be employed to achieve price stabilisation were: a) adjusting the demand and supply of domestic goods and export/import goods, and/or adjusting the allocation of commodities between regions or localities in the country; b) purchasing or selling out of the state reserve; c) controlling goods in stock; d) setting maximum prices, minimum prices or price brackets; e) controlling price components; f) subsidising farm produce prices when the market prices drop too low, thus causing damage to the producers; and g) subsidising prices of other important and essential commodities and services. While changes were made in relation to framed prices, the state continued setting fixed prices for electricity, postal fees, etc.

Agricultural-related products on the price stabilisation list included inputs, outputs and final consumer products (Table 2.3). Specific conditions that defined **abnormal fluctuations** were set by regulation.[19] For paddy this was set at a 15% or greater fall in the purchase price over a 30 day period; for coffee beans, seed cotton and sugarcane it was a fall of 20% or more. An abnormal fluctuation for rice was considered to be a rise in the retail price of 25% or more over a 30 day period. For urea it occurred if the price of one kg of urea fertiliser exceeded the price of two kg of paddy within 30 days.

Table 2.3. **Agricultural-related products subject to price stabilisation**

Market level	1993-99[1]	2004-08[2]	2008-14[3]	2014 onwards[4]
Inputs	Fertilisers	Urea	Chemical fertilisers Plant protection chemicals Certain veterinary products[5] Certain animal feeds	Nitrogenous fertiliser, urea, NPK fertiliser Plant protection chemicals Prophylactic vaccines for livestock and poultry
Farm level output	Paddy Sugarcane Coffee beans Rubber Cashew nuts	Paddy Sugarcane Coffee beans Cotton seed	Paddy	Paddy
Final consumption	Rice	Rice Ginned cotton	Rice Edible sugar Milk products	Rice Edible sugar Milk formula for children under 6 years of age

1. Decision No. 151/QD-TTg dated 12 April 1993.
2. Decree No. 170/2003/ND-CP dated 25 December 2003.
3. Decree No. 75/2008/ND-CP dated 9 June 2008.
4. Decree No. 177/2013/ND-CP dated 14 November 2013.
5. Foot-and-Mouth Disease vaccine; Avian Influenza vaccine: and antibiotics Oxytetracycline. Ampicilline Tylosin and Enrofloxacin.

A number of **amendments** were made in 2008.[20] First, changes were made to the price stabilisation list. Sugarcane, coffee beans and cotton seed were removed, leaving only paddy; the range of agricultural inputs covered increased; and edible sugar and milk products were added to join rice while ginned cotton was removed. Second, alternations were made to the specific conditions defining abnormal fluctuations. For all goods and services covered, it reduced the period over which prices were allowed to vary from 30 to 15 days, allowing greater market flexibility. For the new goods subject to price stabilisation, a 15% or greater increase was considered abnormal for animal feeds and plant protection drugs and a 20% or greater limit was set for chemical fertilisers and certain veterinary drugs. Finally, a third list of goods and services subject to price registration was established. Producers and traders of goods and services on this list were required to register their prices with the relevant state management agencies. Permission to change a registered price must be obtained from the relevant state management agencies before doing so, showing evidence as to why the change is required. Agricultural-related products on the price registration list were very similar to those subject to price stabilisation. It covered specific chemical fertilisers (urea, DAP, NPK and phosphate), specific plant protection products, certain veterinary drugs, table sugar (white sugar and refined sugar), the same types of animal feeds, rice and formula milk for children under 6.

The obligation to **register prices** was initially limited to just SOEs. In 2010 this obligation was extended to all private enterprises producing, importing, distributing and/or selling goods on the price registration list.[21] Moreover, the conditions under which price stabilisation measures were to be implemented also changed. The specific conditions were removed and greater discretion was given to government officials. State management agencies could now introduce measures when the price of a good increases or decreases faster than changes in production costs as determined by the agencies, or if price fluctuations are considered "groundless", or because a producer or trade abuses their market position. No changes were made to the lists of goods and services subject to price stabilisation or price registration.

Effective 1 January 2014 certain animal feeds and veterinary medicines have been removed from the list of products subject to price stabilisation while milk products has been refined to milk products for children under 6 years old.[22] To simplify matters, goods and services subject to price stabilisation also became those subject to price registration. Further, price registration requirements have been restricted to the time when price stabilisation is enacted which in turn has been limited to no more than six months. However, the conditions for implementing price stabilisation measures remain at the **discretion of officials**. These can be imposed when a market price increase or decrease is "unreasonable" compared to the change in production costs, or is "unreasonable" in the case of natural disasters, epidemics, economic crises, etc., or negatively affects economic and social stability. In addition to price registration, other policies that can be implemented include buying into or selling goods from the State Reserve, financial support (e.g. tax concessions and interest rate subsidies) and price support in accordance with international commitments, and regulating demand and supply of domestic goods and exports and imports. A price stabilisation fund has also been re-established, but can only be used to stabilise the prices of a limited range of the goods and services subject to price stabilisation: gas and oil products for domestic consumption; electricity retailing and rice.

Aside from these broader price policy instruments, two specific policies have been introduced to support the **farm gate price for paddy**. Since 2009 the government has subsidised the temporary storage of rice during harvest for the purpose of increasing demand and avoiding price reductions. Under these interventions, the government subsidises all the interest payment on loans taken out by exporting enterprises to purchase rice for temporary storage (usually 3 to 4 months). Enterprises must procure rice at the target paddy price introduced in 2011 to receive the subsidy. MOF and SBV are responsible for allocating funds from the state budget to support these interventions. These interventions are made annually based on the changes in market price of rice, often at the point of lowest price. These procurement policies are a popular intervention by the government and have been used a number of times in recent years (Table 2.4).

In order to meet the commitment made under Resolution No. 63/NQ-CP on food security dated 23 December 2009 to ensure the farm gate price for rice is at a level that provides growers with a profit of more than 30%, **target paddy prices** have been established

Table 2.4. **Temporary storage procurement policy timing and volume, 2009-13**

Decision	Time for rice purchase	Time for storage	Volume (tonnes)
Decision No. 1518/2009/QD-TTg	20 September to 20 November 2009	20 September 2009 to 20 January 2010	0.5 million
Decision No. 993/2010/QD-TTg	15 July to 15 September 2010	15 July to 15 November 2010	1 million
Decision No. 287/2012/QD-TTg	15 March to 30 April 2012	15 March to 15 June 2012	1 million
Decision No. 311/2013/QD-TTg	20 February to 31 March 2013	20 February to 20 May 2013	1 million
Decision No. 850/2013/QD-TTg	15 June to 31 July 2013	15 June to 15 September 2013	1 million

since 2011.[23] Under this policy, production survey data is collected for each province and for each rice season. From this data MOF and MARD calculate an average production cost for eight regions that cover the whole country. Provincial People's Committees use these average costs to determine and proclaim rice paddy floor prices at which enterprises are encouraged to buy rice from farmers, i.e. target paddy prices. Separate target paddy prices (production cost plus 30% profit) are therefore set for each region and for each season (MARD, 2014b).

These interventions have had a positive impact on rice prices, with the average farm gate price lifting above the average export price when measured on an equivalent basis (Figure 2.2). However, the impact has been limited because of the short duration of the subsidised loan coupled with the variation in harvesting time among provinces. A number of factors further **weaken the effectiveness** of this policy measure. First, exporting enterprises often buy rice from local traders or assemblers so they have no direct control over the price farmers receive. Second, there can be large differences in production costs within each of the eight regions, particularly in relation to land lease fees and loan interest rates. Consequently, the target price based on the average will benefit some and disadvantage others. Finally, MOF/MARD are often late in determining production costs, which is supposed to be given at the beginning of each crop season (MARD, 2014b). Consequently, many farmers and state officials agree that most of the **benefits of the interest subsidy is captured by rice exporters**, mainly state-owned enterprises (SOEs), rather than by farmers.[24]

Figure 2.2. **Comparison of different types of rice prices in Viet Nam, 2000-13**

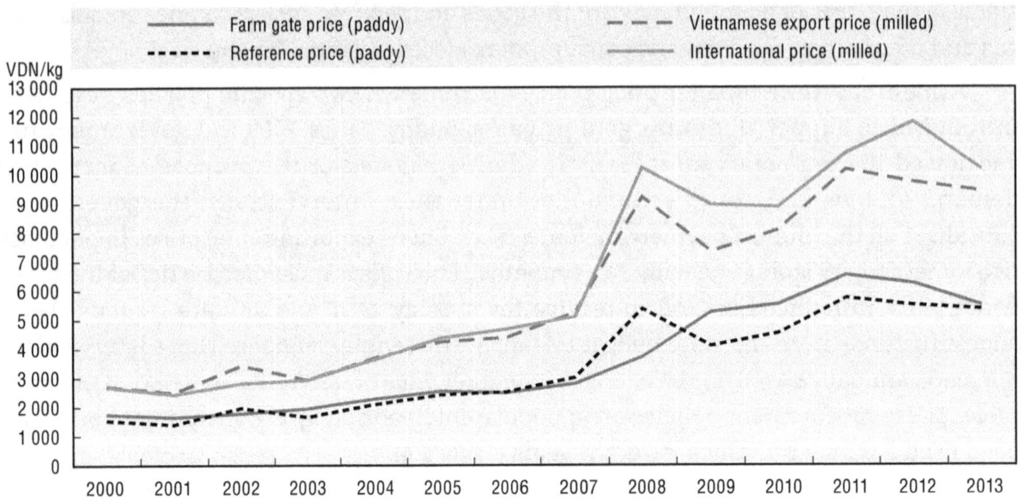

Note: International price is Thailand 15% broken, f.o.b. Bangkok.
Source: Agroinfo, Ministry of Agriculture and Rural Development; USDA Rice Year Book 2014, *www.ers.usda.gov/data-products/rice-yearbook-2014.aspx*.

StatLink http://dx.doi.org/10.1787/888933223757

For example, to support the temporary procurement of one million tonnes of rice in 2013 the government spent at least VND 200 billion (USD 9.5 million) to cover the 7 612 loans that had been taken out at interest rates of 10-10.5%. However, the impact of this was to increase farm gate prices by only VND 100-200 kg (less than USD 0.01 kg). To improve the effectiveness of these policies, MARD is drafting a new policy on paddy/rice procurement in which subsidised loans will be provided to both farmers and enterprises who signed contracts to buy paddy from farmers. With these changes, farmers are expected to have greater ability to store paddy at home and sell whenever they want (MARD, 2014b).

Irrigation service fee exemption

An **irrigation service fee** (ISF) has been collected from farmers and other water users since 1963 to contribute to the expense of managing, maintaining and protecting irrigation works above the "canal gate", i.e. the upper-level systems comprising diversions or pump stations, and primary and secondary canals that lead into tertiary canals. The ISF is collected from farmers on behalf of the irrigation and drainage management companies (IDMC) that manage the upper-level system by water user groups (WUG), which are responsible for administrating the distribution of water to farmers and maintaining the infrastructure below the upper-level system. It was originally collected in the form of paddy production, but from 1995 onwards it has been collected in monetary units. For land in paddy production, it is a fixed change per ha that varies by region and method of irrigation (motor, gravity or combination). Other users pay a volume charge (VND per m^3) that varies according to the output produced, e.g. animal husbandry, permanent crops, aquaculture and industrial.[25]

Since 2009 individuals and households in agricultural, forestry, salt and aquaculture production have been **exempt from payment of the ISF**.[26] For individuals and households in areas subject to socio-economic difficulties, the exemption covered all areas of land and water used; for all others the exemption covered only the land and water used for which they had a land use right certificate (LURC). Funding from central and local government to IDMCs has been increased to offset the fall in ISF revenue, which previously covered about half their costs (Baker et al., 2004). While the exemption applied to the ISF, farmers are still responsible for supporting the management of the tertiary and field canals under the responsibility of WUG through the provision of labour, in-kind contributions and finance. The rationale behind the exemption for farmers is that irrigation canals, bunds and dykes perform a range of public good functions and are widely used for transport. Furthermore, a large portion of the maintenance requirements such as dredging are caused by impacts upstream of particular irrigation systems, and their costs should not necessarily be borne by farmers.

A review of the exemption decision identified a number of positive and negative **outcomes** (Cook et al., 2013). Farmers gained on average a VND 400 000 (USD 20) increase in annual net farm income. Using an average agricultural household income of VND 1.458 million (USD 70) per month, the saving represents 2% of annual income. IDMCs gained from having a more consistent source of funding as previously the collection of the ISF from farmers had varied considerably from commune to commune. However, total central and local government expenditure on supporting operations and maintenance increased from VND 3.3 trillion (USD 203 million) in 2008 to over VND 6.2 trillion (USD 350 million) from 2009 onwards (Figure 2.3). Further, it has weakened the link

Figure 2.3. **Expenditure on supporting irrigation operations and maintenance, 2000-13**

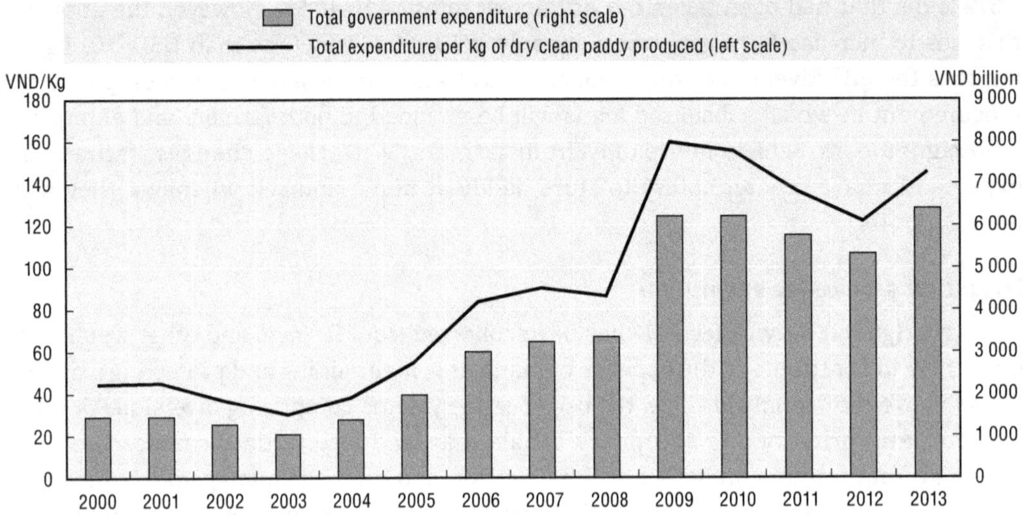

Source: Own tabulation based on OECD (2014), "Producer and Consumer Support Estimates", *OECD Agriculture Statistics Database*.

StatLink http://dx.doi.org/10.1787/888933223763

between farmers, WUGs and IDMCs in managing the resource. Finally, removing the ISF reduces the incentive for farmers to use water efficiently and for IDMCs to provide a quality irrigation service.

Plant genetic and livestock breeding support

There are many programmes through which plant genetic or livestock breeding material are provided to farmers at **subsidised rates**. Many of these are conducted at the provincial level and thus are difficult to quantify. However, the limited information available suggests that the amount of money spent on doing so is not that large. For example, in the two year period 2003-04 the provincial government of Tay Ninh spent VND 150 million (USD 9 500) on providing 80 000 cashew seedlings to growers (Que and Manh, nda). These forms of support have historically been introduced to encourage product diversification or to improve the quality of production. In more recent years, they have been used for disaster recovery to support farmers in response to natural disasters and disease outbreaks.

In particular, the frequency, intensity and diversity of **livestock disease outbreaks**, that has accompanied the expansion of the livestock industry, has created a number of policy challenges. Since 2003 the country's livestock sector has experienced multiple rounds of avian influenza, H5N1 bird flu virus, foot-and-mouth disease (FMD) and Porcine Reproductive and Respiratory Syndrome (PRRS), also known as blue ear disease. Over 49 million birds were culled between 2003 and 2010 due to avian influenza. In 2010 a further 36 272 birds were culled as a result of avian influenza, 371 animals were culled due to FMD, and 77 158 pigs were culled in response to PRRS (OECD, 2012). The domestic livestock industry comprises mainly small-scale or backyard farm operations, which have poor hygiene standards and are susceptible to epidemics. Disease outbreaks will constantly feature as a challenge to the industry as long as it remains fragmented and low in technology and health standards (BMI, 2011).

Following the onset of avian influenza in 2003, the *Emergency Centre for Transboundary Animal Diseases (ECTAD)* was established (Box 2.1). Aiming to control and eradicate these outbreaks, the government established emergency support policies. Following the initial relevant government decisions in 2004 and 2005, several decisions on **compensation** were approved.[27] The rationale is to encourage farmers to declare disease outbreaks at an early stage so that it can be contained among the livestock population. A key component of the support policy is a compensation level that encourages farmers to cull animals rather than sell them illegally on the market. The government's compensation rate for birds culled during stamping-out procedures was raised from 10-15% of the market value of the destroyed/slaughtered livestock in 2004 to 50% in June 2005 and 70% in June 2008 (OECD, 2012).

Box 2.1. **Involvement of the FAO in Vietnamese agriculture**

The FAO first started working in Viet Nam in 1978 and a representation was established in Hanoi one year later. FAO quickly became an important partner and the main contributor of technical assistance in the agricultural sector. In the early 1980s, FAO's programme in Viet Nam was its third largest in the world after India and China. A large part of FAO's initial assistance was in the field of institutional and capacity development. Help was provided to establish a number of new institutions, including the Institute of Agricultural Science, the Soils and Fertiliser Institute and the National Plant Protection Service, as well as to strengthen existing organisations. Through such technical assistance projects, Viet Nam gained access to up-to-date technologies, equipment and techniques.

The focus of FAO's involvement changed during the 1990s towards a concentration on the provision of policy advice. FAO contributed its knowledge and expertise to policy development and planning, including the formulation of key policy and programme documents such as the National Plan of Action for Nutrition, the National Strategy for Rural Development and the National Strategy for Agriculture towards the year 2010. At the same time, FAO contributed to the formulation of legislation aimed at enhancing the ability of the agricultural sector to respond to the challenges and opportunities posed by the new market environment.

During the 2000s, a major focus was on ECTAD. The programme works under four major multi-sectoral initiatives, including the Joint *Government-UN Programme*, the World Bank's VAHIP Project, the One UN Initiative with the largest contribution to FAO coming from United States Agency for International Development (USAID) projects. FAO has worked closely with Viet Nam's government to develop robust, coherent disease control strategies including outbreak detection, investigation, response, laboratory support, epidemiological investigation and capacity building. FAO has also assisted with laboratory support to encourage the transition from detection of agents to diagnosis of diseases. FAO has continued and expanded the ECTAD programme for the prevention and control of highly pathogenic avian influenza (HPAI), including the UN Joint Programme and two USAID projects. FAO is also collaborating with the World Bank on the Viet Nam Avian and Human Influenza Control and Preparedness Project.

Over the five-year period 2006-11 supporting activities to animal health (control of HPAI and other disease control) received the majority of FAO funding (USD 24.5 million out of USD 43.5 million). The second highest was food safety which received USD 3.8 million for 13 projects. Other areas receiving USD 3 million or more were pesticide risk reduction and irrigation and water management and rural development (including strengthening farmer organisations, nutrition, gender aspects and pro-poor risk reduction). Within these areas, the most common activities within projects are capacity building, including training extension

Box 2.1. **Involvement of the FAO in Vietnamese agriculture** *(cont.)*

and technical assistance. This is usually delivered to farmers and local authorities, followed by provincial, district and central government staff. Other activities include analysing and assessing emerging problems, projects and policy interventions. The third area of activities in the FAO projects are supporting state management at all levels such as: developing legal documents, strategies, action plans, policy mechanisms, governance mechanisms as well as agricultural product standards, quality management and food safety, as well as introducing and complying with international norms and standards.

The four priorities in the current five-year Country Programming Framework 2012-16 are: support for effective policies and legal framework on rural livelihood, food and nutrition security and food safety; support for climate change adaptation and mitigation; support for improving the provision of goods and services from agriculture, forestry and fisheries in a sustainable manner; and support for enabling more inclusive and efficient agricultural and food system for the rural vulnerable groups. The financial requirement for implementing CPF 2012-16 is USD 62.5 million.

Source: FAO (2011) and FAO (2013).

In addition to compensating animals, the government has introduced two policies to control disease outbreak and to reduce the costs of preventing and curing diseases affecting porcine production.[28] These **provide veterinary medicines**, for example blue-ear pig and hog cholera vaccines, in response to animal diseases. MARD provides vaccines to deal with diseases with the funds of MOF. Provincial People's Committees implement the support at local level (Tran and Dinh, 2014a).

Since 2009 the government has also taken steps to broaden its support for facilitating the **recovery of agricultural production** after disease outbreaks and natural disasters.[29] The level of support provided is based on the damaged planted area, the number of damaged livestock and the extent of damage. Central government provides up to 80% of this for farmers in mountainous provinces and the Central Highlands, and up to 70% for farmers in other provinces. The remaining part is supplied from provincial budget. MARD is responsible for determining the natural disasters and diseases that are supported; MOF provides funding from the state budget; and Provincial People's Committees implement the support and actively use local funds (Tran and Dinh, 2014a).

Input subsidies have recently been introduced to **support the development of paddy land.**[30] Since 1 July 2012 the government has offered assistance to cover:

- 70% of the expenses for reclaiming and improving unused land into rice-farming land and improving other paddy planting land to specialised land for wet-paddy
- 100% of expenses for paddy seeds in the first year to produce paddy in land that is reclaimed and 70% of expenses for paddy seeds in the first year to produce paddy in other paddy planting land, which is improved to specialised paddy planting land
- 70% of expenses for fertiliser and plant protection products for over 70% of damage and 50% of expenses for fertiliser and plant protection products for 30-70% of damage.

Programme 135

The Socio-economic Development Programme for Ethnic Minorities and Mountainous Areas (known as Programme 135) is the largest and most important **poverty reduction**

programme targeted on ethnic minorities and remote areas. When initially approved in 1998 it was scheduled to last seven years.[31] The programme was extended in 2006 for a further five years (Phase II) and has recently been extended into a third phase from 2011 to 2015.[32] The initial phase consisted of four major components: infrastructure development, e.g. roads, health centres, irrigation systems, water supply systems, markets, etc. at both the village and commune level and at the inter-commune level; training for commune/ village staff in remote and mountainous areas, agricultural and forestry extension (linked to processing industries) and relocation planning. Emphasis was given to developing village, communal and inter- communal infrastructure, with over VND 9 142 billion (USD 600 million) spent during Phase I from 1998 to 2005.

In Phase II, the programme gave greater emphasis to **supporting agricultural production**, capacity building and improved livelihood. Activities in support of market-oriented agricultural production and income generation included: providing agriculture, forestry and fishery extension; establishing demonstration models; distributing agricultural inputs; and delivering equipment and extension services for post-harvest and processing activities. A total of VND 14 trillion (USD 832 million) was spent over the five years on building 4 125 demonstration models of agricultural development and fisheries production, purchasing 42 000 machines for production and post-harvest processing, and running 12 000 capacity building training projects for local officials among other things. Phase III is building on this by improving access for poor and disadvantaged communes to preferential loans for investment in production. The programme will classify each commune according to the difficulties they face, which have different coefficients for the allocation of capital.

Credit policies for farmers

Up until early 2000s, the government controlled credit availability and interest rates in all sectors of the financial market through the activities and regulation of the State Bank of Viet Nam (SBV), and regulations controlling access to credit. The ability for farm households to access **commercial credit** commenced in 1993; previously loans had only been available to households through institutions such as co-operatives.[33] This was supported by the 1993 Land Law, which allocated LURCs to households and gave them the right to use these as collateral for bank loans, and the establishment of the state-owned Viet Nam Bank for Agriculture and Rural Development (VBARD), also known as Agribank. VBARD has been strongly supported by the government through the provision of statutory capital and operating facilities. To further expand access to credit, the concept of "trustable mortgage" *(tin chap)* was introduced in 1999, allowing farm households to borrow up to VND 10 million (USD 700) without collateral.[34] This was quickly raised to VND 20 million (USD 1 400) in 2000. Major changes in interest rate policy were implemented in May 2002 allowing banks to determine interest rates based on the supply and demand of capital, and the level of trust or confidence they have in the customer or customer group (Marsh, Ahn and MacAulay, 2006).[35]

A further financial reform directly affecting the agricultural sector was the establishment of the stand alone, non-profit **Viet Nam Bank for Social Policies** (VBSP), which commenced operation in January 2003.[36] The primary objective of VBSP is to support the government's poverty alleviation efforts through the provision of credit. It achieves this by providing accessible financial services and low interest loans to people living in remote areas, members of ethnic minority groups, students, etc. Loans from the VBSP are subsidised with low interest rates (ranging from 0.0-0.8% per month) and are generally for small amounts. The maximum loan depends on the particular programme but is typically VND 30 million

(USD 1 400). Prior to the establishment of VBSP, these activities were provided by the Viet Nam Bank for the Poor (VBP), which operated through VBARD. Subsequent to the establishment of the VBSP, the provision of preferential credit has been completely removed from VBARD's remit.

As part of the economy-wide demand stimulation package, the government introduced in 2009 a policy to provide agricultural producers with short-term **concessional interest rate** loans to purchase machines, mechanical equipment, facilities and materials.[37] The objective was to reduce investment costs, improve production capacity and promote industrial development in rural areas. For machines, the loans could be equal to 100% of the value of goods, but not exceeding VND 5 million (USD 293) in the case of computers. These loans are exempt from interest payment for at most 24 months. With regard to fertilisers and pesticides, the loans could also be equal to 100% of the value of goods, but not exceeding VND 7 million (USD 410) per ha. The interest rates for these loans were 4% lower than that of commercial loans. The preferential support lasts for at most 12 months (Tran and Dinh, 2014a). These preferential interest rates were available on loans taken out in 2009 and 2010. Just over one million farmers borrowed VND 776 billion (USD 40.8 million) under the programme during 2009 and another VND 147 billion (USD 7.7 million) was lent to 6 424 farmers in the first four months of 2010.

To spur agriculture and rural development and implement the *Tam Nong* Resolution, in 2010 the government increased the limits for **loans without asset security:** to VND 50 million (UDS 2 686) for individuals and households engaged in agricultural activities; VND 200 million (USD 10 745) for households carrying out business or production activities or providing services for agriculture and rural areas; and VND 500 million (USD 26 863) for co-operatives and farm owners.[38] These loans can be used for production costs in the field of agriculture, forestry, fishery and salt production; development of rural production and business lines; construction of rural infrastructure; processing and consumption of agriculture, forestry, fishery and salt products; and trading in products and services for agriculture, forestry, fishery and salt production.

Under a separate policy initiative, access to subsidised credit has been provided since 2010 for the purposes of **mitigating losses** in agricultural production.[39] Post-harvest losses are high in Viet Nam: 11-12% for paddy, 13-15% for maize, 20-22% for vegetable and fruits, 15% for coffee and 18-20% for cassava (MARD, 2014b). Preferential loans can be used to buy machinery and equipment to reduce post-harvest losses including dryers; machines used for the cultivation and harvest of rice, coffee tea and sugarcane; and machines used for aquatic production and cold storage. The machines have to be new, legally standard and have a local content value of at least 60%. The loans could be up to 100% of the cost. The state subsidises 100% of the interest rates for these loans in the first two years and 50% from the third year onwards. They also offer preferential loans to develop projects of production and storage facilities for such purpose. These loans could be up to 70% of project value and last at most 12 years. The financial support provided for these loans is the difference of payment between interest rate of commercial loans and that of state credit for development (currently 10.8% per annum). These preferential loans are channelled through five designated state-owned commercial banks, namely VBARD, Mekong Housing Bank, Joint Stock Commercial Bank for Investment and Development of Viet Nam, Viet Nam Joint Stock Commercial Bank for Industry and Trade and Joint Stock Commercial Bank for Foreign Trade of Viet Nam (MARD, 2014b).

As at April 2014 the total preferential interest loans reached only VND 1 340 billion (USD 64 million). The policies are **not as effective as anticipated** for many reasons. Many farmers cannot access the credit support because of complicated and inconsistent procedures such as the requirement of submitting an invoice. The minimum local content requirement for machines is not realistic. To access the concession loans, enterprises have to sign contracts for production linkages, consumption of agricultural products and usage of agricultural mechanical services with co-operatives, households and individuals (Tran and Dinh, 2014a).

Direct payments

As part of a broad policy package to protect and support the development of paddy land, the government introduced for the first time a **direct per ha payment** to rice farmers in 2012.[40] Over the period 2012-15 the following annual payments will be made:

- VND 500 000 (USD 24) per ha to associations, households and individuals cultivating paddy on wet-paddy farming land (defined as land currently under wet-paddy cultivation or having the conditions for growing two or more wet-paddy crops a year)
- VND 100 000 (USD 5) per ha to associations, households and individuals cultivating paddy on other paddy farming land (land for growing only one wet-paddy crop a year and land for growing upland rice), except upland fields not under paddy planting land-use plans.

This policy has raised the level of transparency and clarity of support for rice because farmers know the exact value of the transfer they are receiving as it is provided directly and not through any intermediate stakeholders. However, there are some **challenges to implementing and monitoring** the programme because of the large number of farmer households – approximately 10 million. There are also difficulties with identifying whether the upland rice field is in planning area or not. However, approximately 95% of current paddy land meets the wet-paddy land definition. Using an average agricultural household income of VND 1.458 million (USD 70) per month and assuming the household farms 2 ha of wet-paddy land, the payment represents almost 6% of annual income. MARD officials wonder whether this is enough to meet the government's objective to stop the switch to other crops or to non-farm activities (MARD, 2014b).

Prior to the introduction of the per ha payment for paddy land, direct support was provided through the "661 Programme" (also called the "5 Million Hectares Reforestation Programme").[41] This programme, which began in 1998, had the objective to **reforest 5 million ha** of land by 2010: 2 million ha of special use and protected forests including natural regeneration, and 3 million ha of production forests (2 million to produce raw materials for wood processing and 1 million to be planted in fruit trees and other perennial crops). A further objective of the programme was to create employment for 2 million people and increase incomes of people in forest areas as a contribution to poverty alleviation, hunger eradication and the development of rural mountainous areas. A number of policy measures were implemented under this programme including loans to large SOEs involved in forestry and direct support to households to establish forest plantations. This latter measure took the form of small grants to households planting trees at a nationwide cost norm of VND 2 million (USD 104) per ha, out of which the costs of seedlings, fertiliser and extension services, etc. had to be paid. Households did not need a LURC to be eligible, but they did have to prepare the land before they could receive support. The minimum plot size was 0.5 ha; the minimum area per household was 1 ha (Sikor, 2011).

Agricultural insurance

Agricultural insurance markets are undeveloped in Viet Nam. For example, premiums from agricultural producers, mainly large-scale rubber plantations and dairy operations, make up only 1% of revenue for the state-owned Bao Viet Group (Tran, 2014c). The high risk of natural disasters and epidemics mean that insurance premiums are quite high, out of reach for many farm households. At the same time insurance offers an attractive policy option for dealing with risk and thereby encouraging production. To overcome this dilemma, a three-year **pilot insurance programme** was introduced in 2011 and provided through two state-owned insurance companies, Bao Viet and Bao Minh.[42] The established insurance premium is subsidised by the government on a progressive scale: 100% for poor farming households; 80% for farming households just above the poverty line; 60% for other farming households; and 20% for agricultural production organisations. In 2013 the pilot period was extended out to 30 June 2014 and the subsidy for farming households just above the poverty line was raised to 90%.[43]

The pilot programme was implemented in **21 provinces**, opened to paddy, livestock (buffalo, cows, pigs and poultry) and aquaculture (catfish and shrimps) producers, and covered a specific list of risks (e.g. storms, foods, blue-ear pig disease and foot-and-mouth disease). In the case of paddy, the programme avoided the complexity associated with ascertaining individual loss by using an index system. A committee of county officials establishes the natural disaster conditions and extent of loss based on area wide surveys. Livestock and fishery coverage is based on assessment of individual loss. Payments of subsidies are handled directly by MOF with the insurance company and loss coverage payments are handled directly by insurance company with the insured party (JICA, 2012).

Just over 304 000 households and one agricultural production organisation participated in the pilot insurance programme (Table 2.5). Poor households accounted for 77% of participants with near poor households accounting for a further 15%. The vast majority of participants (78%) were involved in paddy production. A total of VND 7 748 billion (USD 370 million) was insured under the programme with a total direct insurance fee revenue of VND 394 billion (USD 19 million). Just over VND 700 billion (USD 33 million) had been paid out to producers, 95% of which had been received by aquaculture producers. While a full evaluation of the pilot programme is underway, a number of problems have contributed to the **relatively low take up** of the scheme. These include the fact that many common diseases are not covered, the process of disease certification is confusing and the loss coverage payments are not high enough (Tran, 2014c).

Table 2.5. **Outcomes from the pilot agricultural insurance programme as at June 2014**

	Number of participants[1]	Value insured	Insurance fee paid by government	Compensation provided
Total	304 018	VND 7 748 billion (USD 370 million)	VND 394 billion (USD 19 million)	VND 702 billion (USD 33 million)
		By type of agricultural production (%)		
Paddy	78	28	23	3
Livestock	20	35	21	2
Aquaculture	2	37	55	95

1. The number of participants comprises 304 017 households and one agricultural production organisation (involved in paddy production).

Source: Ministry of Agriculture and Rural Development, 2014.

Income support measures

In July 1993 an **agricultural land use tax** was introduced as the main mechanism of central government to tax farm income, replacing the previous agricultural output tax that dated from 1983. The per hectare tax liability varies according to the land class category (based on land fertility, location, topography, climatic conditions and irrigation) and the type of production (annual versus long-term) (Table 2.6). The liability is set in terms of a fixed quantity of rice per ha and is collected in cash by using a rice price determined by the provincial government based on local market prices. An additional 20% is charged on land held over the land use limit to dissuade land accumulation (Le, 2006).

Table 2.6. **Agricultural land use tax**

Land class category	Land for annual harvest and aquaculture (kg rice/ha)	Land for long-term production (kg rice/ha)
1	550	650
2	460	550
3	370	400
4	280	200
5	180	80
6	50	n.a.

n.a. not applicable.
Note: The annual harvest relates to short-term seasonal agricultural production. Long-term production is related to recurrent crops that do not require replanting annually such as palm oil.
Source: Hong-Loan and McClusky, 2012.

In order to encourage agricultural production and support farmers, **exemptions and reductions** to the agricultural land use tax were introduced in 2003.[44] The rationale was to remove the difficulties and inequities associated with a tax based on a standard rice quantity and where payments in cash are determined by the rice price (Le, 2006 and Tran et al., 2013). For example, in poor-yield years or when the rice price is high, the tax collection is high relative to those years with a good harvest or a lower rice price. Further, for the same land class, the price of rice may be higher in some poor regions than in richer regions, resulting in a higher absolute tax amount for the poor.

The exemptions and reductions were initially provided for a seven-year period but were extended in 2010 out to 2020.[45] Exemptions from the agricultural land use tax is provided for: agricultural land under the land limits assigned by the government to both farm households and individuals; agro-forestry land under the land limits allocated to households from state-owned enterprises; and agricultural land, both under and above the land limits, for poor households and households located areas classified as having "special difficulties". Reductions in the agricultural land use tax by 50% is provided for: organisations which manage and use agricultural land; and land holdings in excess of the land limits which are used for agriculture and forestry by households and individuals, including land allocated by state-owned enterprises. Reductions in land use taxes have also been used to encourage commercial investment in agriculture and incentivise infrastructure development.[46] The 2003 policy change resulted in most farm households and organisations either being exempt from paying agricultural land use tax or having the amount they pay reduced (Figure 2.4).

Figure 2.4. **Revenue from agricultural land use tax, 2000-12**

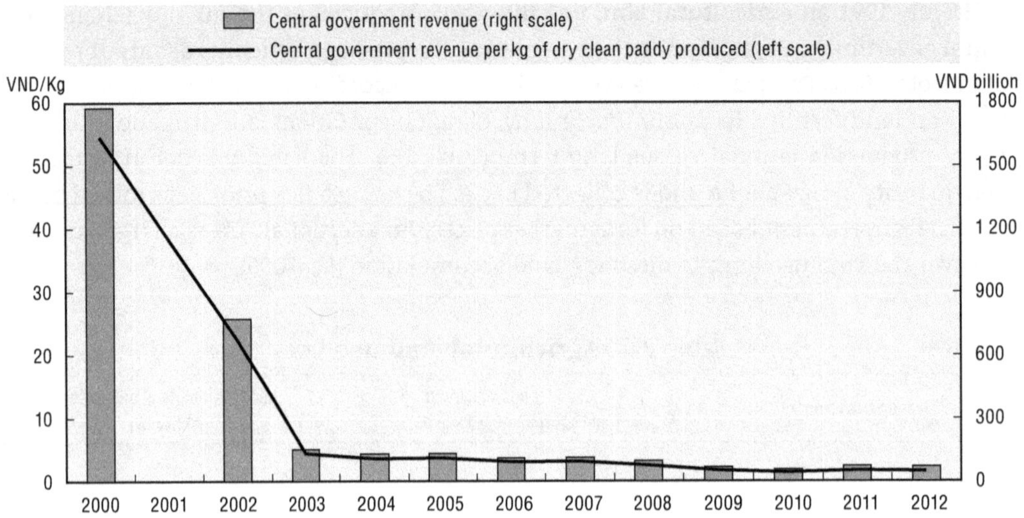

Source: General Statistic Office, National Accounts, State budget revenue final accounts, *www.gso.gov.vn/default_en.aspx?tabid=468&idmid=3&ItemID=15443.*

StatLink http://dx.doi.org/10.1787/888933223776

Extension services

The government's agricultural extension system was officially **established in 1993.**[47] The main objectives of the current system are to: raise producers' awareness through training in production and business knowledge and skills; provision of services to assist farmers in carrying out effective production and business activities adapted to ecological, climate and market conditions; contribute to restructuring the agricultural economy towards commodity production, higher productivity and quality as well as food hygiene and safety; and accelerate agricultural and rural industrialisation, ensuring national food security, socio-economic stability and environmental protection.[48]

The public extension system is organised into 5 levels: a central-level National Agriculture Extension Centre (NAEC) within MARD; provincial agricultural extension centres within their respective DARDs; district agricultural extension stations under the control of the provincial extension centre or the district peoples committee); commune agricultural extension cadres; and village-level agricultural extension collaborators and clubs. According to the current regulations, NAEC has the following responsibilities: developing policies and mechanisms of management for extension in agriculture, forestry, fishery, rural industry; developing economic-technical cost-norms for extension works; leading, organising and guiding the transfer of advanced techniques through setting up demonstration models, disseminating information, training, providing services and international collaboration in related fields.[49]

All 63 provincial governments have their own extension centres with a total of 2 694 staff, an average of 43 persons per centre (Table 2.7). Extension stations exist in 641 out of 703 districts. Almost all of the districts without a station are completely urban. There are 3 335 people employed at the district level, an average of five persons per station. At the commune level, there are 7 804 extension workers, about one people per commune. In total there are about 30 000 people working below the national level, making on average **one public extension worker per 300 farming households** or three workers per

Table 2.7. **Government extension system by region, 2013**

Variable	Region						Total
	Red River Delta	Northern midlands and mountainous areas	North Central and Central coastal areas	Central Highlands	South East	Mekong River Delta	
Agricultural land (000 ha)	771	1 596	1 882	2 000	1 355	2 607	10 211
Agricultural households (000)	1 750	1 789	2 215	743	537	1 833	8 867
Districts in region	129	141	172	61	69	131	703
Communes in region	1 946	2 254	2 416	593	468	1 270	8 947
Stations at the district level	107	135	158	58	53	130	641
Investment (million VND)	220 058	113 537	99 849	38 197	81 360	153 506	706 507
Central budget	19 753	27 697	16 797	5 717	3 052	18 346	91 362
Local budget	198 519	69 652	78 828	27 222	77 496	123 907	575 624
ODA	1 786	16 188	4 224	5 258	812	11 253	39 521
Staff	2 828	14 624	4 635	5 269	511	2 141	30 008
Provincial	799	376	535	180	222	582	2 694
District	440	1 032	647	413	178	625	3 335
Commune	1 589	3 102	1 403	701	111	934	7 840
Village/Hamlet	n.a.	10 114	2 050	3 975	n.a.	n.a.	16 139
Investment per household (USD)	5.97	3.01	2.14	2.44	7.19	3.97	3.78
Central budget per household (USD)	0.54	0.73	0.36	0.37	0.27	0.47	0.49
Local budget per household (USD)	5.38	1.85	1.69	1.74	6.85	3.21	3.08
ODA per household (USD)	0.05	0.43	0.09	0.34	0.07	0.29	0.21
Households per staff	619	122	478	141	1,050	856	295
Staff per commune	1.5	6.5	1.9	8.9	1.1	1.7	3.4

1. Agricultural land in 2011.
2. Agricultural households taken from GSO (2012), Results of the 2011 Rural, Agricultural and Fisheries Census.
3. Investment includes central and local government budget plus ODA.
4. Expenditure and staff include those involved with forestry and fisheries extension.

Source: MARD (2014), *Statistical Yearbook of Agriculture and Rural Development*, 2013.

StatLink http://dx.doi.org/10.1787/888933223782

commune. Expenditure on extension services amounts to about USD 3.80 per farming household, of which 80% comes from the local government.

Since 2001 central government funding for agricultural extension has been allocated through an **open bidding process**. The Science, Technology and Environment Department of MARD is responsible for administering the bidding process for national level projects, and provincial People's Committees' for local level projects. Consequently, in addition to the government extension service, a large number of research institutions, universities, enterprises and NGOs are also involved in the provision of public extension activities. For example, of the 20 national level projects awarded in 2014 only five were awarded to NAEC; the remaining 15 were awarded to universities, research institutes, etc. In many cases, these other providers' sub-contract government extension workers, particularly at the provincial level, to implement projects awarded to them. NAEC supervises the implementation of the projects awarded through this process.

Agricultural extension projects have a strong production focus, which can be described as **top-down, supply-driven** model. They are usually associated with the introduction of new varieties (e.g. hybrids of rice, corn, cotton, sugarcane) or special production techniques (e.g. changing cropping pattern, integrated pest and nutrient management). The extension system also provides farmers information related to new policies and market prices. Projects are delivered through the use of demonstration sites and field days, training farmers, and organising science-technology forums in the fields of crops, livestock, veterinary care, forestry, water resource management, agro-forestry processing and engineering. When participating in extension events, farmers engaged in small-scale production or from poor households are given free access to materials, travel, accommodation and meals. Other farmers receive free materials and are funded for half the cost of travel, accommodation and meals. The future direction for extension in Viet Nam is to promote "socialising the extension program". The intention is to encourage two-way information flow and build farmer-led and demand-driven extension (Nguyen, 2012).

Research and development

The impressive increases in agriculture production since the mid-1980s have been supported by national research efforts that have resulted in scientific solutions to help **improve agricultural production**. Research has contributed to the introduction of new plant breeds, diversification of crops, and improved pest and disease management (JICA, 2012).

Prior to 2005 there were 30 different research agencies within MARD (28 research institutes and 2 universities), each with their own budget and often with overlapping mandates. In order to achieve **greater co-ordination** these institutes were merged and reorganised in 2005 into 16 agencies, comprising 12 research institutes and four universities:

- Vietnamese Academy of Agricultural Sciences (VAAS), including 18 institutes and centres
- Vietnamese Academy of Forest Sciences (VAFS), including 13 institutes and centres
- Vietnam Academy for Water Resources (VAWR), including 15 institutes and centres
- National Institute of Animal Sciences (NIAS)
- National Institute of Veterinary Research (NIVR)
- Vietnam Institute of Agricultural Engineering and Post-harvest Technology (VIAEP)
- Research Institute of Aquaculture I (RIA1)
- Research Institute of Aquaculture II (RIA2)
- Research Institute of Aquaculture III (RIA3)
- Research Institute of Marine Fisheries (RIMF)
- National Institute of Agricultural Planning and Projection (NIAPP)
- Vietnam Institute of Fisheries Economics and Planning (VIFEP)
- Vietnam National University of Agriculture (VNUA)
- Vietnam Forestry University (VFU)
- Water Resource University (WRU)
- Bac Giang Agriculture and Forestry University (BAFU).

VASS is the biggest research institute under MARD.[50] The three-fold purpose of VAAS is to provide a comprehensive vision, strategic direction and oversight of agriculture research and development programmes; to conduct basic and applied research and foster the transfer of new technologies; and to provide post-graduate and professional training. In terms of **research strategies**, VAAS is focusing on the following eleven areas: develop basic research approaches that conserve and effectively utilise plant and animal genetic and other agricultural resources; efficiently apply agricultural biotechnology; select and develop animal breeds and crop varieties with high productivity, good quality, high resistance to biotic and abiotic stresses; implement Integrated Crop Management, Good

Agricultural Practice and Hi-tech production technologies for the major cropping systems; ensure appropriate solutions to meet society's demand for food safety and food security; reduce postharvest losses; effectively use natural resources – soil, water and biodiversity; improve the agricultural environment; study on mitigation and adaptation to climate change; develop suitable agrarian systems based on integrated socio-economic approaches, household farming production, and appropriate policies; and study rural development.

Government funding for agricultural research is provided through MARD, MOST and provincial governments. Between 2000 and 2012 government expenditure on agricultural research increased from VND 153 billion (USD 10 million) to VND 822 billion (USD 40 million), an annual average increase of 11% per annum in USD terms. Despite the impressive increase, funding as a percentage of GDP remains relatively low at around 0.03% of GDP. The limited funding means that much of the research has not met the practical requirements of farmers, business and science. Further, the rigid application of common policies and technologies across the country without considering local circumstances has wasted financial and human resources and discouraged product diversification. Finally, the lack of autonomy for and within research institutes does not create incentives for scientific research personnel, leading to a serious brain drain of agricultural scientists in Viet Nam (Tran and Dinh, 2014a).

In recent years, Viet Nam has introduced policies to develop research and development activities in agriculture that are **consistent with the goal of modernising** the sector. First of all, the NA enacted the Resolution No. 26/2012/QH13 on continuously raising the effectiveness and efficiency of public investment for agriculture, farmers and rural areas. Its main focus is on identifying the prioritised agricultural investment portfolio. To be more concise, the priority is science and technology in biotechnology, post-harvest processing, crop seeds, livestock and fishery breeds (Box 2.2).

Box 2.2. **Biotechnology in Viet Nam**

The commercialisation of agricultural biotechnology has been a goal of the government for many years and is an integral part of the restructuring agenda that seeks to increase the utilisation of high technologies in agriculture and reduce the country's dependence on maize imports. All of the necessary regulations required for commercialising agricultural biotechnology were completed in early 2014.

Decision No. 14/2008/QD-TTg on approving the master plan on biotechnology development and application in Viet Nam up to 2020, dated 22 January 2008, entered into force on 17 February 2008. Two of the key aims of the plan are to research, develop and apply biotechnology in a wide and effective manner to production and life; and to concentrate resources on and diversify forms and the effectiveness of investment in biotechnology. This was followed in 2010 by the first-ever biotech regulation: Decree No. 69/2010/ND-CP dated 21 June 2010 and effective 10 August 2010. This was revised by Decree No. 108/2011/ND-CP dated 30 November 2011, which changed the Ministry responsible for food use certification from MOH to MARD. Together these decrees provide the legal framework for both food and bio-safety management of genetically modified organisms (GMO), genetic specimens, and products derived from GMOs. As the next step in the process, implementation circulars have been issued by MONRE and MARD.

Circular No. 8/2013/TT-BTNMT, dated 16 May 2013 and effective 1 July 2013 provides the procedure for granting and revoking Certificates of Biosafety, and the regulatory structure for evaluating the biosafety of agricultural traits derived from biotechnology.

Box 2.2. **Biotechnology in Viet Nam** *(cont.)*

Circular No. 2/2014/TT-BNNPTNT, dated 24 January 2014 and effective 10 March 2014 provides the procedure for granting and revoking Certificates for Food and Feed Safety for GMO plants. As part of the process MARD has established a Feed and Food Safety Committee, consisting of 11 experts and scientists representing different Ministries including MONRE, MARD, MOH, MOIT, the Vietnam Academy of Sciences and Technology, VASS, and Ho Chi Minh City's Biotechnology Centre, to review and evaluate applications.

Bio-tech developers submitted applications for registration as soon as these Circulars entered into force. On 11 August 2014 the Minister of Agriculture and Rural Development signed the first four Certificates for Food and Feed Safety for four genetically modified maize traits; one of which is glyphosate tolerant and the remaining three are insect resistant. A number of additional traits are currently under review at MARD for feed/food use approval, including one soybean trait. On 27 August 2014, following just over a year review by Vietnam's Biosafety Committee, the Minister of Natural Resources and Environment issued the first Certificate of Biosafety for one of the insect resistant maize traits previously approved by MARD for food and feed use. The additional three traits have been subsequently received a Certificate of Biosafety. These traits will be able to be commercially grown in Vietnam following variety registration that will take one season to complete. All four genetically modified maize traits went through confined and multi-location field trials conducted by MARD during 2010-12.

Source: GAIN-VM4047 (2014), "Vietnam: Agricultural Biotechnology Annual", USDA FAS, 6 August and GAIN-VM4047 (2014), "GVN approves first biotech traits for cultivation and feed food use", USDA FAS, 5 September.

In December 2012 MARD set in place a strategy for the development of science and technology for agriculture and rural development over the period 2013-20.[51] It sets **specific targets** for science and technology to become a key driving force for the industrialisation and modernisation of agriculture and rural development; contributing 40% to the value-added agriculture in 2015 and 50% in 2020; high technology products of will represents 15% of the agricultural product value in 2015 and 35 % by 2020. The programme of activities to support these strategies was approved in 2013 including research and development of staple crops, livestock husbandry and animal health, agricultural engineering and post-harvest technology, irrigation technology and research of policies on agriculture and rural development.[52]

Irrigation and flood protection

Irrigation has played an important part in the success of Viet Nam's efforts to raise agricultural production. According to FAO, the irrigation potential in Viet Nam is 9.4 million ha, of which close to 50% (4.6 million ha) has been developed. Investment in irrigation and flood protection has been a **major focus of the government** since the 1970s, with some 80% of the capital investment funds available to the agriculture sector allocated to improving and expanding irrigation, and protecting flood prone areas from damage. The rapid expansion in agricultural production in response to the *Doi Moi* reforms was enabled by earlier investments in irrigation systems (Tsukada, 2011).[53] Total central and local government expenditure on capital development increased from around VND 3 trillion (USD 200 million) in the early 2000s to over VND 13.5 trillion (USD 700 million) in 2010 before reducing to just under VND 10 trillion (USD 460 million) in recent years (Figure 2.5).

Figure 2.5. **Expenditure on irrigation capital development, 2000-13**

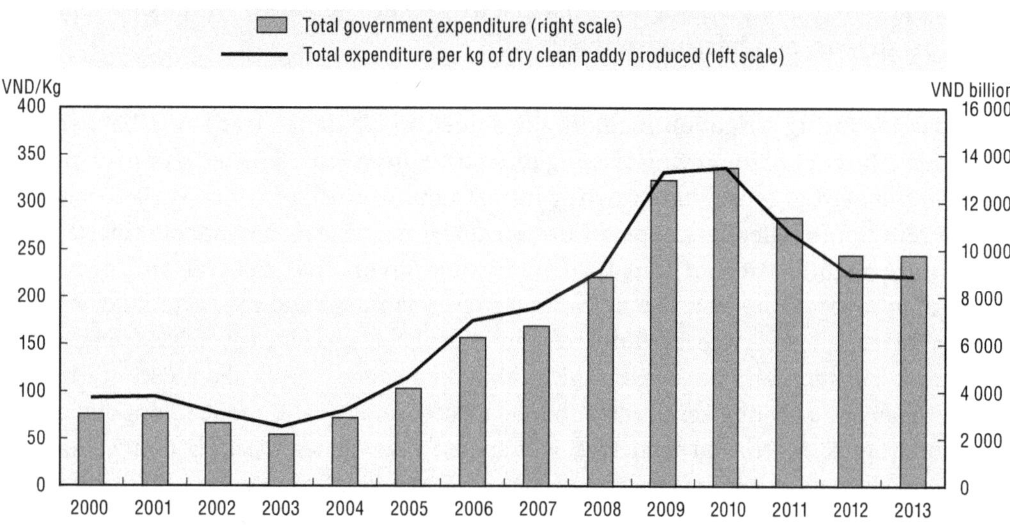

Source: Own tabulation based on OECD (2014), "Producer and Consumer Support Estimates", *OECD Agriculture Statistics Database*.

StatLink http://dx.doi.org/10.1787/888933223795

The actual area currently equipped for irrigation, the state of repair of the irrigation systems and other **statistics on irrigation are uncertain** because of confusion surrounding definitions and weaknesses in data collection. To remedy this, the government has commenced a new nationwide census of irrigation but results are not yet available. There are 1 014 separate irrigation schemes throughout the country, of which the vast majority (904) service less than 2 000 ha. FAO estimate that 1.6 million ha are serviced by small irrigation systems (< 5 000 ha), 1.2 million ha by medium irrigation schemes (5 000-50 000 ha) and 1.7 million ha by large irrigation schemes (> 50 000 ha). Around half of the total irrigation area is located in the MRD with a further 16% in the RRD. Supporting these irrigation networks are an estimated 5 600 reservoirs which store and supply water when needed, supplementing water diverted directly from rivers. About 2.1 million ha is pump irrigated, with over 11 500 pumps lifting water to higher ground when water levels are too low to reach fields (JICA, 2012).

MARD has the primary responsibility for **irrigation management**. In carrying out this role, MARD works with the Ministry of Natural Resources and Environment (MONRE) and the National Water Resources Committee (NWRC). Since its establishment in 2002 MONRE has taken over responsibly from MARD for the management of water resources in general including the allocation of interprovincial water resources. The NWRC was established in 2000 to solve conflicts over water resource management between ministries and between ministries and provinces. In addition to irrigation, MARD is responsible for dykes, flood and storm management, and rural water supply. Funding of large capital projects, including investment for main canals of large irrigation and flood-control projects, is largely carried out by the central government.

Provincial People's Committees are responsible for the public irrigation systems within their boundaries. Under the guidance of the respective PPCs, provincial DARDs are administratively responsible for operating, maintaining and repairing public irrigation, drainage and flood-control systems, and for survey, design and construction of minor new works within their respective provinces. IDMCs are responsible for the actual operation and maintenance of irrigation and flood control systems. Small-scale structures (such as dams and reservoirs or pumping stations) that irrigate or drain areas within one commune or co-operative are administered at that level.

The 2009 Viet Nam Water Sector Review launched by a group of donors led by ADB lists many **issues facing irrigation** including the need to i) balance trade-offs between the economic efficiency of improving existing infrastructure versus expansion of new irrigated areas; ii) the need to rehabilitate existing infrastructure, much of which is 30-40 years old and suffers from inadequate expenditure on O&M; iii) achieve sustainable financing for O&M and rehabilitation of irrigation facilities, given that central and provincial government budgets are insufficient for major refurbishment and the current government subsidy for irrigation service fees limits the ability of IDMCs to fully cover O&M; iv) the need to improve irrigation service coverage, given that an average of only about 68% of irrigation design areas are currently serviced; v) manage water quality and reduce nonpoint source pollution from fertilisers and pesticides, which pose a public health risk in many areas; and vi) effectively manage multipurpose reservoirs for irrigation, hydropower, and water supply (JICA, 2012).

In 2014 an **irrigation restructuring scheme** was introduced with the following objectives: improving efficiency in the irrigation sector to contribute to agricultural restructuring towards greater added value and sustainable development; meeting the development requirements of socioeconomic sectors; building capacity for disaster prevention and response to climate change; and contributing to the modernisation of agricultural and rural infrastructure and new rural development.[54] The major solutions that are being implemented are innovations in planning, including irrigation planning associated with agricultural restructuring, the application of water-saving irrigation technologies, and the reorganisation of agricultural production. Further solutions are innovations in technology, including research on equipment integration; forecast capacity enhancement; flood warnings, drought and saltwater intrusion, as well as research on hydrological regimes and flows to improve the quality of reservoir operation processes, particularly in emergency situations.

In addition to irrigation, an extensive system of dykes provides flood protection. There is over 8 000 km of dykes in Viet Nam, of which roughly three-quarters are river dykes and one-quarter sea dykes. In order to provide greater protection from forecast sea level rises associated with climate change, the government has embarked on a programme of maintaining and upgrading the MRD sea dyke system.[55]

Storage infrastructure

To support the temporary storage policy, the government has encouraged the expansion of **rice storage capacity** of enterprises. Considering its importance both to national food security and export revenues, rice warehouse infrastructure is poor. In 2009 the warehouse storage capacity was only 1.5 million tonnes compared with annual paddy production of over 20 million tonnes. Of this capacity, less than one-third was regularly used because it was often decrepit and inconveniently located (BMI, 2011). To enhance the warehouse system, the government set a target of building a four million tonne storage system in the MRD by the end of 2011: upgrading the existing 1.5 million tonne storage system and building new warehouses with capacity for 2.5 million tonnes.[56] The objective of these improvements was to increase the returns from rice exports by giving enterprises more flexibility as to the timing of shipment and lifting the quality of stored rice. To stimulate the USD 400 million investment required a variety of incentives were provided by the government including low interest loans and business income tax exemptions (MARD, 2014b). By the end of 2011 less than 50% of the new warehouse capacity target had been constructed. And this was primarily due to the actions of the two largest SOEs, Vinafood 1 and Vinafood 2, who had completed 66% and 58% of their proposed investments (Tran, 2014a). Consequently, the target date for completion of the expansion was pushed back until the end of 2013. As at 31 July 2013 the warehouse storage capacity had reached 5.4 million tonnes, consisting of 4.4 million tonnes for rice and 1 million tonnes for paddy.

Consumer support

State Reserves of rice are used to provide **direct food support** to households through a number of different programmes. Poor households involved in forestation and forest protection are provided 15 kg rice per capita per month during the period when they are not able to provide themselves with staple food (for not longer than seven years).[57] Poor households in the border areas are granted 15 kg rice/person/month until they can be self-sufficient in food. The government also uses rice from the State Reserve to support households in food-deficit provinces before the harvest and in provinces suffering natural disasters. This is one of the direct support policies for the poor but the government lack of measures to identify the right beneficiaries and provide the necessary quantity (MARD, 2014b).

In addition to direct food aid, the government also intervenes to prevent sharp rises in the price of essential food products purchased by households. Support is implemented through tax concessions and interest rate subsidies for retail enterprises. Consequently, it is mainly food that is distributed through more forma marketing channels such as the supermarket system that receive this support.

2.4. Trade policies affecting agro-food trade flows and agricultural commodity prices

Strongly connected to the reforms in domestic agriculture policy was the gradual integration of domestic markets with the global economy. This section summarises **key developments in trade policy** since the mid-1970s. It details the important trade measures currently affecting imports and exports of agro-food, including price based instruments (e.g. tariffs and other import duties, and export taxes), quantitative restrictions (e.g. import quotas and export bans) and regulatory requirements (e.g. licensing and quarantine arrangements). Multilateral, regional and bilateral trade relations are also discussed.

Overall reforms of the trade system

Prior to the late 1980s, Viet Nam was a **relatively closed country**. Foreign trade was severely constrained: controlled by central decision makers and carried out by a small number of SOEs with monopoly rights. Exports were discouraged by an overvalued exchange rate and the use of export duties. Obligations to partners within the Council for Mutual Economic Assistance had to be met before product could be sold to the convertible currency area. Imports had to proceed through an extensive system of quotas and licenses. Additionally, Viet Nam faced a trade embargo with the United States that was only lifted in 1994 (McCaig and Pavcnik, 2013).

Market oriented **trade reforms began in 1989** as part of the broader *Doi Moi* reforms that commenced in 1986. Initial reforms included the unification and devaluation of the exchange rate, relaxations on import and export quotas, simplification of licensing procedures for import and export shipments, and delisting items from export duties and reducing the rates for remaining products. Permission for private as well as SOEs to establish direct links with foreign markets was given in 1991. Import tariffs were first introduced on 1 January 1988. The original tariff schedule covered only 130 commodity categories with tariff rates of 0% to 60%. This was replaced in 1992 by a detailed, consolidated schedule based on the Harmonised System (HS) of tariff nomenclature. It provided tariff coverage for all commodities and established both normal and preferential tariffs. Preferential rates (50% of the normal tariff) were applied to exported/imported goods to/from countries that signed bilateral trade agreements with Viet Nam.

Although these initial reforms were extensive, they did not necessary go far enough and were sometimes reversed. While the private sector was granted the ability to engage directly in international trade, in order to obtain an import or export licence, enterprises were required to have a foreign contract and a shipping license, sufficient working capital, and trade experience. With the removal of trading right monopolies, other forms of protection were sometimes provided to **support domestic industries**. For example, the tariff on meat increased from 10% in 1992 to 30% in 1999, and for sugar from 10% in 1992 to 45% in 1999 (Nguyen, 2006). Similarly, non-tariff barriers (NTBs) such as quantitative restrictions and foreign exchange management were used to balance "supply and demand of the economy" to protect domestic production and regulate consumption. The number of goods under this form of quantitative control increased from five in 1996, to eight in 1997 and 1998, 16 in 1999, and 12 in 2000 (petroleum, fertiliser, steel, cement, paper, sugar, liquors, motorcycles, passenger cars, ceramic and granite tiles, and refined vegetable oil) (Vo, 2005).

A further series of trade reforms occurred towards the end of the 1990s and early 2000s. The monopoly position of SOEs in foreign trading activities was gradually weakened and the **abolishment of trade licenses** in 1998 was a most significant step forward in trade liberalisation.[58] Since then all domestic enterprises have been allowed to freely trade in commodities except those prohibited or under specialised management (Vo, 2005). In April 2001 a trade policy roadmap for the five-year period 2001-05 was announced for the first time.[59] This replaced the practice of announcing one-year regimes, making a more transparent and predictable export-import environment. Effective 1 May 2001 most quantitative restrictions on imports remaining in place were removed (Vo, 2005). Those remaining, including for products such as dairy, birds eggs and sugar, have been subsequently eliminated and replaced in some cases with tariff quotas.

Reforms to trade policy were complemented by a concerted effort to enter into **international trade agreements and partnerships**. In 1995 Viet Nam became a member of the Association of Southeast Asian Nations (ASEAN) and its associated ASEAN Free Trade Area (AFTA). Viet Nam was formally admitted as a member of the Asia Pacific Economic Community (APEC) in November 1998. In December 2001 the US-Viet Nam Bilateral Trade Agreement[60] came into effect. This led to a huge increase in Vietnamese exports to the US, predominantly in light manufactured products such as clothing, textiles, and footwear (McCaig and Pavcnik, 2013). Viet Nam used these bilateral and regional agreements to promote exports, cement in reforms and prepare for engagement with world (Abbott et al., 2006).[61] The culminating act was obtaining WTO membership in 2007. The WTO's binding and transparent rules are considered by Viet Nam to be the most efficient mechanism to guard against protectionism and to address global trade issues (WTO, 2013). Viet Nam is working to bring its regulatory environment into conformity with international rules, e.g. bringing its sanitary and phytosanitary (SPS) requirements into compliance with CODEX Alimentarius (Tran and Dinh, 2014a).

Import policy measures

Tariffs

During the 1990s and through to the mid-2000s, agricultural production was protected with an **average applied tariff** of around 25%. Since WTO accession in 2007 the simple average Most Favoured Nation (MFN) applied tariff on agricultural products (WTO definition) has declined by one-third to reach 16.2% in 2013 (Figure 2.6). While tariff protection for agricultural products has fallen, it is higher than the overall simple average MFN applied tariff of 9.5% and almost double the 8.3% average for non-agricultural products. Viet Nam agreed to bind all tariff lines as part of its WTO accession process. Further, all agricultural tariffs are *ad valorem*, making the regime very transparent.

Figure 2.6. **Average MFN applied tariffs for agricultural and non-agricultural commodities, 2003-13**

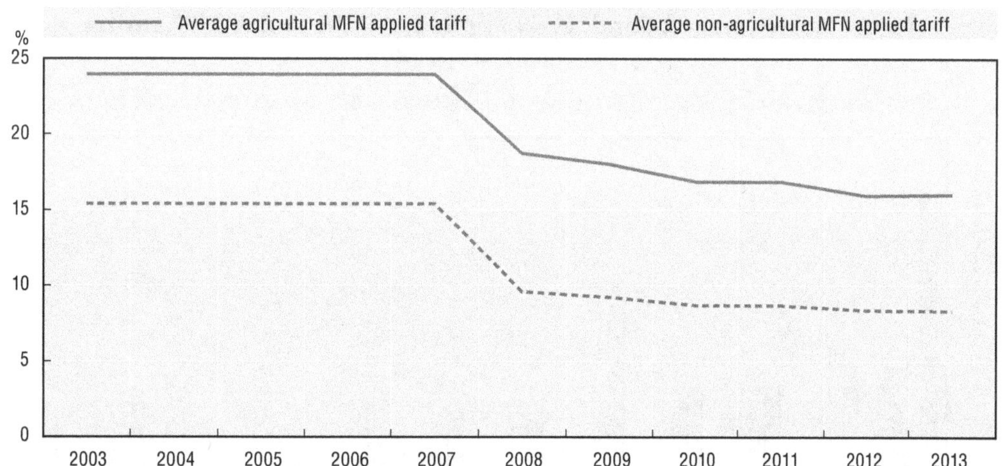

Note: Simple average tariffs based on pre-aggregated HS six digit averages. Pre-aggregated means that duties at the tariff line level are first averaged to six digit subheadings. Subsequent calculations are based on these pre-aggregated averages.

Source: WTO Tariff Download Facility, *http://tariffdata.wto.org/Default.aspx.*

StatLink http://dx.doi.org/10.1787/888933223801

One-third of agricultural MFN applied tariffs fall within the range of 0-5% (Figure 2.7). Almost two-thirds of agricultural imports enter Viet Nam at these **low tariff rates**, with almost 40% entering duty-free. At the other extreme, around one-quarter of MFN applied and final bound agricultural tariffs are 25% or more. Only 6% of imports in 2012 entered Viet Nam through these higher tariff lines.

Among the agricultural product categories attracting the **highest MFN applied tariffs** are beverages and tobacco; coffee, tea and cocoa; and fruit, vegetables, plants (Figure 2.8). Within these product groups, MFN applied tariffs are highest for cigarettes and cigars (100-135%) and wine and spirits (45-55%). An MFN applied tariff of 40% applies to a range of commodities including meat of poultry, turkey and duck, tea (green and black), grapefruit, milled rice, refined sugar, and many types of prepared or preserved fruits and vegetables.

Figure 2.7. **Frequency distribution of agricultural final bound and MFN applied tariff lines and imports by tariff rates, 2013**

Source: WTO Tariff Profile of Viet Nam 2014, *http://stat.wto.org*.

StatLink http://dx.doi.org/10.1787/888933223814

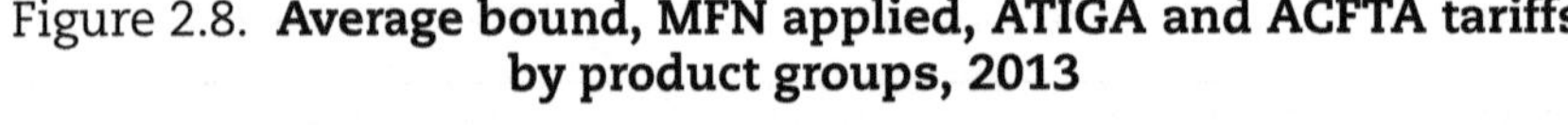

Figure 2.8. **Average bound, MFN applied, ATIGA and ACFTA tariffs by product groups, 2013**

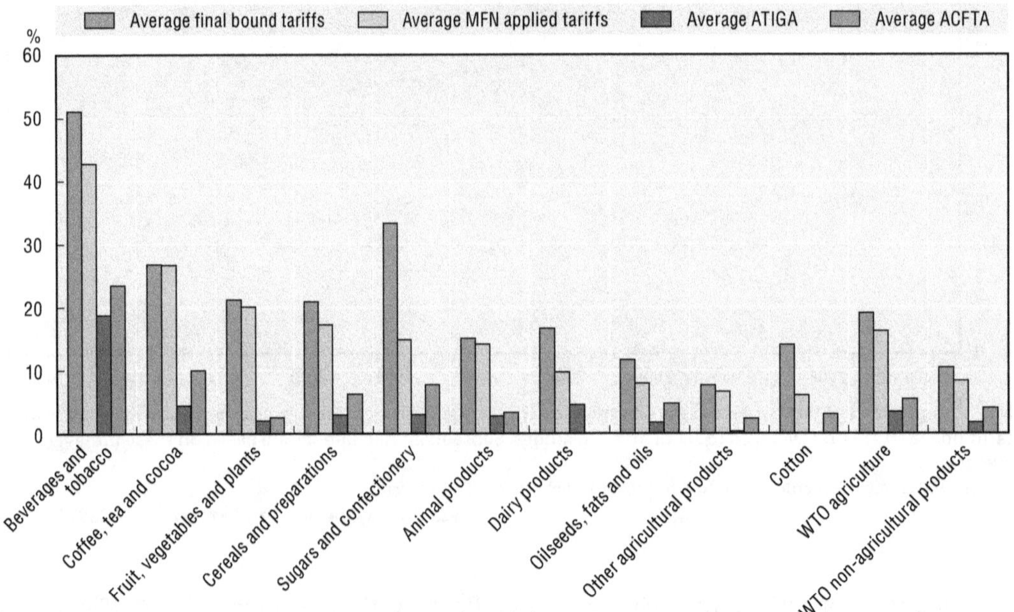

ATIGA – ASEAN Trade in Goods Agreement; ACFTA – ASEAN China Free Trade Agreement.

Note: Simple average tariffs based on pre-aggregated HS six digit averages. Pre-aggregated means that duties at the tariff line level are first averaged to six digit subheadings. Subsequent calculations are based on these pre-aggregated averages. Product groupings are ordered according to average MFN applied tariff.

Source: Final bound and MFN applied tariffs: WTO Tariff Profile of Viet Nam 2014, *http://stat.wto.org*; ATIGA and ACFTA: World Trade Organisation (2013) Trade Policy Review of Vietnam: Report by the Secretariat, WT/TPR/S/287/Rev.1.

StatLink http://dx.doi.org/10.1787/888933223823

Viet Nam accords tariff preferences under regional and bilateral preferential trade agreements, i.e. to its ASEAN partners, Korea, China, Australia and New Zealand, India, and Japan. The **preferential tariff treatment** in Viet Nam's market is generally substantial for ASEAN and China, and somewhat less pronounced for Viet Nam's other FTA partners

(Table 2.A1.3). The simple average applied tariff on agricultural commodities imported from ASEAN members and China is just 3.4% and 5.4% respectively.

The average MFN applied tariffs for agriculture and industrial goods are lower than the average WTO bound rates of 19% and 10% respectively. The difference between bound and currently applied MFN rates leaves some **scope for flexibility** in Viet Nam's tariff policy. In some instances, Viet Nam has implemented tariff reductions ahead of the committed timetable. Tariff cuts may also have been employed on occasion to reduce inflationary pressures in the domestic economy or to mitigate fluctuations in domestic energy prices. However, a number of tariff rate increases since 2008 seem primarily motivated by a willingness to afford higher protection to certain domestic sectors, e.g. meat producers. Although all tariff rate increases have been within the WTO bound limits set by Viet Nam's tariff commitments, frequent changes in the applied tariff introduce uncertainty and may undermine the predictability of access to the Vietnamese market (WTO, 2013).

The import tariff is 0% for most materials and inputs associated with agricultural production, such as fertilisers, corn and rice seeds. The MFN applied tariff rate for most agricultural machinery (tractors, harvesting machines, seeders, manure spreaders, machines for cleaning, sorting or grading, etc.) is 5%, with the exception of ploughs and harrows (20%). All agricultural machinery is duty-free from ASEAN members. The government has implemented these **low import taxes for agricultural inputs** with the purpose of supporting farm production.

Tariff-rate quotas

Under its WTO accession commitments for agriculture, Viet Nam provides **tariff-rate quotas** (TRQs) for eggs, sugar (raw and refined), and unmanufactured tobacco and tobacco refuse (Table 2.8).[62] The TRQ for sugar was first introduced in 2007 to replace an import licensing regime. TRQs for eggs and unmanufactured tobacco and tobacco refuse have been in place since 2003 when they were established along with TRQs for salt, cotton, dairy products and maize.[63] These had been introduced as a step in the process of increasing market access, replacing non-tariff measures such as prohibitions, licenses or import quotas affecting the same products. TRQs on dairy products, cotton and maize were eliminated on 1 April 2005 (WTO, 2006).[64]

Table 2.8. **Tariff-rate quota commitments for eggs, sugar and tobacco**

Product	Initial quota quantity 2007	Volume commitment	Final quota quantity	In-quota tariff rate	Out-of quota final tariff rate	Administration method
Birds' eggs, in shell, fresh, preserved or cooked[1]	30 000 dozen	5% increase per year	Infinity	40%	80%	Quotas allocated to firms that own a business registration certificate and who have actual import demand
Cane or beet sugar and chemically pure sucrose, in solid form[2]	55 000 tonnes	5% increase per year	Infinity	25% for cane; 50% for beet sugar and 60% for sucrose	85% for cane; 100% for beet and 85% for sucrose	Allocated to end-users based on past performance; portion of TRQ allocated to new importers
Unmanufactured tobacco, tobacco refuse[3]	31 000 tonnes	5% increase per year	Infinity	30% except 15% for tobacco stems	80-90%	Quotas allocated to end-users who have cigarette-producing permits

1. 04070091; 04070092; 04070099.
2. 17011100; 17011200; 17019100; 17019911; 17019919; 17019990.
3. 24011010; 24011020; 24011030; 24011090; 24012010; 24012020; 24012030; 24012050; 24012090; 24013010; 24013090.

Source: World Trade Organisation (2013), *Trade Policy Review of Vietnam: Report by the Secretariat*, WT/TPR/S/287/Rev.1.

According to Viet Nam's WTO Goods Schedule, the TRQ volumes for all three groups of products must increase by 5% annually without any upper time limit. The size of the annual import TRQ is fixed by MOIT, whereas the tariff rates for out-of-quota imports are determined by MOF. TRQs are allocated to end-users (WTO, 2013). The most recent notification on imports under TRQs for 2010 shows no imports of eggs under the TRQ and only about half the tobacco TRQ was used.[65] According to the authorities, there are no imports of eggs due to a lack of demand. The TRQ volume for sugar in 2010 was set at 250 000 tonnes, well above the accession commitment level.

In addition to its WTO commitments, Viet Nam has opened **TRQs for preferential imports** of rice and dried tobacco leaves from Cambodia and Lao People's Democratic Republic (PDR) (Table 2.9). The regime is regulated in accordance with bilateral memoranda concluded in 2005 (Lao PDR) and 2007 (Cambodia). TRQ volumes are subject to 0% import duty. The TRQs, which are announced by MOIT, may be stipulated for one or two years at a time (WTO, 2013).

Table 2.9. **Preferential tariff-rate quotas for Cambodia and Lao PDR, 2008-13**

Product	Country	2008	2009	2010	2011	2012	2013
Rice[1]	Cambodia	150	150	250	250	300	300
(000 tonnes)	Lao PDR	40	40	40	40	70	70
Dried tobacco[2]	Cambodia	n.a.	n.a.	3	3	3	3
(000 tonnes)	Lao PDR	3	3	3	3	3	3

1. 1006100090; 1006301900; 1006303000.
2. 2401101000; 2401102000; 2401103000; 2401109000 for both Cambodia and Lao PDR plus 2401201000; 2401204000 and 2401301000 for Lao PDR.

Source: World Trade Organisation (2013), Trade Policy Review of Vietnam: Report by the Secretariat, WT/TPR/S/287/Rev.1.

VAT rates and other duties on imports

Viet Nam passed its first Law on Value Added Tax (VAT) in 1999.[66] There are **three VAT rates**: 0%, 5% or 10%, with 10% being the standard rate.[67] The zero rate applies to exports of goods and services, and international transportation. The reduced (5%) rate applies to a select group of "essential" goods and services including many related to agricultural production: clean water; fertilisers, insecticides and pesticides; feed for cattle, poultry and poultry; special purpose machinery for agricultural production such as ploughs, harrows and harvesters; unprocessed products of cultivation, husbandry and forestry; sugar and sugar by-products; semi-processed cotton; preliminary processed rubber latex; and fresh food (World Bank, 2014).[68] VAT is applied on the duty-paid value of imports, and is due at the same time as the payment of import duties. For domestic producers, VAT is collected monthly and settled at the end of the calendar year. VAT constitutes almost one-third of the total tax revenue, while trade taxes account for about 10% (WTO, 2013).

A number of agricultural related goods and service are **VAT exempt** including the main outputs from farming, i.e. products of cultivation, husbandry, fishery or aquaculture which have not yet been processed into other products or which have only been semi-processed by organisations or individuals self-producing and selling such products, and products at the stage of importation. Domestic raw and semi-processed agricultural production has been VAT exempt since the tax was first introduced in 1999. As part of its WTO commitments, Viet Nam extended the exemption to include imports from 1 January 2006, i.e. prior to this date imports of these products were subject to 5% VAT (WTO, 2006). Other goods and service not subject to VAT of relevance to agriculture include certain inputs, specifically animal breeds and plant variety products, including breeding eggs, breeding animals, seedlings, seeds, sperm, embryos and genetic materials; irrigation and drainage; soil ploughing and harrowing; dredging of intra-field canals and ditches for agricultural production; services of harvesting farm produce (World Bank, 2014).

Import licensing and state trading

The importation of various goods is subject to "line management", i.e. import licences are issued by MOIT but other **ministries regulate imports**.[69] In such cases, neither the value nor the quantity of imports is restricted. The purpose of this system is, inter alia, to enforce minimum quality or performance standards for goods related to human, animal, or plant health; local network compatibility (telecommunications equipment); monetary security; or cultural sensitivities. According to Vietnamese authorities, the "line management" system includes automatic and non-automatic licensing procedures. Under this system, MARD regulates the importation of veterinary medicines, pesticides, plant and animal strains, animal feeds, fertilisers and genetic sources of plants, animals and micro-organisms used for scientific purposes (Table 2.A1.4).

Prior to the current arrangements, various agricultural commodity groups including dairy products, refined vegetable oil and sugar as well as fertiliser had been subject to quantitative restrictions through **import licensing**. These had been introduced at various stages during the 1990s and removed at various points during the 2000s (Table 2.10).

Some restrictions remain. Goods identified as subject to "**state trading**" may only be imported by designated enterprises. Thus, cigars and cigarettes; crude oil and petroleum products; newspapers, journals and periodicals; recorded media for sound or pictures; and aircraft and spacecraft, may only be brought into Viet Nam by designated importers. The Viet Nam Tobacco Import Export Company (VinatabaIMEX), a subsidiary of the Viet Nam National Tobacco Corporation (Vinataba), is the leading importer of machinery, tobacco and materials for the tobacco industry in Viet Nam, and is an exporter of tobacco leaf, cut rag, and cigarettes (WTO, 2013).

In 2008 Viet Nam introduced what it considers to be an **automatic licensing system** for a wide range of consumer products and agricultural items.[70] Licences are valid for 30 days from the date of issue, and are not transferable among importers. Importers must apply for a new licence when the old licence has expired. The measure was introduced to gather more detailed statistics and trade data for import assessment. In 2010 the product coverage was extended to include additional agriculture and food items, textiles, and clothing.[71] At that time, the measure affected imports of meat and meat products; certain fish and fish products; sugar confectionary, including chocolate; certain vegetable, fruit, cereal, and flour preparations; beverages, spirits and vinegar; plastic products; textiles and apparel; footwear; cosmetics; home electrical appliances; motor vehicles and motorcycles; furniture; toys; and steel products. The product coverage was subsequently reduced somewhat in 2011 and a temporary suspension of this "automatic" licensing arrangement has been in place since late 2012 for all products except certain steel products (WTO, 2013).[72]

Prior to WTO accession, Viet Nam had eliminated foreign exchange restrictions on "dispensable and non-essential" import items and consumer goods, and "payment method" restrictions. However, in April 2010 MOIT promulgated a long list of "non-essential" imported commodities and consumer goods **not encouraged** for import.[73] The SBV subsequently

Table 2.10. **Agricultural related products subject to quantitative import restrictions and phasing out schedule**

Product description	HS tariff lines	Established	Phasing out
Milk and dairy products	0401, 0402, 0403, 0404	1994[1]	Replaced with a TRQ in May 2003[2] which was abolished on 1 April 2005
Live animals for breeding	Chapter 1	1996[3]	Removed 1 May 2001[4]
Birds' eggs, in shell, fresh, preserved or cooked	040700	1996[3]	Replaced with a TRQ in May 2003[2]
Human hair, unworked, whether or not washed or scoured; waste of human hair	050100	1996[3]	Removed 1 May 2001[4]
Pigs', hogs' or boars' bristles and hair; badger hair and other brush making hair; waste of such bristles or hair	502000		
Bulbs, tubers, tuberous roots, corms, crowns and rhizomes, dormant, in growth or in flower; chicory plants and roots other than roots of heading No. 12.12	060100	1996[3]	Removed 1 May 2001[4]
Other live plants (including their roots), cuttings and slips; mushroom spawn	060200		
Maize (corn) for sowing	10050010	1996[3]	Removed 1 May 2001[4]
Rice in the husk (paddy of rough) for sowing	10061010		
Seeds, fruit and spores, of a kind used for sowing	120900	1996[3]	Removed 1 May 2001[4]
Vegetable materials of a kind used primarily for plaiting (for example, bamboos, rattans, reeds, rushes, osier, raffia, cleaned, bleached or dyed cereal straw, and lime bark)	140100	1996[3]	Removed 1 May 2001[4]
Residues and waste from the food industries; prepared animal fodder	Chapter 23	1996[3]	Removed 1 May 2001[4]
Sugar	1701	1997[5]	Replaced with a TRQ on 1 January 2007[6]
Fertiliser	3102 3103 3105	1997[5]	Removed 1 May 2001[4]
Refined vegetable oil	15079010 15089010 15119090 15131910 15155090	1999[7]	Removed 1 January 2002[4]

1. Circular No. 04/TM-XNK dated 4 April 1994 guiding the implementation of Decision No. 78/TTg dated 28 February 1994 on the management of importation and exportation.
2. Decision No. 91/2003/QD-TTg dated 9 May 2003 on the application of tariff quota on imports to Viet Nam.
3. Decree No. 89/ND-CP dated 15 December 1995.
4. Decision No. 46/2001/QD-TTg dated 4 April 2001 on the management of importation and exportation during the period of 2001-05.
5. Decision No. 28/QD-TTg dated 13 January 1997 on the management of importation and exportation in 1997.
6. Decision No. 19/2006/QD-BTM dated 20 April 2006.
7. Decision No. 242/1999/QD-TTg dated 30 December 1999 on the management of importation and exportation in 2000.

Source: WTO (2004).

instructed credit institutions to consider carefully or restrict the provision of foreign currency loans to finance imports of the listed items. In 2011 the list was expanded.[74] Among the discouraged items are live animals, dairy products, sugar confectionary, fish and crustaceans, and table salt (WTO, 2013).

Food safety and quarantine measures

Viet Nam undertook to comply with the requirements of the **Sanitary and Phytosanitary** (SPS) Agreement upon its accession to the WTO without recourse to any transitional arrangements. The national enquiry point for SPS matters is the SPS Office located within the International Cooperation Department of MARD, although several government agencies are responsible for SPS-related matters. Decision No. 147/2008/QD-TTg sets out the process Viet Nam will follow to meet its obligations under the SPS Agreement, such as harmonising its food hygiene and safety and sanitary and phytosanitary standards with those of international organisation such as Codex Alimentarius, the World Organisation for Animal Health (OIE) and the Commission on Phytosanitary Measures. This Decision also sets out objectives for risk assessment, control measures, and institutional capacity, and states that the same standard should be used for exports and domestic consumption.

Since joining the WTO, Viet Nam has imposed different SPS measures on meats, fresh fruits and vegetables, feed, dairy, processed foods, and other food imports from different countries. Further, while Viet Nam has agreed to equivalence of foreign food safety measures, it has not fully adopted standards provided by international organisations. For example, Viet Nam maintained protective measures against bovine spongiform encephalopathy (BSE) on beef imports from the United States that contained stricter restrictions than the science-guided measures recommended by the OIE (Arita and Dyck, 2014).

The umbrella law guiding **food safety** provides organisations and individuals with rights and obligations to ensure food safety; conditions for food safety; food production and trading; food import and export; food advertisement and labelling; food testing; food risk analysis; prevention and dealing with food safety incidents; information, education and communication on food safety; and state management of food safety.[75] Responsibility for state management of food safety rests with MARD, MOIT and MOH. Together they oversee: i) the conformity to technical regulations for food safety regulations, ii) safety requirements for genetically modified foods, iii) granting and withdrawing food safety certificates for establishments that meet food safety requirements, iv) state inspection on food safety for imported and exported foods, and v) labelling of food products.[76] The three ministries are currently developing Circulars and Technical Regulations to enforce sections of the Decree. Among those developed to date, is one that provides additional clarification regarding which Ministry is responsible for what set of food products. Responsibility covers both domestically produced food as well as imported food products.[77]

The National **Strategy on Food Safety** for 2011-20 sets a general objective of implementing master plans on food safety from production to consumption by 2015, and controlling food safety over the entire food supply chain by 2020. Specific objectives of the strategy are: awareness raising and food safety practicing for target groups; capacity building for food safety management system; significantly improvement of food safety assurance in manufacture, processing, selling facilities; actively prevention of acute food poisoning. One of the key programmes is assigned to MARD. In collaboration with the provincial people committees, MARD will draft and lead the "Development on safe food supply chain model".

Despite these efforts to set in place a legal framework and structure for quarantine and food safety that conforms to international obligations, further work is required to effectively implement the regulatory regime. There are issues associated with the system of legal documentation and technical regulations, the organisational system and human resources (Dao T.A., 2014). The large number of legal documents relating to food safety, about 400 documents issued by the central government and ministries and about 1 000 documents issued by local governments, result in overlap and lack a clear focus. There is poor co-ordination between agencies, risk analysis and identification systems, both at the central government level and between central and local government (Tran and

Dinh, 2014a). The capacity of testing agencies is limited, leading to inconsistent enforcement that adds to uncertainty for foreign producers (Arita and Dyck, 2014).

In addition to quarantine and food safety documentation, importers of food and agricultural products are often required to provide authorities with **other supporting documents**. Depending on the product, these may include a certificate of free sale (CFS) from the competent authorities of the exporting country, stating that the product is produced and freely sold in the country of origin. In terms of agriculture, the goods affected are mostly non-food agriculture and fishery products but all products containing genetically modified materials, products that were irradiated, and products that were produced by new technologies require a CFS when imported to Viet Nam (WTO, 2013).

Standards and labelling

Viet Nam undertook to comply with the obligations of the Technical Barriers to Trade (TBT) Agreement from the date of its accession to the WTO. Since WTO's accession, Viet Nam has sent over 50 TBT notifications of regular measures covering a variety of products. At the end of 2012 there were 6 800 Vietnamese standards, of which 40% were harmonised with international, regional and foreign standards (up from 25% in 2005) (Tran and Dinh, 2014a).

Under Law No. 05/2007/QH12 on Products and Goods Quality, products and goods are divided into two groups: Group 1 are goods which are "incapable of causing unsafety" and Group 2 are goods which are "capable of causing unsafety" (defined as "those products and goods which, under rational conditions of transportation, storage, preservation and use for proper purposes, can latently cause harms to humans, animals, plants, assets or the environment"). Producers control products in Group 1 on the basis of applicable standards. The competent state agency controls products in Group 2 on the basis of relevant technical regulations as well as by producers on the basis of applicable standards. Different ministries are responsible for the control of quality of products and goods under their responsibility, and for issuing lists of products and goods that can cause "unsafety" and are subject to mandatory inspection. A wide range of goods are subject to **mandatory inspection and quality control by MARD**: plants, animals, fertilisers, animal feeds, plant protection drugs, veterinary drugs, bio-products for use in agriculture, forestry and aquaculture, and irrigation works and dykes (WTO, 2013).[78]

Export policy measures

Export licences and quotas

Prior to 1995 several key agricultural products, namely rice, tea and coffee beans, were subject to export quotas. These were removed for tea and coffee in 1995. However, coffee beans remained subject to line management by MARD requiring their approval before export. **Export quotas on rice** were introduced in 1992 to ensure food security and price stability. The total quantity of rice exports as well as the allocation of export quotas was determined by the government in two phases on the basis of rice production and consumption estimates: an initial allocation for the period up to September and then for the rest of the year after the September crop (CIE, 1998). Initially the government directly assigned an export quota to each company. While the number of companies approved to export rice varied year-by-year (15-40), they were all SOEs, either established at the national level, e.g. Vinafood 1 and Vinafood 2 or by provincial governments. In order to increase competition, from 1997 the government allocated 60-70% of rice export quotas to provincial governments, based on their province's share of national commercial paddy production, for allocation to enterprises. The export quota for rice rose from less than one million tonnes in 1992 to 4.5 million tonnes in 1998 (Nguyen and Grote, 2004). The intense lobbying by these enterprises to share in the export quotas during the early years of rice exports suggests that the quotas were binding and ensured that the domestic price was below the relevant border price (Athukorala et al., 2009).

In 1998, partly in response to growing concerns about trade deficits and a slowdown in foreign financing, export licensing controls were abolished (WTO, 2006).[79] This gave permission for all domestic private companies, SOEs and co-operatives to export their production without requiring further licensing or approval. This initiative applied to all but a narrow list of commodities. An even greater relaxation of controls was extended earlier that year to foreign owned enterprises, which were now authorised to export products above and beyond those identified in their investment license.[80] Prohibited exports, quota controlled exports and commodities subject to specialised export regulation were excluded from this change. For agriculture this meant rice and coffee (CIE, 1998). One consequence of these changes was that a share of the rice export quota was allocated to private firms, but they accounted for just 4% of total rice exports in 1999 (Nielsen, 2002).[81]

The rice export quota system was formally abandoned in 2001.[82] Enterprises were from then on permitted to export rice provided they had obtained general business licenses to trade in rice or other agricultural products. However, a **flexible control system** has developed in its place (WTO, 2006). This has been formalised most recently in Decree No. 109/2010/ND-CP on rice export business dated 4 November 2010. The role of various agencies in rice export management is divided as follows: MARD is responsible for forecasting the quantity of rice available for commercial export based on domestic production and consumption estimates and reserve volumes; MOIT for seeking markets and negotiating government-to-government (G2G) food exporting agreements; and the Viet Nam Food Association (VFA) is in charge of operating contract registration of rice exporting enterprises and allocating G2G contracts among exporters (Box 2.3).

Box 2.3. **Viet Nam Food Association**

The Viet Nam Food Association (VFA), formerly known as the Viet Nam Food Import and Export Association, was established in 1989 (Decision No. 727/KDDN-QD dated 13 November 1989). VFA is a social organisation of enterprises involved in producing, processing and trading of agricultural produce, food and other processed food products. It is organised and operates under a charter ratified by the Minister of the Interior and under state management of MARD. Members of VFA work together to co-ordinate food trading activities for the protection of legitimate interests of its members; to contribute to food security; to import, export food in the international market in compliance with the state policies. It receives no government funding to carry out these activities.

In recent years, there have been more than 200 registered exporters with VFA. Nevertheless, the trade remains highly concentrated, with the ten largest exporters accounting for 70% or more of the total trade. VINAFOOD I and II are the two main SOEs and account for 44% of the volume and 53% of the value of Viet Nam's rice trade over the 2007 to 2009 period. Each of these is affiliated – through full or partial ownership – to a range of other companies many of which are themselves specialised rice milling and trading companies. Interactions among these companies involve some combination of competition

Box 2.3. **Viet Nam Food Association** *(cont.)*

and co-operation. In 2008 majority or fully state-owned enterprises accounted for 79% of the value of the trade, the private sector and companies with a minority state ownership stake accounted for 19%, and co-operatives for 2%. Most private companies are very small, but there are now a few which can export in excess of 100 000 tonnes per year. Many companies that are eligible to join VFA and export are also often denied membership.

Source: VFA website (*www.vietfood.org.vn*) and Tran et al. (2013).

At the beginning of each year, the government announces an indicative total export volume available to all enterprises; there is no indicative export volume of rice allocated to individual enterprises. All traders having legally registered their business are free to sign rice export contracts at their own discretion. However, companies must first submit the export contract to the VFA for approval. In the usual case, this approval is automatic. However, when the government believes that export restrictions are needed, the VFA is asked by the government to stop approving new rice-export contracts. Without the approval from the VFA, enterprises cannot export. Through this mechanism the government is able to adjust indirectly the progress of total rice exports whenever needed. The government used this instruction mechanism to limit rice exports during 2007 and 2008 (Tran et al., 2013; Nguyen and Talbot, 2013).

On 1 January 2011 two important modifications to this system occurred. First, the right to export rice from Viet Nam, which had been reserved up till that point for Vietnamese individuals with registered business and enterprises, was extended to foreigners in line with Viet Nam's WTO commitment. However, to enhance the competitiveness and bargaining power of Vietnamese rice exporters in comparison to foreign partners and competitors, effective from the same date, exporters have been required to meet stricter requirements regarding storage and processing facilities. According to Decree No. 109/2010/ND-CP dated 4 December 2010 **rice exporters require a certificate from MOTI.**[83] To qualify for a certificate an enterprise must own at least one specialised warehouse with a minimum capacity of 5 000 tonnes of paddy and a rice milling facility with a minimum capacity of 10 tonnes of paddy per hour, and meet other technical requirements intended to improve the value added of rice exports. Enterprises are also required to maintain a minimum volume of reserves equivalent to 10% of their rice exports in the preceding six months. The government is able to request traders to sell product from these reserve volumes into the domestic market to stabilise any sudden price increase (Tran and Dinh, 2014a).

Minimum export prices for various grades of rice have been announced since commercial trade resumed in 2008 to limit price declines. Decree No. 109 provides the current legal basis for setting these prices.[84] The VFA again plays a key role in the administration of this policy. Specifically, rice-exporting enterprises determine their individual floor prices based on purchase costs, taxes, etc. and report this to VFA.[85] VFA uses the submitted floor prices of enterprises to determine the average floor prices of export rice nationwide at the beginning of each season.

While there is a greater degree of competition among export companies under the current arrangements in comparison to the former export quota system, the rice export market is far from being competitive. The VFA exerts a **large degree of control** over the export market and largely favours SOEs, in particular Vinafood 1 and Vinafood II and their subsidiaries, with credit and G2G contract allocation that results in unfair trading privileges (Tran et al., 2013). The recent capacity requirements have pushed small-scale exporters out of the market. The current system also creates an element of market uncertainty because an export company cannot predict precisely when total exports will hit the policy target. This creates the incentive for export companies to submit their export contracts to the VFA as soon as they can to avoid facing the possible suspension of rice exports. Since the strategic motivation for exporting earlier than others is valid for all the export companies, this eventually leads to the actual early suspension of rice exports. The strategic uncertainty may result in suboptimal timing and quantity of rice exports for an individual company, since the quantity and timing of rice exports should be ideally determined based on global and domestic market conditions rather than on the strategic motivation induced by the first-come, first-served basis (Tsukada, 2011). It also reduces the incentive to develop or expand market opportunities (Tran, 2014a). Finally, the sudden termination of contracts when the export volume is reached reduces the prestige of Viet Nam enterprises with their partners. All these factors mean that the current system keeps Viet Nam in the vicious cycle of supplying low-quality rice, and the market expects this from Viet Nam.

State trading

Before 1989 the **state held the monopoly position** in foreign trade.[86] In the period 1975-80 the Ministry of Foreign Trade established Import and Export Companies, and only these companies were allowed to trade. Major partners were the Soviet Union and centrally-planned economies of central and eastern Europe. During 1981-88 foreign trade was decentralised. As a result not only import and export companies which belonged to the Ministry of Foreign Trade were allowed to import and export, but also those belonging to other ministries or local governments. In 1989 the monopoly of the SOEs was broken. Private trading companies were allowed to engage in trade but their activities were severely impeded because import and export licenses were required. Private companies that produced exports were allowed to choose state-owned exporting companies as entrustees while those with annual export revenues above USD 5 million could apply for export licenses.[87] Since 1991 all private companies with licences were allowed to export directly, not through entrustees. In 1998 the licensing requirements for trading were largely abolished, and since 2001 private companies as well as SOEs have been allowed to export most products without any licence.[88]

While the legal control of SOEs over exports may have been eliminated, SOEs still **exert a high degree of influence** over the export of important agricultural commodities like rice, coffee, rubber and tea. Although there are about 200 registered rice exporters, most export less than 1 000 tonnes per year. Eleven companies account for 70% of the rice trade. SOEs, which are responsible for G2G transactions, account for about 80% of exported rice and distribute the exported rice through concessional government programmes in the Philippines, Indonesia, and Cuba (Phan et. al., 2013). Viet Nam National Coffee Corporation (Vinacafe), through its subsidiaries, member companies, and associated companies, has interests in all stages of the coffee chain (WTO, 2013). Subsidiaries of the Viet Nam Rubber Group (VRG) control about 270 thousand ha of total rubber plantations in 2009, corresponding to 40% of total nationwide area and 85% of export production (VietCapital Securities, 2011). SOEs at the national level still account for 60% of tea export with SOEs at the provincial level, private enterprises and joint-venture enterprises account for the remaining 40% (Nguyen and Grote, 2004).

Export taxes and charges

Export taxes were first introduced as part of the early market-orientated reforms through the Law on Export and Import Duties dated 29 December 1987 (Nguyen and Grote, 2004). The duties were justified at the time by the need to raise revenue, protect the environment, conserve natural resources and retain inputs for domestic production (Athukorala et al., 2009). In terms of agricultural products, an initial export tax of 10% was levied on rice, peanut, cashew nut, coffee, tea, rubber and raw hides and skins. The tax rates were gradually reduced over time. For example, they were reduced in 1989 to 5% on rice, 4% on rubber and 3% on cashew nut, tea, coffee and pepper.[89] Rates on some products, such as rice and coffee, were changed quite frequently (CIE, 1998). For example, the export tax on rice was revised twice in 1995, once in September (from 0% to 2%) and again in October (from 2% to 3%). Export taxes applying to most agricultural products were removed in the late 1990s/early 2000s. In accordance with the current Law on Export and Import Duties, in effect since 1 January 2006, export taxes are levied on just a few agricultural related products (WTO, 2013).[90] Cashew nuts in shell are currently zero-rated, while duty rates of 0-10% apply to raw hides and skins, and 0-5% to rubber.

During the period of sharply rising international prices for rice in 2008, the government imposed **export taxes on rice** for a six-month period as part of broad range of policy measures to limit price increases on the domestic market. On 21 July 2008 the government announced the introduction of a progressive export tax, ranging from a minimum VND 500 000 (USD 30) tonne for an export price of USD 600 tonne to a maximum VND 2.9 million (USD 160) tonne for export prices of USD 1 300 tonne and higher.[91] This announcement, made in the context of increasing world rice prices, served to slow purchases by exporters and lower domestic prices (Pham, 2010). As prices fell, the minimum taxable price was raised to USD 800 tonne on 15 August. When domestic prices returned to pre-crisis levels, the export tax on rice was removed in November 2008.

In addition to export taxes, **customs surcharges** had historically applied to the export of certain products. For agricultural-related products, these affected unprocessed cashew nut, unprocessed rubber latex and coffee. An export surcharge applied to coffee during the early 1990s but was removed in 1995 (WTO, 1996). For unprocessed cashew nut, an export surcharge had applied since 1995 and from 2001 in the case of rubber latex. Revenue from customs surcharges was used to finance first the Price Stabilisation Fund (PSF) and then the Export Promotion Fund (EPF). Viet Nam removed these surcharges from the date of its WTO accession as part of negotiated commitments.

Export subsidies and promotion

Viet Nam has not provided direct **export subsidies** to agricultural products since the mid-2000s. In 1998 export subsidies were first provided to canned pineapple, and in addition, an Export Reward Fund (ERF) was established. It provided financial support and preferential loans to enterprises exporting fruits and vegetables as well as meat products. In 1999 the ERF together with the PSF was transformed into the EPF.[92] The EPF provided subsidised interest payments relating to agricultural exports when their international prices decline, assisted some exports which faced losses due to their weak competitiveness or other reasons, and rewarded exporters who promoted new exports, accessed new foreign markets or enlarged their exports to foreign markets. Bonuses contingent on export performance were paid to enterprises exporting rice, coffee, pork, canned fruit and canned vegetables in 2001. The export bonus programme was extended in 2002 to also cover beef and poultry meat; fresh, dried and semi-processed fruit and vegetables; tea; peanuts; pepper; and cashew nuts (WTO, 2006). However, as Viet Nam joined the WTO with a commitment to not maintain agricultural export subsidies from the date of accession, these types of direct payments were discontinued. The EPF was closed down in 2008, with the remaining funds made available for general trade promotion activities (WTO, 2013).[93]

A national **trade promotion programme** has been in operation since 2005.[94] The programme provides funds for a wide range of trade promotion activities, such as the hiring of domestic and foreign experts for advice and assistance on export development or product quality improvements; the organisation of trade fairs, exhibitions; sponsorship to participate in trade events; and to carry out surveys or market investigation (Tran and Dinh, 2014a). In addition, the Viet Nam Development Bank (VDB), established in 2006, provides export credits, investment credit guarantees, and export project performance security along with implementing state policies with respect to the financing of investments, post-investment assistance, and investment credit guarantees (WTO, 2013).

In recent times the government has launched some **agricultural specific measures** to facilitate greater access to commodity markets and trade promotion. Since 2011 the government has provided credit facilities for exporters of tea, pepper, cashew nuts, processed vegetables (box, fresh, dried, pre-processing, fruit juice), sugar, meat, poultry, coffee, seafood and handicrafts such as rattan, bamboo and wicker products.[95] Effective from 1 January 2014 the government supports 50% of the cost of advertising on the mass media; 50% of the cost for fair exhibitors in the country; and 50% of the cost for obtaining market information and other services from the state promotion agencies.[96]

Trade relations

Bilateral trade agreements

Prior to WTO accession, Viet Nam had concluded bilateral trade agreements with 40 partners. The principal aim of these agreements was to establish trade relations based on reciprocal MFN treatment.[97] The agreements were typically short documents with fairly standard text. The exception was the **US-Viet Nam Bilateral Trade Agreement**, signed on 13 July 2000 and entered into force on 10 December 2001. Viet Nam was granted MFN trade status, providing it with substantially better access to the United States: average tariffs fell from 40% to less than 3%. In return, Viet Nam agreed to open up some of its services sectors (banking, insurance, and telecommunications), enhance protection of intellectual property rights and improve its foreign investment regime (WTO, 2013). The Agreement was prepared on the basis of WTO principles and was regarded as a very important step towards WTO membership (Vo, 2005).

Bilateral negotiations have continued following Viet Nam's accession to the WTO. In June 2007 Viet Nam and the United States signed a Trade and Investment Framework Agreement. While this did not introduce additional specific concessions or commitments, it established the US-Viet Nam Council on Trade and Investment, *inter alia,* for the monitoring of implementation of obligations under the WTO Agreement and the bilateral trade agreement. The Viet Nam-Japan Economic Partnership Agreement entered into force on 1 October 2009. It is a comprehensive agreement covering goods, services and investment, as well as issues such as business environment, labour mobility and co-operation on technical standards. A bilateral Free Trade Agreement (FTA) with Chile was signed in November 2011 and entered into force on February 2014. The Agreement, which is focused on market access, will provide tariff free entry for 90% of Chilean exports to Viet Nam including beef, pork, dairy products and fruits. Negotiations for bilateral FTAs with Korea and the customs union of Belarus, Kazakhstan and the Russian Federation, which began in 2012 and 2013 respectively, were concluded in December 2014. Both are expected to come into force during 2015. Bilateral negotiations are continuing with EFTA (Iceland, Liechtenstein, Norway, and Switzerland) and the EU, both of which began in 2012.

Regional trade agreements

Viet Nam became the **seventh member of ASEAN** on 28 July 1995, joining the ASEAN-6 comprising the five founding members: Indonesia, Malaysia, Philippines, Singapore and Thailand, plus Brunei, which joined in 1984.[98] As a requirement of membership, it signed the ASEAN Free Trade Area (AFTA) Agreement that the ASEAN-6 had concluded on 28 January 1992. Viet Nam began granting preferential treatment for goods to its ASEAN partners under the Common Effective Preferential Tariff (CEPT) scheme on 1 January 1996.[99] Reduction commitments were completed in 2013, although commitments were finished on 96% of tariff lines in 2006.

Under CEPT there were four product lists, each with a different tariff reduction period (Table 2.11). Members could decide which tariff lines were included in each list. New ASEAN members were given the same time period for tariff reduction as the ASEAN-6 with the starting point determined by the date of joining. Products on the Inclusion List (IL) had tariffs reducing to 0-5% over ten years under the Normal Track and seven years under the Fast Track; those on the Temporary Exclusion List (TEL) had tariffs reducing to 0-5% by the same end date but with a delayed start; while those on the Sensitive List (SL) had a longer delayed start period, a ten-year implementation period and higher end rates. Once a product is included in the CEPT, quantitative restrictions were to be eliminated immediately and other NTBs were to be removed within five years (Tantraporn and Tuchinda, 2012). Agricultural products on Viet Nam's sensitive list included bird's eggs, certain citrus (grapefruit and lemons), rice (paddy and husked), and sausages and other prepared or preserved meat. Unmanufactured tobacco, cigarettes and other products manufactured from tobacco are on the General Exception List (GEL) not subject to reduction commitments.

Table 2.11. **CEPT time frame for ASEAN member states**

Country	Manufactured and processed agricultural goods			Unprocessed agricultural goods		
	Inclusion List Fast Track	Inclusion List Normal Track	Temporary Exclusion List	Inclusion List	Temporary Exclusion List	Sensitive List
ASEAN-6	1993-2000	1993-2003	1996-2003	1996-2003	1997-2003	2001-10
Viet Nam	1996-2003	1996-2006	1999-2006	1999-2006	2000-06	2004-13
Lao PDR and Myanmar	1998-2005	1998-2008	2001-08	2001-08	2002-08	2006-15
Cambodia	2000-07	2000-10	2003-10	2003-10	2004-10	2008-17

Note: CEPT stands for Common Effective Preferential Tariff.
Source: Information gathered from the official ASEAN website, *www.asean.org*.

At the ninth ASEAN Ministerial Summit held from 7-8 October 2003 in Bali, the ten members of ASEAN signed an ambitious accord to establish an **ASEAN Economic Community** (AEC). The agreement, called the "Bali Concord II", aims to create a community in Southeast Asia based on three pillars: economic co-operation, political and security co-operation, and socio-cultural co-operation. The ultimate goal of the agreement is to create a competitive region with a free flow of investment, goods, services and skilled labour, combined with a freer flow of capital, stable and equitable economic development, and reduced poverty and socio-economic disparities by the year 2020. As a step to achieving this goal, ASEAN members signed the ASEAN Trade in Goods Agreement (ATIGA) in February 2009, consolidating all existing ASEAN initiatives, obligations, and commitments on trade in goods into a single document. ATIGA entered into force on 17 May 2010 (WTO, 2013). Under ATIGA, end tariff rates on the Sensitive List products were reduced to a common standard of 5%.

In the terms of food and agriculture, the underlying objectives of **co-operation between ASEAN countries** have been to strengthen food security and ensure food safety in the region. ASEAN Ministers of Agriculture and Forestry have established a Ministerial Understanding (MU) on ASEAN Cooperation in Food, Agriculture and Forestry, signed in October 1993, to facilitate and promote trade in the region. In response to the high fluctuation of food prices coupled with the global financial crisis that started in 2008, ASEAN Leaders adopted the ASEAN Integrated Food Security (AIFS) Framework and the Strategic Plan of Action on ASEAN Food Security (SPA-FS) for 2009-13 at the 14th ASEAN Summit in 2009. Its goals were to strengthen and expand existing regional initiatives in the areas of food security, sustainable food trade, integrated food security information and agricultural innovation. One of major activities was the establishment of the ASEAN Plus Three Emergency Rice Reserve (APTERR), which entered into force on 12 July 2012.[100] Member countries have pledged to stockpile 787 000 tonnes of rice, of which Viet Nam has committed 14 000 tonnes, for disposal under the collective scheme. Research suggests that this volume is not large enough to alone deal with a 5% production shock in China and Indonesia, with governments having to also rely on domestic measures (Briones et al., 2012). A revised AIFS and SPA-FS covering the period 2015-18 has been developed.

In addition to trade liberalisation among its own members, ASEAN has also been actively negotiating trade agreements with its major regional trading partners in what are **termed ASEAN+ agreements** (Table 2.12). The coverage of these agreements has evolved from traditional measures (e.g. tariffs on goods) to non-traditional elements (e.g. trade in services, and investment) and more recently (e.g. AANZFTA) to measures including rules on investment and competition that go beyond its WTO commitments (Vu, 2014). All of these agreements acknowledge the differences in the levels of development within ASEAN by allowing the four non-ASEAN-6 member states extended dates for total compliance

Table 2.12. **Viet Nam's tariff reduction commitments under ASEAN+ agreements**

FTA	Date signed[1]	Implementation begins[2]	Full implementation deadline	Tariff lines liberalised
ASEAN-China Free Trade Agreement (ACFTA)	29 November 2004	1 January 2005	NT 1: 2015 NT 2: 2018	90% tariff lines (HS-6 digit level)
ASEAN-Japan Comprehensive Economic Partnership Agreement (AJCEP)	1 April 2008	1 December 2008	NT 1: 2018 NT 2: 2023 NT 3: 2024	88.6% tariff lines (HS 10-digit level)
ASEAN-Korea Free Trade Agreement (AKFTA)	24 August 2006	1 January 2010	NT 1: 2015 NT 2: 2018	90% tariff lines (HS-6 digit level)
ASEAN-India Free Trade Agreement (AIFTA)	13 August 2009	1 January 2010	NT 1: 2018 NT 2: 2021	80% tariff lines (HS 6-digit level)
ASEAN-Australia-New Zealand Free Trade Agreement (AANZFTA)	27 February 2009	1 January 2010	NT: 2018 NT: 2020	90% tariff lines (HS 8-digit level)

1. Date on which the agreement on trade in goods signed by Viet Nam.
2. Normal Track (NT) reductions commence for Viet Nam.
Source: Vu (2014); *WTO RTA Database, http://rtais.wto.org/UI/PublicAllRTAList.aspx.*

(usually five years) and requiring a commitment to liberalise fewer tariff lines. Negotiations on a Regional Comprehensive Economic Partnership (RCEP) between the ten ASEAN member states and the six states with which ASEAN has existing FTAs were formally launched in November 2012. The aim is to conclude an agreement establishing an open trade and investment environment in the region to facilitate the expansion of regional trade and investment and contribute to global economic growth and development by the end of 2015.

Having initially participated as an "associate member", Viet Nam announced its decision to be a full participant in the negotiations on a **Trans-Pacific Partnership (TPP) Agreement** in November 2010. While the final scope of the agreement is still being negotiated, trade gains for Vietnamese agriculture are likely to be limited to smaller export sectors such as cassava starch, processed food and honey. This is because its major exports, e.g. coffee, rubber, cashews and pepper, already benefit from existing low or duty-free rates with many of the TPP partners with the exception of rice to Japan and Korea (Arita and Dyck, 2014).

WTO

Viet Nam became the **WTO's 150th Member** on 11 January 2007. This was the result of eleven years of preparation, including eight years of negotiations (OECD, 2009). Viet Nam agreed to comply with key WTO Agreements such as the Agreements on Trade-Related Aspects of Intellectual Property Rights, Technical Barriers to Trade, the Application of Sanitary and Phytosanitary Measures, and customs valuation from the date of accession without recourse to any transitional period (WTO, 2013). Anchoring domestic reforms to the requirements of WTO membership has created a uniform basis for important advances in many areas of the national economy (OECD, 2009). As part of its accession commitments, Viet Nam agreed to bind 100% of its entire tariff schedule including all agricultural tariff lines.[101] Other key agricultural commitments were to:

- implement tariff reductions over a 3-5 year period
- eliminate all non-tariff barriers at accession and establish import quotas for birds eggs, sugar, unmanufactured tobacco and salt
- grant trading rights (the right to import and export) for all goods to all foreign individuals and organisations at accession with an adjustment period for fertiliser (1 January 2010) and rice (1 January 2011)
- maintain domestic support within the *de minimis* ceiling for developing countries, i.e. not above 10% of production values
- eliminate export subsidies immediately at accession.

In the current WTO **Doha Round** of negotiations, Viet Nam has been stressing the importance of the development dimension. In the rules area, anti-dumping and countervailing measures in the fisheries sector are of particular interest (WTO, 2013). Viet Nam joined the Recently Acceded Members (RAMs) group immediately on accession and became the 20th member of the Cairns Group on 2 December 2013.

2.5. Evaluation of support to agriculture

This section presents a **quantitative evaluation of support** provided to agriculture in Viet Nam through the domestic and trade policies discussed in detail in the previous sections of this chapter. The evaluation is based on the indicators of agricultural support developed by the OECD, including the Producer Support Estimate (PSE), Consumer Support

Estimate (CSE), General Services Support Estimate (GSSE), Total Support Estimate (TSE) and others (Box 2.4). Evaluation of agricultural support for Viet Nam covers the period between 2000 and 2013.

Box 2.4. OECD indicators of support to agriculture

INDICATORS OF SUPPORT FOR PRODUCERS

Producer Support Estimate (PSE): The annual monetary value of gross transfers from consumers and taxpayers to agricultural producers, measured at the farm gate level, arising from policy measures that support agriculture, regardless of their nature, objectives or impacts on farm production or income.

Percentage PSE (%PSE): PSE transfers as a share of gross farm receipts (including support).

Producer Nominal Assistance Coefficient (producer NAC): The ratio between the value of gross farm receipts (including support) and gross farm receipts valued at border prices (measured at farm gate).

Producer Nominal Protection Coefficient (producer NPC): The ratio between the average price received by producers at farm gate (including payments per tonne of current output), and the border price (measured at farm gate). The producer NPC is also available by commodity.

Producer Single Commodity Transfers (producer SCT): The annual monetary value of gross transfers from consumers and taxpayers to agricultural producers, measured at the farm gate level, arising from policy measures directly linked to the production of a single commodity such that the producer must produce the designated commodity in order to receive the transfer.

Producer Percentage Single Commodity Transfers (producer %SCT): The commodity SCT expressed as a share of gross farm receipts for the specific commodity (including support).

INDICATORS OF SUPPORT TO CONSUMERS

Consumer Support Estimate (CSE): The annual monetary value of gross transfers from (to) consumers of agricultural commodities, measured at the farm gate level, arising from policy measures that support agriculture, regardless of their nature, objectives or impacts on consumption of farm products. If negative, the CSE measures the burden (implicit tax) on consumers through market price support (higher prices), that more than offsets consumer subsidies that lower prices to consumers.

Percentage CSE (%CSE): CSE transfers as a share of consumption expenditure on agricultural commodities (measured at farm gate), net of taxpayer transfers to consumers.

Consumer Nominal Assistance Coefficient (consumer NAC): The ratio between the value of consumption expenditure on agricultural commodities (at farm gate) and that valued at border prices (measured at farm gate).

Consumer Nominal Protection Coefficient (consumer NPC): The ratio between the average price paid by consumers (at farm gate) and the border price (measured at farm gate).

Consumer Single Commodity Transfers (consumer SCT): The annual monetary value of gross transfers from (to) consumers of agricultural commodities, measured at the farm gate level, arising from policy measures directly linked to the production of a single commodity.

INDICATORS OF SUPPORT TO GENERAL SERVICES FOR AGRICULTURE

General Services Support Estimate (GSSE): The annual monetary value of gross transfers to general services provided to agricultural producers collectively (such as research, development, training, inspection, marketing and promotion), arising from policy measures

Box 2.4. **OECD indicators of support to agriculture** *(cont.)*

that support agriculture regardless of their nature, objectives and impacts on farm production, income, or consumption. The GSSE does not include any transfers to individual producers.

Percentage GSSE (%GSSE): GSSE transfers as a share of Total Support Estimate (TSE).

INDICATORS OF TOTAL SUPPORT TO AGRICULTURE

Total Support Estimate (TSE): The annual monetary value of all gross transfers from taxpayers and consumers arising from policy measures that support agriculture, net of associated budgetary receipts, regardless of their objectives and impacts on farm production and income, or consumption of farm products.

Percentage TSE (%TSE): TSE transfers as a percentage of GDP.

A detailed description of the **OECD methodology** to estimate agricultural support (the "PSE Manual") and a comprehensive database for OECD and selected non-OECD countries including Viet Nam are available from *www.oecd.org/tad/agricultural-policies/producerandconsumersupportestimatesdatabase.htm*. The methodology applied in this study is fully consistent with that used for other countries as presented in OECD reports that monitor and evaluate agricultural policies (OECD, 2014). Box 2.5 provides basic information on how this methodology has been applied in the case of Viet Nam.

Box 2.5. **Viet Nam's PSEs: What and how?**

Period covered: 2000-13

Products covered: Rice, natural rubber, coffee (green), maize, cashew nuts (with shell), sugar cane, pepper, tea, beef and veal, pigmeat, poultry and eggs. These 12 commodities account for 81% of the total value of gross agricultural output (GAO) in Viet Nam during the entire thirteen-year period 2000-13 and 83% in 2011-13. The eight crop products account for 80% of the value of total crop production in 2011-13 while the four livestock products represent on average 95% of total livestock production. For the purposes of calculating market price gaps, six are treated as exportables: rice, natural rubber, coffee, cashew nuts, pepper, pigmeat and tea. The remaining five are considered importables.

Market Price Support

Producer prices: Average prices received by producers, sourced from MARD.

Price gap estimates: For all the above listed products with the exception of pork, relevant data have been collected and price gaps calculated. For pork, which is a marginally exported commodity subject to import tariffs, the price gap has been set to zero.

External reference prices: For the six exportable commodities, the average export unit values registered at the Vietnamese border are used. The average import unit value at the Vietnamese border is used for maize. For the remaining four commodities, a variety of alternative reference prices are used because of the limited volume of imports into Viet Nam: the average Thailand FOB price for refined sugar; the average Australian FOB price for beef and veal; the average Chinese FOB price for poultry and the Chinese farm gate price for eggs.

Marketing margins: The marketing margin indicates processing, handling and transportation costs for a given commodity. For all but one product, coffee, margins were calculated as a fixed percentage of the farm gate price based on discussion with MARD.

Box 2.5. **Viet Nam's PSEs: What and how?** *(cont.)*

These ranged from 8% in the case of beef to 32% in the case of natural rubber. A fixed price of USD 17 per tonne was used for coffee.

Quality adjustments: No quality adjustments were made.

Budgetary Support

Budgetary information for the period 2000-13 originates from MARD and covers budgetary expenditure undertaken by MARD and MOF. It incorporates transfers to provincial governments for agricultural programmes and where possible local government expenditure. However, the value of local government expenditure is underrepresented in the budgetary data. The cost to the government of subsidising fertiliser production in Viet Nam is not included as evidence suggests that this support is not passed on to farmers in terms of lower fertiliser prices.

Support to agricultural producers

Level of producer support

The **percentage Producer Support Estimate** (%PSE) is the OECD's key indicator to measure the level of support provided to the agricultural sector. It expresses the monetary value of support transfers to agricultural producers as a share of gross farm receipts. Because it is not affected by inflation or the size of the sector, it allows comparisons in the level of support to be made over time and between countries. The level of support is important because it provides insights into the burden that agricultural support policies place on consumers (MPS) and taxpayers (budgetary transfers).

Viet Nam's %PSE averaged 7% in the three-year period 2010-13 indicating that less than one tenth of gross receipts of agricultural producers were generated by support policies (Table 2.13 and 2.14). Producer support was estimated to be at a similar level in 2000-02. No distinct long term-term trend in producer support can be observed over the period 2000-13 (Figure 2.9). However, the %PSE fluctuated considerably, reaching a low of minus 21% in 2008 and a peak of 16% in 2009. With the exception of 2002 and 2008, fluctuations in support were within a positive range, indicating that overall policies were **supportive of domestic producers**.

Like a lot of other countries, changes in the level of support in Viet Nam are driven by fluctuations in MPS. For example, the 55% decrease in producer support between 2012 and 2013 is due to a 67% fall in MPS. This is because transfers from consumers are relatively large as compared to transfers from taxpayers. The share of budgetary transfers in the total PSE was 20% on average in 2011-13, a similar level as in 2001-03. These swings are relatively greater in Viet Nam and often produce negative values because of the government's efforts to **balance the interests between producers and consumers**. On the one hand, the government wishes to increase prices received by producers to encourage production and improve farmer incomes. On the other, it wants to keep prices paid by final consumers at an affordable level to help alleviate poverty and avoid social tension. A similar pattern of support is observed in Indonesia.

In comparison with OECD and selected non-OECD countries, the average level of producer support in Viet Nam of 7% measured over 2011-13 is considerably **lower than the OECD** average of 18% (Figure 2.10). It is the lowest of the five Asian economies for which

Table 2.13. **Estimates of support to agriculture in Viet Nam, VND million**

	2000-02	2011-13	2011	2012	2013p
Total value of production (at farm gate)	**128 610 574**	**749 037 505**	**776 920 800**	**735 134 000**	**735 057 714**
***of which** share of MPS commodities (%)*	*81.6*	*83.5*	*82.5*	*88.2*	*79.7*
Total value of consumption (at farm gate)	**112 148 440**	**628 416 550**	**653 749 282**	**605 641 081**	**625 859 286**
Producer Support Estimate (PSE)	**8 878 569**	**53 663 299**	**60 990 064**	**68 762 051**	**31 237 781**
Support based on commodity output	7 057 707	43 104 503	51 970 395	58 388 303	18 954 812
Market Price Support[1]	*7 057 707*	*43 104 503*	*51 970 395*	*58 388 303*	*18 954 812*
Payments based on output	*0*	*0*	*0*	*0*	*0*
Payments based on input use	1 510 528	6 462 651	6 347 335	5 950 149	7 090 469
Based on variable input use	1 510 528	6 448 671	6 326 365	5 939 664	7 079 984
with input constraints	*0*	*0*	*0*	*0*	*0*
Based on fixed capital formation	*0*	*13 980*	*20 970*	*10 485*	*10 485*
with input constraints	*0*	*0*	*0*	*0*	*0*
Based on on-farm services	*0*	*0*	*0*	*0*	*0*
with input constraints	*0*	*0*	*0*	*0*	*0*
Payments based on current A/An/R/I, production required	0	4 096 144	2 672 333	4 423 600	5 192 500
Based on Receipts/Income	0	131 111	83 333	176 000	134 000
Based on Area planted/Animal numbers	0	3 965 033	2 589 000	4 247 600	5 058 500
with input constraints	0	3 965 033	2 589 000	4 247 600	5 058 500
Payments based on non-current A/An/R/I, production required	0	0	0	0	0
Payments based on non-current A/An/R/I, production not required	0	0	0	0	0
With variable payment rates	0	0	0	0	0
with commodity exceptions	0	0	0	0	0
With fixed payment rates	0	0	0	0	0
with commodity exceptions	0	0	0	0	0
Payments based on non-commodity criteria	310 333	0	0	0	0
Based on long-term resource retirement	310 333	0	0	0	0
Based on a specific non-commodity output	0	0	0	0	0
Based on other non-commodity criteria	0	0	0	0	0
Miscellaneous payments	0	0	0	0	0
Percentage PSE (%)	**7.3**	**7.1**	**7.8**	**9.2**	**4.2**
Producer NPC (coeff.)	**1.09**	**1.10**	**1.12**	**1.12**	**1.06**
Producer NAC (coeff.)	**1.08**	**1.08**	**1.08**	**1.10**	**1.04**
General Services Support Estimate (GSSE)	**4 235 433**	**15 401 235**	**16 138 827**	**14 830 699**	**15 234 180**
Agricultural knowledge and innovation system	349 070	1 641 978	1 545 605	1 680 665	1 699 665
Inspection and control	51 601	73 477	72 298	74 354	73 780
Development and maintenance of infrastructure	3 735 025	13 043 445	13 888 071	12 467 880	12 774 385
Marketing and promotion	18 429	26 242	25 821	26 555	26 350
Cost of public stockholding	81 308	616 093	607 032	581 245	660 000
Miscellaneous	0	0	0	0	0
Percentage GSSE (% of TSE)	***n.a.***	**23.8**	**20.9**	**17.7**	**32.8**
Consumer Support Estimate (CSE)	**-9 376 321**	**-69 540 643**	**-85 176 047**	**-74 080 534**	**-49 365 347**
Transfers to producers from consumers	-9 459 449	-62 424 742	-78 155 141	-70 495 069	-38 624 017
Other transfers from consumers	-292 172	-9 415 262	-9 246 835	-8 692 452	-10 306 499
Transfers to consumers from taxpayers	0	0	0	0	0
Excess feed cost	375 300	2 299 361	2 225 929	5 106 987	-434 831
Percentage CSE (%)	**-8.9**	**-11.0**	**-13.0**	**-12.2**	**-7.9**
Consumer NPC (coeff.)	**1.11**	**1.13**	**1.15**	**1.15**	**1.08**
Consumer NAC (coeff.)	**1.10**	**1.12**	**1.15**	**1.14**	**1.09**
Total Support Estimate (TSE)	**13 114 002**	**69 064 534**	**77 128 890**	**83 592 750**	**46 471 961**
Transfers from consumers	9 751 621	71 840 004	87 401 976	79 187 520	48 930 516
Transfers from taxpayers	3 654 553	6 639 791	-1 026 251	13 097 681	7 847 944
Budget revenues	-292 172	-9 415 262	-9 246 835	-8 692 452	-10 306 499
Percentage TSE (% of GDP)	**2.8**	**2.2**	**2.8**	**2.6**	**1.4**
GDP deflator (2000-02 = 100)	**100**	**305**	**280**	**310**	**325**

p. provisional, n.a.: not available.

NPC: Nominal Protection Coefficient.

NAC: Nominal Assistance Coefficient.

A (area planted), An (animal numbers), R (receipts), I (income).

1. MPS commodities for Viet Nam are: rice, natural rubber, coffee, maize, cashew nuts, sugar cane, pepper, tea, beef and veal, pigmeat, poultry and eggs. Market Price Support is net of producer levies and Excess Feed Cost.

Source: OECD (2014), "Producer and Consumer Support Estimates", *OECD Agriculture Statistics Database*.

StatLink http://dx.doi.org/10.1787/888933223844

Table 2.14. **Estimates of support to agriculture in Viet Nam, USD million**

	2000-02	2011-13	2011	2012	2013p
Total value of production (at farm gate)	**8 719**	**35 695**	**37 611**	**34 604**	**34 870**
***of which** share of MPS commodities (%)*	*81.6*	*83.5*	*82.5*	*88.2*	*79.7*
Total value of consumption (at farm gate)	**7 600**	**29 949**	**31 649**	**28 509**	**29 690**
Producer Support Estimate (PSE)	**614**	**2 557**	**2 953**	**3 237**	**1 482**
Support based on commodity output	490	2 055	2 516	2 748	899
Market Price Support[1]	*490*	*2 055*	*2 516*	*2 748*	*899*
Payments based on output	*0*	*0*	*0*	*0*	*0*
Payments based on input use	103	308	307	280	336
Based on variable input use	103	307	306	280	336
with input constraints	*0*	*0*	*0*	*0*	*0*
Based on fixed capital formation	*0*	*1*	*1*	*0*	*0*
with input constraints	*0*	*0*	*0*	*0*	*0*
Based on on-farm services	*0*	*0*	*0*	*0*	*0*
with input constraints	*0*	*0*	*0*	*0*	*0*
Payments based on current A/An/R/I, production required	0	195	129	208	246
Based on Receipts/Income	0	6	4	8	6
Based on Area planted/Animal numbers	0	188	125	200	240
with input constraints	0	188	125	200	240
Payments based on non-current A/An/R/I, production required	0	0	0	0	0
Payments based on non-current A/An/R/I, production not required	0	0	0	0	0
With variable payment rates	0	0	0	0	0
with commodity exceptions	0	0	0	0	0
With fixed payment rates	0	0	0	0	0
with commodity exceptions	0	0	0	0	0
Payments based on non-commodity criteria	21	0	0	0	0
Based on long-term resource retirement	21	0	0	0	0
Based on a specific non-commodity output	0	0	0	0	0
Based on other non-commodity criteria	0	0	0	0	0
Miscellaneous payments	0	0	0	0	0
Percentage PSE (%)	**7.3**	**7.1**	**7.8**	**9.2**	**4.2**
Producer NPC (coeff.)	**1.09**	**1.10**	**1.12**	**1.12**	**1.06**
Producer NAC (coeff.)	**1.08**	**1.08**	**1.08**	**1.10**	**1.04**
General Services Support Estimate (GSSE)	**288**	**734**	**781**	**698**	**723**
Agricultural knowledge and innovation system	24	78	75	79	81
Inspection and control	4	4	4	4	4
Development and maintenance of infrastructure	254	622	672	587	606
Marketing and promotion	1	1	1	1	1
Cost of public stockholding	6	29	29	27	31
Miscellaneous	0	0	0	0	0
Percentage GSSE (% of TSE)	***n.a.***	**23.8**	**20.9**	**17.7**	**32.8**
Consumer Support Estimate (CSE)	**-646**	**-3 317**	**-4 123**	**-3 487**	**-2 342**
Transfers to producers from consumers	-652	-2 978	-3 784	-3 318	-1 832
Other transfers from consumers	-20	-449	-448	-409	-489
Transfers to consumers from taxpayers	0	0	0	0	0
Excess feed cost	26	109	108	240	-21
Percentage CSE (%)	**-8.9**	**-11.0**	**-13.0**	**-12.2**	**-7.9**
Consumer NPC (coeff.)	**1.11**	**1.13**	**1.15**	**1.15**	**1.08**
Consumer NAC (coeff.)	**1.10**	**1.12**	**1.15**	**1.14**	**1.09**
Total Support Estimate (TSE)	**902**	**3 291**	**3 734**	**3 935**	**2 205**
Transfers from consumers	672	3 427	4 231	3 728	2 321
Transfers from taxpayers	249	313	-50	617	372
Budget revenues	-20	-449	-448	-409	-489
Percentage TSE (% of GDP)	**2.8**	**2.2**	**2.8**	**2.6**	**1.4**
GDP deflator (2000-02 = 100)	**100**	**305**	**280**	**310**	**325**

p. provisional, n.a.: not available.

NPC: Nominal Protection Coefficient.

NAC: Nominal Assistance Coefficient.

A (area planted), An (animal numbers), R (receipts), I (income).

1. MPS commodities for Viet Nam are: rice, natural rubber, coffee, maize, cashew nuts, sugar cane, pepper, tea, beef and veal, pigmeat, poultry and eggs. Market Price Support is net of producer levies and Excess Feed Cost.

Source: OECD (2014), "Producer and Consumer Support Estimates", *OECD Agriculture Statistics Database.*

StatLink http://dx.doi.org/10.1787/888933223852

Figure 2.9. **Level and composition of Producer Support Estimate in Viet Nam, 2000-13**

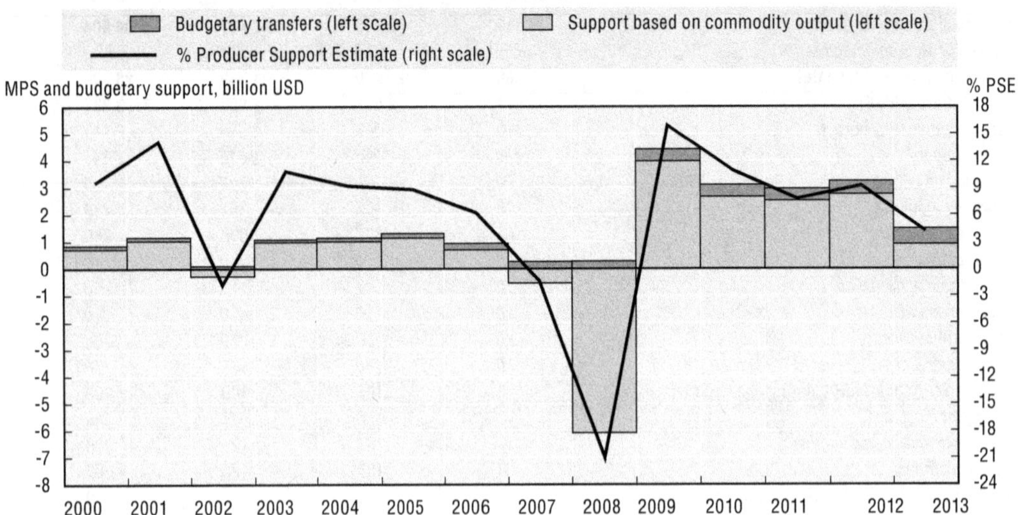

Note: Percentage PSE (%PSE) is the monetary value of support transfers to agricultural producers as a share of gross farm receipts.

Source: OECD (2014), "Producer and Consumer Support Estimates", *OECD Agriculture Statistics Database.*

StatLink http://dx.doi.org/10.1787/888933223832

Figure 2.10. **Producer Support Estimate in Viet Nam and selected countries, 2011-13 average**

Per cent of gross farm receipts

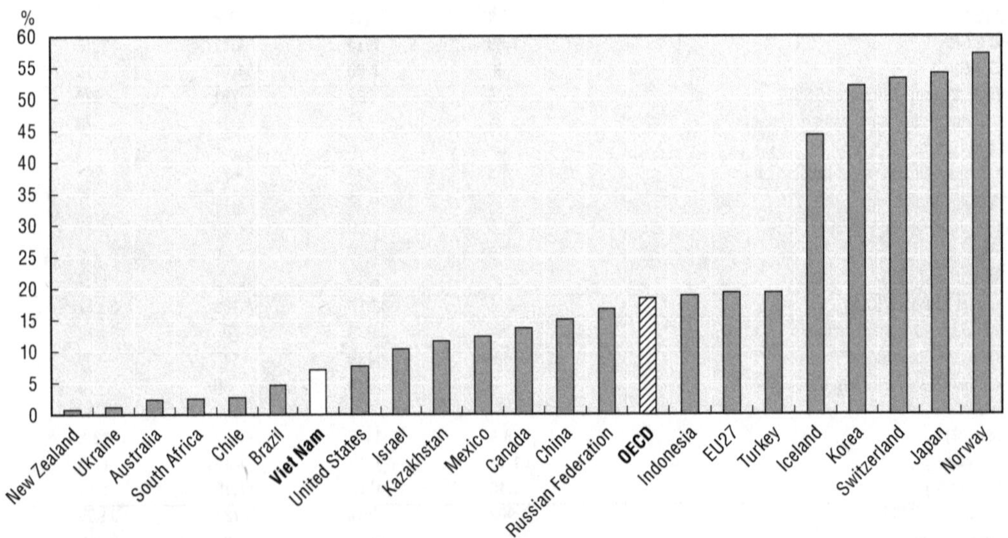

1. The statistical data for Israel are supplied by and under the responsibility of the relevant Israeli authorities. The use of such data by the OECD is without prejudice to the status of the Golan Heights, East Jerusalem and Israeli settlements in the West Bank under the terms of international law.
2. The OECD total does not include the non-OECD EU member states.
3. 2010-12 for Brazil, China, Indonesia, Kazakhstan, Russia, South Africa and Ukraine.

Source: OECD (2014), "Producer and Consumer Support Estimates", *OECD Agriculture Statistics Database.*

StatLink http://dx.doi.org/10.1787/888933223865

indicators of support are calculated: less than half the level of support provided to producers in China and Indonesia, and much lower than the two OECD members Korea (52%) and Japan (54%).

Composition of producer support by policy category

In addition to the level of support, it is also necessary to analyse the way in which that support is provided to producers. The **composition of support** is important because how support is provided determines its impact on the agricultural sector and the distribution of benefits to society as a whole. For example, market price support can have a large effect on production and trade, but it imposes additional and regressive costs on domestic consumers, is not effective in improving farm income and can have negative effects on the environment. On the other hand, income support not based on current commodity production is much more effective at improving farm income with less spill-over effects. Policies that directly target non-commodity criteria such as landscape elements, environmental performance or traditional breeds of animals are also typically more effective at reaching these societal objectives. While targeted policies are likely to be more politically sustainable as they can be clearly explained, higher implementation costs (the costs associated with designing, implementing, monitoring and evaluating policy measures) make the move towards targeted policies more challenging (van Tongeren, 2008; Martini, 2011).

MPS is the dominant component of producer support in Viet Nam. Its aggregate value is the outcome of implicit taxation through negative price gaps for some commodities (a negative MPS) and price support of others (a positive MPS) (Figure 2.11). Annual variations depend on movements in world prices, domestic prices and exchange rates, as well as changes in production levels. For example, at the aggregate level, the 61% decrease MPS between 2012 and 2013 was caused by a narrowing of the price gap between domestic and border prices which more than offset the 3% increase in total production. In general, producer prices fell faster than the decrease in border prices.

Figure 2.11. **Level and composition of Market Price Support in Viet Nam, 2000-13**

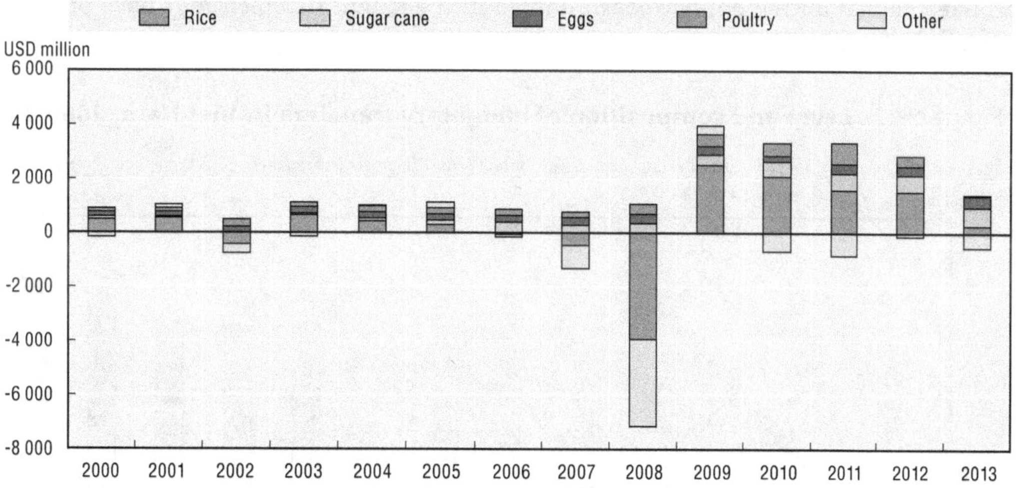

Source: OECD (2014), "Producer and Consumer Support Estimates", *OECD Agriculture Statistics Database*.

StatLink http://dx.doi.org/10.1787/888933223878

Because rice represents around one-third of the total value of production, changes in the **MPS for rice** have a significant influence on changes in total MPS. The gap between domestic and border prices for rice was kept fairly close during the period 2000-08; sometimes positive, sometimes negative. In 2008 international prices for rice and other grains rose dramatically. Concerned about the impact of rising prices on poor households, the Vietnamese government responded with a number of policies to increase supply on the domestic market, e.g. tightened export controls and imposed an export tax. Farm gate prices rose but not as fast as border prices, leading to significant negative MPS for rice. Following this event, additional policies were put in place to support farm gate prices for rice. With the subsequent fall in international prices, a large positive gap emerged. However, as a major net exporter of rice this positive gap could only be sustained for a period of time and it has subsequently fallen year by year.

Sugar cane and livestock products, illustrated in Figure 2.11 by eggs and poultry, are the two principal agricultural commodities/sectors that receive market price support. These sectors produce **import-competing commodities** and are protected through import tariffs and stringent food safety import regulations. In comparison, export-competing commodities, such as natural rubber, coffee, cashew nuts and tea, generally have negative market price support, i.e. farm gate producer prices are lower than border prices measured at the farm gate. The presence of SOEs involved in the processing of these products may explain this outcome.

Budgetary support to agricultural producers has increased since 2000, rising from just over USD 131 million in 2000 to USD 583 million in 2013 (Figure 2.12). As a share of gross farm receipts, it has however **remained relatively constant** over the period at about 1.5%. Budgetary support is primarily given in the form of payments based on variable input use. Expenditure associated with subsidising the irrigation fee exemption is the dominant payment in this category. Prior to 2011 the government was providing support through the reforestation programme to take land out of agriculture production (non-commodity criteria). Since 2011 it has been providing a per hectare payment with the objective of keeping about 4 million ha in paddy production. The interest concession programmes that have been introduced since 2010 to assist with the purchase of machinery (payments based on fixed capital formation) have been limited in the extent to which they have provided support to farmers.

Figure 2.12. **Level and composition of budgetary transfers in Viet Nam, 2000-13**

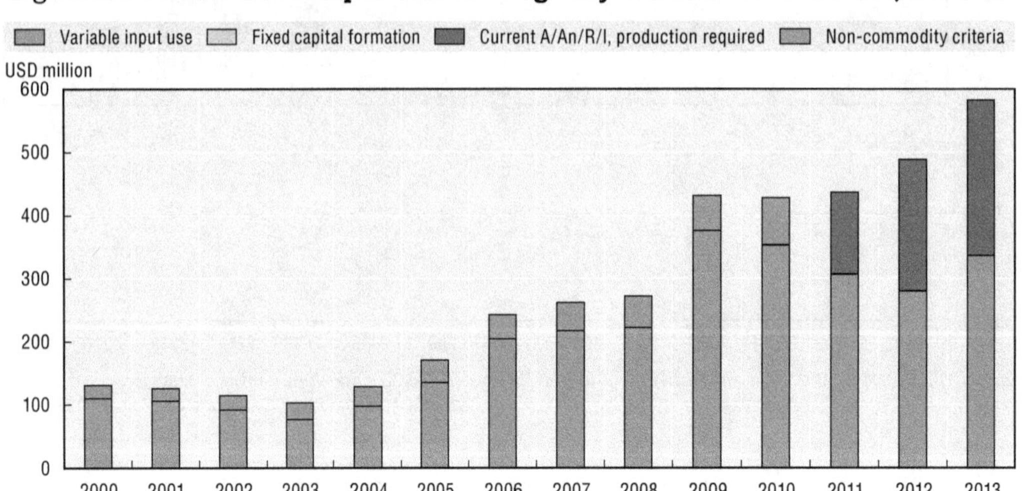

A (area planted), An (animal numbers), R (receipts), I (income).

Source: OECD (2014), "Producer and Consumer Support Estimates", *OECD Agriculture Statistics Database.*

StatLink http://dx.doi.org/10.1787/888933223889

Commodity profile of producer support

Producer Single Commodity Transfers (SCT) is an indicator that shows the extent to which agricultural policies are **commodity specific**. It indicates the flexibility that policies accord to producers in their choices of product mixes. For example, a payment designated for only one specific commodity implies that in order to receive payment, a farmer must produce that commodity. Alternatively, payment may be provided for any commodity in a designated group (for example, any crop within a cereal group), or simply to any commodity without distinction. The latter payments progressively give freedom to those who receive support to define their production mix, and producers become more responsive to market signals. The SCT corresponds to the first type of support and includes MPS and payments provided for the production of only a specified individual commodity. The SCT can be expressed in relative terms as a percentage of gross receipts for a given commodity. A figure of 33%, for example, indicates that the value of transfers that are specific to that commodity is equivalent to one-third of gross farm receipts for that commodity.

Producer SCT as a share of commodity gross farm receipts (%SCT) is highest for certain livestock products (beef and veal, poultry and eggs) and sugar cane, with the value of transfers to these commodities representing **20% or more of gross farm receipts** (Figure 2.13). In Viet Nam these indicators principally reflect the market price support for these commodities as the only other single-commodity payments provided are per hectare payments for rice (Other SCT). These commodities also had the highest %SCT in 2000-02. Although there has been a decrease in the %SCT indicator for eggs and poultry, it has risen marginally for beef and veal, and substantially in the case of sugar cane.

Figure 2.13. **Producer SCTs by commodity in Viet Nam, averages 2000-02 and 2011-13**

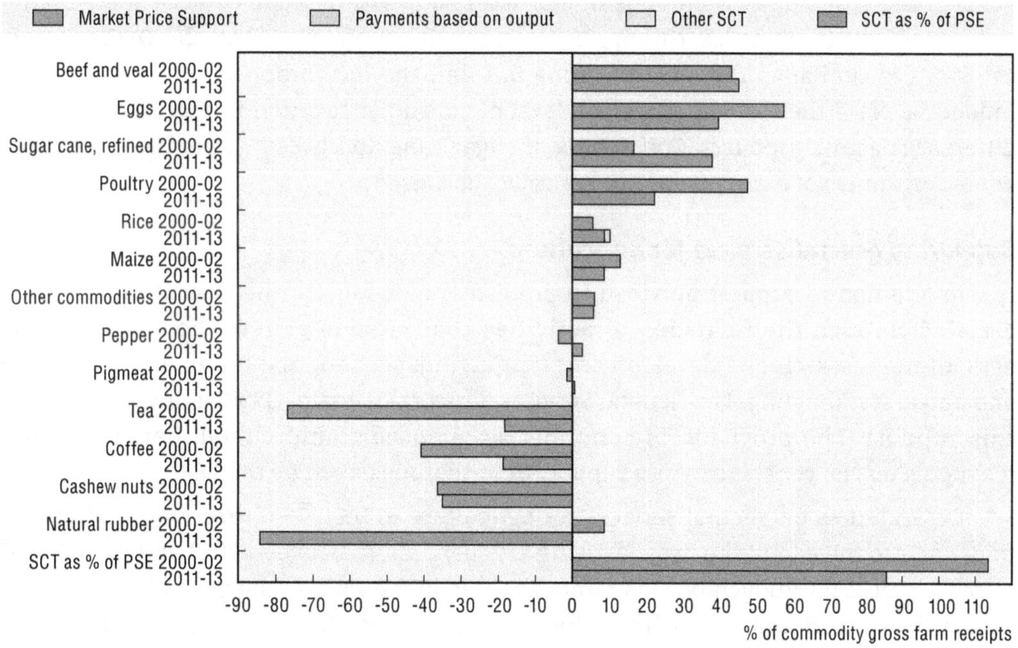

Note: Commodities are ranked according to 2011-13 levels. SCT: Single Commodity Transfers.

Source: OECD (2014), "Producer and Consumer Support Estimates", *OECD Agriculture Statistics Database.*

StatLink http://dx.doi.org/10.1787/888933223890

The **negative producer SCT** for tea, coffee, cashew nuts and natural rubber indicates implicit losses in receipts that producers of these commodities incur because they receive prices that are below world prices. In the case of tea, coffee and cashew nuts, the negative producer SCT has become smaller over the period under review indicating an improvement in their situation. However, natural rubber has switched from being supported in 2000-02 to having the highest negative SCT of any commodity evaluated. It should be noted that in all cases it would be incorrect to interpret implicit taxation of crop products exclusively as a policy outcome. For example, poor infrastructure can impede market adjustment and exacerbate any policy impact on prices, and therefore contributing to the negative results.

Support to consumers of agricultural products

The Consumer Support Estimate (CSE) is a related indicator measuring the **cost to consumers** arising from policies that support agricultural producers by raising domestic prices. In the OECD methodology, the consumer is understood as the first buyer of these products. A negative CSE indicates that consumers are paying more than they would in comparison to border prices (an implicit tax); when it is positive, consumers are able to purchase product cheaper on the domestic market (an implicit subsidy). In the majority of countries monitored by OECD, consumers are taxed but may be partly compensated, e.g. through direct budgetary subsidies to processors, various forms of food assistance. In the absence of consumer support policies, such as in Viet Nam, the CSE mirrors the developments in MPS (Tables 2.13 and 2.14).

Similar to the PSE, the CSE can be expressed in relative terms as a percentage of consumption expenditures (%CSE). In 2000-02, consumers were implicitly taxed through agricultural policies at a moderately high level with a %CSE of -9%, indicating that policies to support agricultural prices increased consumption expenditure by 9% on aggregate. By 2011-13 the cost imposed on consumers had risen, with a %CSE of -11% (Figure 2.14). Comparing across countries, this aggregate tax on consumers is **above the OECD average** of -8%. It is similar to the level in China but half the level imposed on consumers in Indonesia. Nevertheless, low aggregate level of consumer taxation in Viet Nam disguises differences across products. Consumers of sugar cane and livestock products are taxed, while consumers of export crops are typically subsidised.

Support to general services for agriculture

In addition to support provided to producers individually, the agricultural sector is assisted through the financing of activities that provide **general benefits**, such as agricultural research and development, training, inspection, marketing and promotion, and public stockholding. The General Services Support Estimate (GSSE) indicator measures this support. The provision of common, as opposed to individual, benefit is what distinguishes the general services support from that measured by the PSE.

Expenditures on general services for agriculture in Viet Nam **rose sharply** from the mid-2000s to the end of the decade, rising 25% per annum between 2003 and 2009 in USD terms (Figure 2.15). Budgetary constraints imposed in the wake of the global financial crisis resulted in a 20% reduction from the peak. The most important GSSE category is development and maintenance of infrastructure, which is dominated by expenditure on irrigation systems. Over the period from 2000 to 2010 this category represented around 90% of GSSE expenditure but has fallen to around 85% since 2010. The next most important GSSE category is agricultural knowledge and innovation systems. This comprises

Figure 2.14. **Consumer Support Estimate in Viet Nam and selected countries, 2011-13 average**

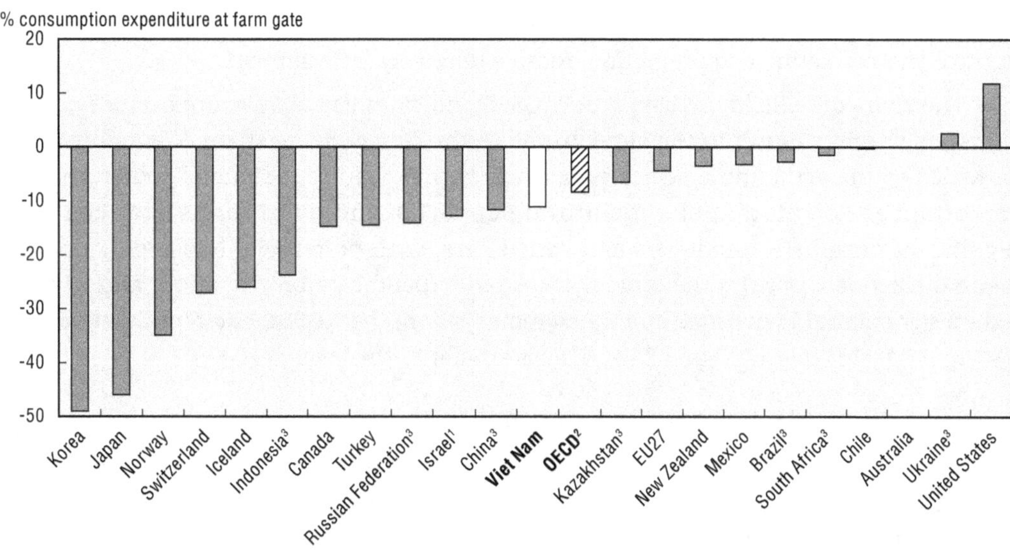

1. The statistical data for Israel are supplied by and under the responsibility of the relevant Israeli authorities. The use of such data by the OECD is without prejudice to the status of the Golan Heights, East Jerusalem and Israeli settlements in the West Bank under the terms of international law.
2. The OECD total does not include the non-OECD EU member states.
3. 2010-12 for Brazil, China, Indonesia, Kazakhstan, Russian Federation, South Africa and Ukraine.

Source: OECD (2014), "Producer and Consumer Support Estimates", *OECD Agriculture Statistics Database.*

StatLink http://dx.doi.org/10.1787/888933223904

Figure 2.15. **Level and composition of General Services Support Estimate in Viet Nam, 2000-13**

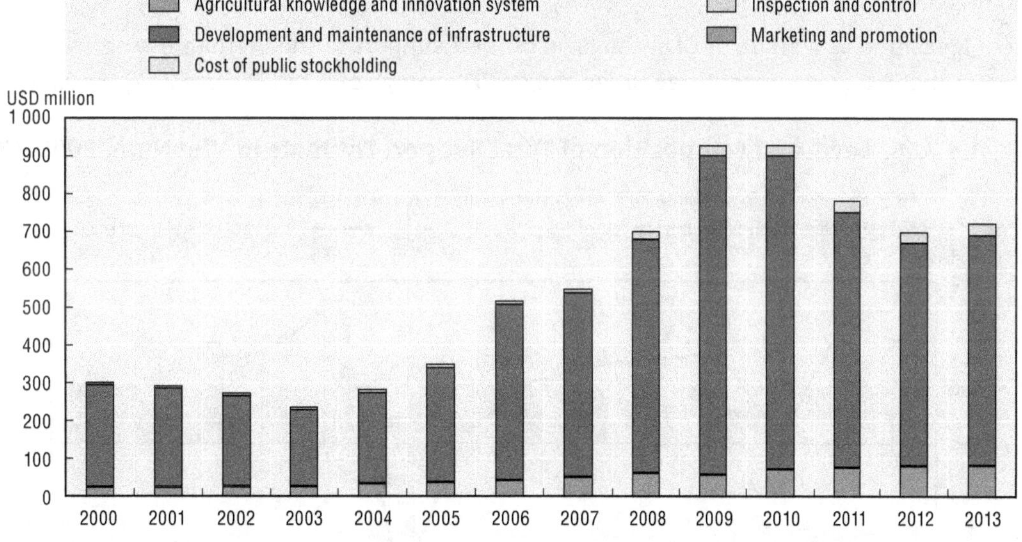

Source: OECD (2014), "Producer and Consumer Support Estimates", *OECD Agriculture Statistics Database.*

StatLink http://dx.doi.org/10.1787/888933223918

expenditure on agricultural research and development, extension and advisory services, and agricultural education. Government spending on these services has grown steadily since 2003. The category experiencing the most rapid increase is the cost of public stockholding. A noticeable increase in expenditure on this category has occurred since the food price crisis in 2008-09 and the issuing of Resolution No. 63/2009/NQ-CP on food security dated 23 December 2009. Some areas that are critical for lifting the quality of agricultural production and per unit returns received by farmers, such as inspection and control and marketing and promotion, receive relatively little support.

The share of GSSE in total support (%GSSE) indicates the relative importance of these transfers within support to the agricultural sector. The growing share of support that is provided to the agricultural sector as a whole rather than to individual producers is an important re-orientation of agricultural support spending to forms that can bring significant benefits to producers and consumers, with potentially less production and trade distortions. Despite the large increase in expenditure on irrigation, and to a less extent agricultural knowledge and innovation system, the **%GSSE has remained small** at 23% in 2011-13 (Tables 2.13 and 2.14). This is similar to the level in 2000-01.

Support to the agricultural sector as a whole

The Total Support Estimate (TSE) is the broadest indicator of support, representing the sum of transfers to agricultural producers individually (PSE) and collectively (GSSE), and direct budgetary transfers to consumers. Expressed as a percentage of GDP, the %TSE provides an indication of the cost that support to the agricultural sector places **on the overall economy**. Its value depends on the degree to which the agricultural sector is supported in a country, the size of this sector and its importance relative to the overall economy.

Viet Nam's TSE averaged VND 69 trillion (USD 3.3 billion) per year in 2011-13, representing 2.2% of GDP. Between 2000 and 2013 the %TSE in Viet Nam fluctuated considerably due to large variations in MPS (Figure 2.16). However, since peaking in 2009 at VND 95 trillion (USD 5.3 billion), equivalent to 5% of GDP, the TSE has **steadily declined** to reach 1.36% in 2013. The fall in MPS, along with budgetary reductions for irrigation, are the main contributors to this decrease.

Measured as a share of GDP, the level of total support to the Vietnamese agricultural sector in 2011-13 is almost **three times the OECD** average of 0.78% (Figure 2.17). At 2.2%, it

Figure 2.16. **Level and composition of Total Support Estimate in Viet Nam, 2000-13**

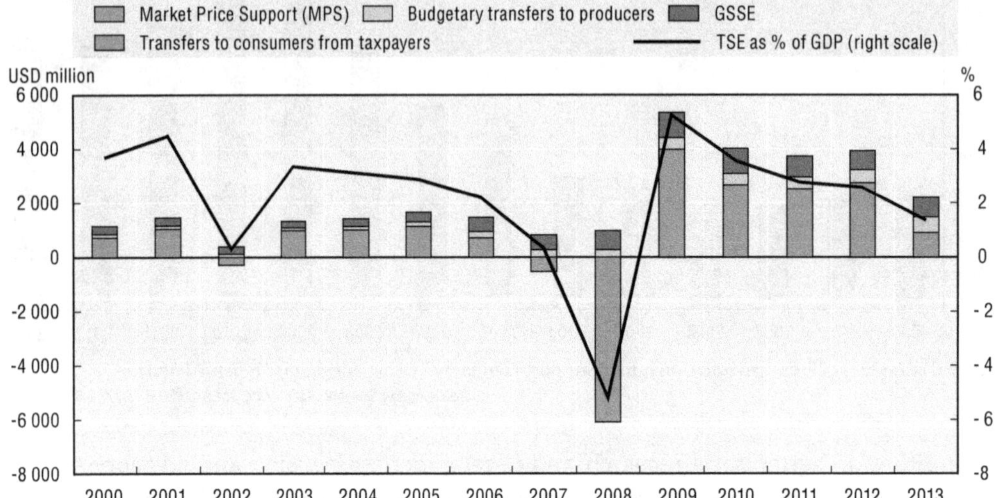

Note: GSSE: General Services Support Estimate; TSE: Total Support Estimate.

Source: OECD (2014), "Producer and Consumer Support Estimates", *OECD Agriculture Statistics Database*.

StatLink http://dx.doi.org/10.1787/888933223928

Figure 2.17. **Total Support Estimate in Viet Nam and selected countries, 2011-13 average**

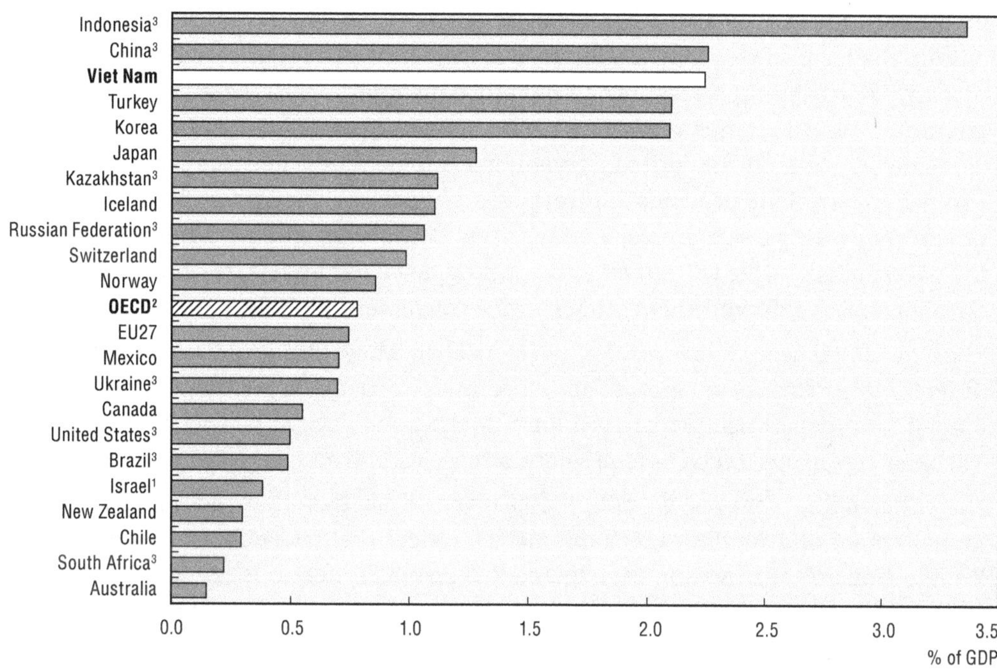

1. The statistical data for Israel are supplied by and under the responsibility of the relevant Israeli authorities. The use of such data by the OECD is without prejudice to the status of the Golan Heights, East Jerusalem and Israeli settlements in the West Bank under the terms of international law.
2. The OECD total does not include the non-OECD EU member states.
3. 2010-12 for Brazil, China, Indonesia, Kazakhstan, Russia, South Africa and Ukraine.

Source: OECD (2014), "Producer and Consumer Support Estimates", *OECD Agriculture Statistics Database.*

StatLink http://dx.doi.org/10.1787/888933223933

is similar to the level of support provided in China, Turkey and Korea but below that of Indonesia. Given that Viet Nam had the lowest %PSE among these five countries, the high %TSE shows that for a developing country with a large agricultural sector and still low GDP, the cost of support to the economy can be relatively high even if the level of agricultural support as measured by the PSE is low. A %TSE of 2.2% highlights the potential burden of the current policy mix and the need to ensure that the money is spent effectively.

2.6. Summary

- The main **priorities** of Viet Nam's agricultural policy are to lift the quality and competitiveness of output, raise the income of the rural population, develop an adequate level of infrastructure, strengthen the capacity of the sector to integrate with international markets, use natural resources in a sustainable manner, and improve the management efficiency of the sector.
- Two **key resolutions** are driving current policy initiatives: Resolution No. 26/2008/NQ-TW on agriculture, farmers and rural areas dated 5 August 2008 and Resolution No. 63/2009/NQ-CP to ensure national food security dated 23 December 2009. The first seeks to increase the market orientation of the sector; the second to guarantee adequate food supplies sourced from domestic production, particularly for rice. There is potential for conflict in achieving both at the same time.

- **Government intervention** is of two general types: long-term policies to increase yields/production and short-term policies to respond to price fluctuations. While investment incentives and other fiscal policies are designed to increase agricultural output, the focus of short-term policies is to influence market prices. The incompatible objectives of keeping prices low to benefit consumers while keeping them high to guarantee rural incomes gives rise to inconsistent price stabilisation policies that combine procurement and price interventions with quotas. When rice prices are low, the central government provides interest rate support to enterprise to buy rice from producers, putting upward pressure on prices. Conversely, when world prices are high, the government is able to limit exports, putting downward pressure on prices, harming farming households while benefiting net rice consumers.
- The agricultural policy framework is characterised by a high degree of fragmentation across different departments in various ministries. As a consequence of **weak co-ordination** among these departments, there is a lot of overlap and gaps in policies and regulations. These emerge in areas such as food safety, disease control and water management.
- Notwithstanding changes to policy objectives, little has changed in terms of the **management and funding mechanisms** for agricultural research and there has been limited transfer of irrigation management responsibilities to local communities and producer groups. These and other changes are required to the function of MARD if it is going to serve a more market-based agricultural sector. Certain functions such as international co-operation, policy analysis, sectoral monitoring and standard setting will need to be done at a much higher level, while others such as undertaking commercial activities and certain licensing practices will need to be reduced.
- While **SOEs** now operate in a more competitive market place and have been opened to private ownership, they still have a considerable degree of influence over certain agricultural sectors, particularly in relation to input supply and further processing. Moving forward the government will have to ensure a level playing field if it expects the private sector to participate fully in the development of the agricultural sector.
- Similarly, the use of **industry associations** such as VFA to implement policy needs to be fully reviewed, as there is a strong possibility for vested interests to limit competition in the market place.
- **Co-operatives** and other forms of "farmer groups" play a relatively small role in the sector, despite various government attempts to revitalise these forms of organisation. Consequently, many farmers are failing to obtain the benefits from acting collaboratively to secure better inputs and outputs.
- The commitment to ensure that farmers receive a **30% profit on rice production** is not sustainable. It does not encourage production efficiency and reduces the incentive for farmers to consider alternative products. As a major trader, export prices are more influenced by world market conditions than by domestic production costs. There is a large possibility that Viet Nam will either lose market share or require government support to make up the difference, at the possible cost to other economic and social priorities.
- The exemption of most individual and households involved in agricultural production from the payment of the **irrigation service fee** is a backward step. It was implemented as simple method to raise farm income with low transaction costs. However, removing the charge for water does not support the objective of improving resource efficiency and supporting environmental sustainability.

- The use of a competitive bidding process in the selection of **extension projects** creates the possibility of efficient allocation. However, there appears to be potential for overlap in projects awarded at the central and local government level. Moreover, extension projects have a strong production focus with less attention paid to meeting demand requirements, e.g. how to market, engage in contracts, meet food safety requirements, etc.
- **Tariff protection** has fallen by one-third over the period as a result of regional and unilateral tariff reduction commitments. The simple average MFN applied tariff on agricultural products was 16% in 2013. The average agricultural tariff is just 3.4% and 5.4% on imports from ASEAN members and China respectively. However, an MFN applied tariff of 40% applies to a range of commodities including meat or poultry, turkey and duck, tea (green and black), grapefruit, milled rice, refined sugar, and many types of prepared or preserved fruits and vegetables.
- There is an on-going need to **strengthen the capacity of policy-making and implementation** to ensure the protection of human, plant and animal health, improve regulatory reputation and support the export of value-added agricultural goods. It is important that import requirements for food safety, quarantine, and standards and labelling purposes are implemented in a transparent manner, consistent with international guidelines and practice.
- The current system for **controlling rice exports** needs to be overhauled. It is where the conflict between the objectives of improving the market orientation of the sector and ensuring food security comes to the fore. The system limits competition in the market place and reduces the incentive to develop long-term market arrangements.
- The **level of producer support** as measured by the %PSE averaged 7% in 2011-13; less than half the level of support provided to producers in China and Indonesia.
- Over the period 2000-13 the level of support was quite variable without revealing any distinct long-term trend. Nevertheless, the **%PSE remained positive** over most of this period, indicating that producers generally received moderate support.
- The variations in producer support were driven mainly by sharp fluctuations of its **market price support** component. Budgetary transfers have remained relatively constant at about 20% of producer support on average over the period 2000-13.
- **Budgetary transfers** supporting producers are mainly in the form of payments based on variable input use. Direct payments have switched from retiring agricultural land from production to maintaining land in paddy production.
- Producer support in Viet Nam is based predominantly on the most **distorting forms** of support, although their share in total support has declined compared to the early 2000s.
- Producers of **import-competing commodities** such as beef and veal, poultry, eggs and sugar cane are highly supported, receiving prices for their outputs above international prices. In contrast, producers of export-competing commodities such as natural rubber, coffee, cashew nuts and tea are implicitly taxed in that are paid prices for their outputs that are lower than international prices.
- There has been considerable variation in the level of **support provided to rice production.** Rice has moved from being a commodity with little support, to being highly taxed, to receiving support, to seeing this fall back to a low level again. This reflects the challenge of trying to support producers while protecting consumers. Further efforts to move away from using the price of rice as the mechanism to achieve both goals are needed.

- **General services** for the agricultural sector have remained relatively constant as a share of total support transfers, suggesting there has been little re-orientation of policies towards those that can benefit both producers and consumers. Expenditure on some general services such as inspection and control and marketing and promotion receive relatively limited support.
- **Total support** to agriculture is high relative to the overall economy and is comparable to that in China, Turkey and Korea, and much smaller than in Indonesia.

Notes

1. Decision 10-2007-QD-TTg dated 23 January 2007 on the system of economic branches in Viet Nam.
2. In 1978 the United States imposed a trade and investment embargo in response to Viet Nam sending troops into Cambodia. Aid from China terminated in 1979 due to border conflict.
3. Based on agricultural initiatives, the same reform also proceeded in the industrial sector. Table 2.A1.1 contains a guide to understanding the hierarchy, content, numbering and coding of Vietnamese legal documents.
4. The momentum for economic reform was further strengthened by the success of economic reforms experienced in China and the Soviet perestroika movement, and the decision of the Soviet Union to cease its economic aid to Viet Nam.
5. As with the previous Directive 100 reform, many villages had already been practising "full contracts" (*khoan trang*), in which paddy fields were leased to households.
6. Vinafood I and Vinafood II were established in 1995 for the purposes of setting up food businesses, buying farmers' produce, assisting with regional food balance and contributing to the stabilisation of food prices.
7. The US-Viet Nam bilateral trade agreement provided substantial new impetus for legal reform in Viet Nam. The commercial code was almost entirely rewritten, with significant new Enterprise, Competition, and Investment Laws all introduced. The final negotiations for Viet Nam's WTO accession were enabled by significant additional legal reforms undertaken particularly in 2005 as the US and other WTO members insisted that Viet Nam implement reforms before accession would be granted (Abbot et al., 2006).
8. The other specific production targets are: increasing corn acreage to 1.3 million ha and the quantity of corn up to 7.5 million tonnes; ensuring fruit trees planted area of 1.2 million ha to yield 12 million tonnes of fruits; 1.2 million ha of vegetables to yield 20 million tonnes of vegetable, producing 8 million tonnes livestock meat; 1 million tonnes of fresh milk, 14 billion units of poultry's eggs.
9. According to Decree No. 199/2013/ND-CP on defining the functions, tasks, powers and organisational structure of MARD dated 26 November 2013 and which entered into force on 15 January 2014. This replaced Decree No. 01/2008/ND-CP of 3 January 2008 and Decree No. 75/2009/ND-CP of 10 September 2009 which amended Decree No. 01/2008/ND-CP by substituting Article 3 with a new article on the organisational structure of the Ministry.
10. There are 63 "provincial" governments (58 provinces and 5 municipalities), 662 "districts" (536 rural districts, 25 provincial cities, 59 district level towns and 42 urban districts) and 10 776 "communes" (9 012 communes, 583 commune-level towns and 1 181 wards).
11. All revenues collected from taxes and fees related to international trade must be transferred to the central budget. Local governments retain 100% of the revenues they collected from land (e.g. renting, tax on land use transfers and land use tax), from natural resource tax, registration fees and from lottery. Another source of revenue for provinces is part of the revenues collected from VAT, corporate income tax, personal income tax, and gasoline fee. There exists a sharing mechanism between central and provincial government for these revenues, but the majority of provinces can retain 100% for their own budget. Richer provinces (HCMC, Hanoi, Quang Ninh, etc.) have to transfer a part of these revenues to the central budget (World Bank, 2012). However, current regulations state that the revenue sources can only be allocated between central state budgets to the local agencies where the revenues are collected, say VAT or corporate income taxes.
12. Law No. 23/2012/QH13 on Co-operatives dated 20 November 2012 came into effect on 1 July 2013, replealing Law No. 18/2003/QH11 on Co-operatives dated 26 November 2003, which entered into force on 1 July 2004. This in turn repealed the Law on Co-operatives dated 20 March 1996.

13. The State Reserve system first began in North Vietnam on 7 August 1956.
14. Decision No. 2091/2012/QD-TTg dated 28 December 2012.
15. Decision No. 137/1992/HDBT of the Committee of Ministers (now known as the government) on prices dated 27 April 1992.
16. Decision No. 151/1993/QD-TTg dated 12 April 1993.
17. Decision No. 195/1999/QD-TTg dated 27 September 1999.
18. Decree No. 170/2003/ND-CP dated 25 December 2003 and in effect from 9 January 2004 annulled Decision No. 137. Decree No. 170 was passed to implement Ordinance No. 40/2002/PL-UBTVQH10 on Prices, dated 26 April 2002, which provided the legal framework for market price regulation.
19. Circular No. 15/2004/TT-BTC dated 9 March 2004 guiding the implementation of Decree No. 170/2003/ND-CP.
20. Decree No. 75/2008/ND-CP dated 9 June 2008 and in effect from 24 June 2008, and Circular No. 104/2008/TT-BTC dated 13 November 2008 and in effect from 28 November 2008 guided the implementation of Decree No. 75 and replaced Circular No. 15/2004/TT-BTC.
21. Circular No. 122/2010/TT-BTC amending and supplementing Circular No. 104, dated 12 August 2010 and in effect from 1 October 2010.
22. Decree No. 177/2013 ND-CP dated 14 November 2013 detailing and guiding the implementation of some articles of the Law No. 11/2012/QH12 on Price dated 20 June 2012 and in effect since 1 January 2013 replaced Ordinance No. 40/2002 which was considered out-dated and inconsistent with certain WTO obligations.
23. Circular No. 430/TTg-KTN on product consumption for farmers dated 12 March 2010 and Decree No. 109/ND-CP on export management dated 4 November 2010.
24. Similarly, although the government has policies in place to impact fertiliser prices, both price stabilisation measures and input subsidies to fertiliser manufacturers (Chapter 1), evidence suggests that domestic prices in Viet Nam are normally 5-10% higher than import prices (Tran, 2014). The subsidy on raw materials therefore does not appear to be passed on to farmers. Instead producers of fertiliser, main SOEs, enjoy the benefits. For this reason, the report does not include an estimate of the value of fertiliser subsidies in the calculation of support to agriculture.
25. Decree No. 143/2003/ND-CP dated 28 November 2003 provides the general provisions for the current system for funding irrigation including the setting of ISFs.
26. Decree No. 115/2008/ND-CP dated 14 November 2008 and effective 1 January 2009 amending and supplementing a number of articles of Decree No. 143/2003/ND-CP. These exemptions were continued by Decree No. 67/2012/ND-CP dated 10 September 2012 to amend Decree No. 143/2003/ND-CP and replace Decree No. 115/2008/ND-CP.
27. Decisions No. 132/2007/QD-TTg (15 August 2007), No. 1037/QD-TTg (15 August 2007), No. 738/QD-TTg (18 May 2006), No. 719/QD-TTg (5 June 2008), No. 80/2008/TT-BTC (18 September 2008), No. 142/2009/QD-TTg (31 December 2009), and, most recently, No. 1442/QD-TTg (23 August 2011).
28. Decision No. 1681/QD-TTg dated 11 September 2010 on supplying germicide from State Reserves for local areas to handle blue-ear pig disease and Decision No. 1791/QD-TTg dated 15 October 2011 on mechanism and policy to support vaccines for blue-ear pig disease and hog cholera in order to boost livestock production, ensure supply and stabilise market prices.
29. Decision No. 142/2009/QD-TTg dated 31 December 2009 and its amendment Decision No. 49/2012/QD-TTg dated 8 November 2012.
30. Decree No. 42/2012/ND-CP dated 11 May 2012 on management and use of paddy land, and in effect since 1 July 2012, and implemented by MOF Circular No. 205/2012/TT-BTC.
31. Decision No. 135/1998/QD-TTg dated 31 July 1998.
32. Decision No. 07/2006/QD-TTg dated 10 January 2006.
33. Decree No. 14/1993/ND-CP.
34. Decision No. 67/1999/QD-TTg dated 30 March 1999.
35. Decision No. 546/2002/QD-NHNN dated 30 May 2002.
36. Decision No. 131/2002/QD-TTg dated 4 October 2002.

37. Resolution No. 12/2009/NQ-CP dated 6 April 2009, and implemented through Decision No. 497/2009/QD-TTg dated 17 April 2009, and its amending Decision No. 2213/QD-TTg dated 21 December 2009, and Circular No. 02/2010/TT-NHNN dated 22 January 2010.

38. Decree No. 41/2010/ND-CP dated 12 April 2010 on credit policies for agricultural and rural development, which entered into force on 1 June 2010.

39. Resolution No. 48/2009/NQ-CP dated 23 September 2009, and implemented by Decisions No. 63/2010/QD-TTg of 15 October 2010, No. 65/2011/QD-TTg of 2 December 2011 and No. 68/2013/QD-TTg of 14 November 2013.

40. Decree No. 42/2012/ND-CP on management and use of paddy land dated 11 May 2012 and in effect since 1 July 2012, and implemented by MOF Circular No. 205/2012/TT-BTC

41. Decision No. 661/QD-TTg dated 29 July 1998.

42. Decision No. 315/2011/QD-TTg on the pilot provision of agriculture insurance during 2011-13, dated 1 March 2011 and which entered into force on the date of its signing. Circular No. 47/2011/TT-NNPTNT lists the natural disasters and diseases that are insured for each type of agricultural product.

43. Decision No. 358/2013/QD-TTg dated 27 February 2013.

44. Resolution No. 15/2003/QH11 on agricultural land use tax exemption and reduction dated 17 June 2003 and implemented by Decree No. 129/2003/ND-CP dated 3 November 2003, effective from 21 November 2003.

45. Resolution No. 55/2010/QH12 on exemption and tax reduction on agricultural land use in the period 2011-20 dated 24 November 2010 and implemented by Decree 20/2011/ND-CP dated 23 March 2011.

46. For example, Decree No. 61/2010/ND-CP on incentive policies for enterprises investing in agriculture and rural development dated 4 June 2010 and Decision No. 57/2010/QD-TTg on exemption from land rents for projects on construction of warehouses to store 4 million tonnes of rice or maize, refrigerated warehouses for preservation of fishery products, vegetables and fruits and coffee temporary reserve warehouses. Difficulties in implementing Decree No. 61 has meant that it has been replaced by Decree No. 210/2013/ND-CP.

47. Decree No. 13/1993/ND-CP on agricultural extension dated 2 March 1993.

48. Decree No. 02/2010/ND-CP on agricultural extension dated 8 January 2010, effective from 1 March 2010. This repealed Decree No. 56/2005/ND-CP of 26 April 2005 on agricultural promotion and fisheries promotion which in turn repealed Decree No. 13/1993/ND-CP of 2 March.

49. NAEC was created in 2005. Prior to this a Department of Agriculture and Forestry Extension (DAFE) within the MARD held the functions of both the state governing organisation managing production as well as the technology transfer body. However, due to difficulties in serving these two assignments, the government divided DAFE into two departments: Department of Crop Production and NAEC.

50. The 18 research institutes and centres that operate under VASS include: Cuu Long Delta Rice Research Institute (CLRRI); Southern Horticultural Research Institute (SOFRI); Institute of Agricultural Science for Southern Viet Nam (IAS); Sugarcane Research Institute (SRI); Plant Resources Centre (PRC); Institute for Agricultural Environment (IAE); Soil and Fertiliser Research Institute (SFRI); Plant Protection Research Institute (PPRI); Agricultural Genetics Institute (AGI); Northern Mountainous Agriculture and Forestry Science Institute (NOMAFSI); Fruits and Vegetables Research Institute (FAVRI); Maize Research Institute (MRI); Viet Nam Sericulture Research Centre (Vietseri); Centre for Technology Development and Agricultural Extension (CETDAE); Field Crops Research Institute (FCRI); Agricultural Science Institute of Northern Central Viet Nam (ASINCV); Agricultural Science Institute of Southern Coastal Central of Viet Nam (ASISOV); and Western Highlands Agro-Forestry Scientific & Technical Institute (WASI).

51. Decision No. 3246/QD-BNN-KHCN dated 27 December 2012, which was based on Decision No. 418/QD-TTg dated 4 November 2012 that set the overall strategy of science and technology in Viet Nam for the period of 2011-20.

52. Decision No. 1259/QD-BNN-KHCN dated 4 June 2013.

53. Although long-running time-series data on the ratio of irrigated rice fields is not available, it is evident that the irrigation of rice fields had expanded to a considerable extent, roughly 60%, by the mid-1980s. This was partly the result of the favourable initial conditions in the RRD, which has a long history of irrigation and water management. In addition to the advantage from history, the government had tried since reunification to renovate and rehabilitate the large-scale irrigation

systems both in the RRD and MRD. These earlier investments can help explain how a series of agricultural reforms could be put into practice without severe constraint from the poorly developed infrastructure (Tsukada, 2011).

54. Decision No. 794/2014/QD-BNN-TCTL on the restructuring scheme of irrigation sector dated 21 April 2014.
55. Decree No. 667/2009/QĐ-TTg dated March 2009.
56. Decree No. 48/2009/ND-CP for storage development in MRD dated 23 September 2009 followed by Decision No. 3242/2010/QD-BNN-CB dated 2 December 2010.
57. Resolution No. 30/NQ-CP.
58. Decree No. 57/1998/ND-CP dated 31 July 1998.
59. Decision No. 46/2001/QD-TTg dated 4 April 2001.
60. Officially titled as the "Agreement between the United States of America and the Socialist Republic of Viet Nam on Trade Relations".
61. Joining ASEAN has been described as a "fitness gym" which allowed Viet Nam to conduct training "work out" in preparation for the serious exercise of competing in the global economy (Vo, 2005).
62. Under Viet Nam's WTO Goods Schedule commitments, it also operates a TRQ for salt, with an initial quota of 150 000 tonnes increasing by 5% annually. While outside the WTO definition of agriculture, salt production is considered an agricultural activity in Viet Nam. Salt production is an important source of income for hundreds of thousands of poor farmers living in coastal areas where it can be difficult to use land for agriculture. The TRQ aimed at securing employment and ensuring income stability for these farmers (WTO, 2006).
63. Decision No. 91/2003/QD-TTg dated 9 May 2003.
64. Decision No. 46/2005/QD-TTg dated 3 March 2005.
65. G/AG/N/VNM/2 of 3 November 2011.
66. At present, VAT is levied on goods and services according to the Law No. 13/2008/QH12 on VAT dated 3 June 2008 and applied since 1 January 2009.
67. A fourth rate of 20% applicable to some specific services was also established as part of the original VAT structure but this was abolished on 1 January 2004.
68. Other categories subject to a 5% VAT rate are medical equipment and medicines; teaching and learning aids; children's toys and books; scientific and technological services; special purpose machinery and equipment for newsprint; products made from jute and bamboo; and cultural exhibits and sports activities (World Bank, 2014).
69. Decree No. 12/2006/ND-CP dated 23 January 2006.
70. Decision No. 24/2008/QD-BCT and Circular No. 17/2008/TT-BCT dated 12 December 2008.
71. Circular No. 24/2010/TT-BCT dated 28 May 2010 replaced Circular No. 17/2008/TT-BCT.
72. Circular No. 32/2011/TT-BCT dated 5 September 2011 and Circular No. 27/2012/TT-BCT dated 26 September 2012 respectively.
73. Decision No. 1899/2010/QD-BCT dated 16 April 2010.
74. Decision No. 1380/2011/QD-BCT dated 25 March 2011.
75. Law No. 55/2010/QH12 on Food Safety dated 17 June 2010 and entered into force on 1 July 2011. It supersedes the Viet Nam Food Ordinance approved in 2003.
76. Decree No. 38/2010/ND-CP detailing implementation of a number of articles of the Law on Food Safety dated 25 April 2012 and effective 11 June 2012.
77. Joint Circular No. 13/2014/TTLB-BYT-BNNPTNT-BCT dated 9 April 2014 and effective 26 May 2014.
78. Circular No. 50/2009/TT-MARD.
79. Decree No. 57/1998/ND-CP dated 31 July 1998.
80. Decision No. 10/1998/QD-TTg dated 23 January 1998.
81. Export quota was offered to private traders on the basis of four criteria: previous experience in rice trade, ownership of milling facilities, capacity to export at least 5 000 tonnes per shipment and proof of financial security (Nielsen, 2002).

82. Decision No. 46/2001/QD-TTg dated 4 April 2001 on the management of goods export and import in the 2001-05 period.
83. Implemented by Circular No. 44/2010/TT-BCT dated 31 December 2010.
84. Implemented by Circular No. 89/2011/TT-BTC dated 17 June 2011.
85. Floor price of export rice = average costs of export rice (for each kind of rice) + expected margin + related taxes and fees, where the average cost of export rice is to be at least equal to the target paddy price plus preliminary processing costs and logistic costs for export.
86. Circular No. 53/BNG-P dated 2 October 1982.
87. Circular No. 10/1989/KTDN/XNK dated 7 August 1989.
88. Decree 57/1998/ND-CP dated 31 July 1998.
89. Decision No. 222/TC-CTN dated 29 December 1989.
90. Law No. 45/2005/QH11 dated 14 June 2005. Metals, including mineral ores and metal scraps, and wood products are the goods most subject to export taxation by Viet Nam.
91. Decision No. 104/2008/QD-TTg dated 21 July 2008.
92. Decision No. 195/1999/QD-TTg dated 27 September 1999.
93. Decision No. 124/2008/QD-TTg dated 8 September 2008.
94. Decision No. 279/2005/QD-TTg dated 3 November 2005 and currently operating in accordance with Decision No. 72/2010/QD-TTg dated 14 November 2010.
95. Decree No. 75/2011/ND-CP dated 30 August 2011.
96. Decree No. 210/2013/ND-CP dated 19 December 2013.
97. Since its WTO accession, Viet Nam has concluded only one further bilateral agreement in the same format, i.e. with Angola (2008).
98. Lao PDR and Myanmar (Burma) joined two years later on 23 July 1997 and Cambodia on 30 April 1999.
99. Despite the word "Common" in the CEPT, it should be noted that AFTA is not a customs union, but merely a free trade agreement, meaning that while ASEAN member states have common effective tariffs among themselves in AFTA, the level of tariffs with non-ASEAN countries will continue to be determined individually. The time frame for tariff reductions was originally set at 15 years commencing 1 January 1993, but two years after its initial implementation ASEAN members agreed to shorten the time period to ten years, i.e. from 2008 to 2003.
100. This reserve builds on the East Asia Emergency Rice Reserve (EAERR), the pilot project of the ASEAN Ministers of Agriculture and Forestry and the Ministers of Agriculture of the PRC, Japan, and the Republic of Korea. The EAERR in turn is a revitalisation and expansion of the ASEAN Emergency Rice Reserve (AERR) that was established in 1979.
101. The full texts of Viet Nam's commitments and related documents can be found on the WTO website *www.wto.org*.

ANNEX 2.A1

Policy tables

Table 2.A1.1. Hierarchy, content, and numbering and coding of legal documents

Hierarchy of legal documents

1. Constitution, laws and resolutions of the National Assembly (NA).
2. Ordinances and resolutions of the Standing Committee of the NA.
3. Orders and decisions of the State President.
4. Decrees of the Government.
5. Decisions of the Prime Minister.
6. Resolutions of the Justices' Council of the Supreme People's Court and circulars of the Chief Justice of the Supreme People's Court.
7. Circulars of the President of the Supreme People's Procuracy.
8. Circulars of Ministers or Heads of Ministry-equivalent Agencies.
9. Decisions of the State Auditor General.
10. Joint resolutions of the Standing Committee of the NA or the Government and the central offices of socio-political organisations.
11. Joint circulars of the Chief Justice of the Supreme People's Court and the President of the Supreme People's Procuracy; those of Ministers or Heads of Ministry- equivalent Agencies and the Chief Justice of the Supreme People's Court, the President of the Supreme People's Procuracy; those of Ministers or Heads of Ministry-equivalent Agencies.
12. Legal documents of People's Councils and People's Committees.

Content of relevant legal documents

- **Laws of the NA** address fundamental issues across a wide range of fields as well as rights and obligations.
- **Resolutions of the NA** focus on socio-economic development tasks and state budget issues.
- **Ordinances of the Standing Committee** contain regulations explaining the constitution and laws.
- **Resolutions of the Standing Committee** provide interpretation of the constitution, laws and ordinances.
- **Decrees by the government** provide guidelines on the implementation of higher legal documents including specific action to implement policy, allocation of specific tasks to ministries and identifying areas which are not mature enough to develop into laws or ordinances.
- **Decisions of the Prime Minister** focus on ways to lead, manage and administer the government's operations and public administration system.
- **Circulars of Ministers** provide detail guidelines on the implementation of higher legal documents, regulations on technical processes and standards and ways to exercise management of the sector/area.

Numbering and coding

Alphabetical letters at the end of each policy's name include two parts connected by the hyphen "-". They represent the abbreviated names for the type of document and the promulgating agency in Vietnamese. For example, in the case of the Circular No. 120/2011/TT-BTC, TT is the acronym for Circular in Vietnamese *(Thông tư)*, and BTC is the acronym for the Ministry of Finance in Vietnamese (*Bộ Tài Chính*). The following is a list of abbreviations for the types of documents and issuers listed in this report:

NQ (Nghị quyết)	Resolution	QH13	National Assembly (in this case the 13th National Assembly)
PL (Pháp lệnh)	Ordinance	UBTV	Standing Committee of NA
ND (Nghị định)	Decree	TW	CPV Executive Committee
QD (Quyết định)	Decision	CP	Government
TT (Thông tư)	Circular	TTg	Prime Minister
TTLB	Joint Circular	BCT	Ministry of Industry and Trade
		BNN	Ministry of Agriculture and Rural Development
		BTC	Ministry of Finance
		BTNMT	Ministry of Natural Resources and Environment
		BYT	Ministry of Health
		NHNN	State Bank of Vietnam

Source: Law No. 17/2008/QH12 on the Promulgation of Legal Documents dated 3 June 2008.

Table 2.A1.2. **Main tasks of units under the Ministry of Agriculture and Rural Development**

Category	Units	Activities
Functional departments	Ministry Administrative Office (Office of the Ministry)	Integrates and co-ordinates the operation of the entities within the Ministry.
	Organisation and Personnel Department	Responsible for staffing and training of public servants.
	Planning Department	Oversees the integrated management of strategies, master plans, plans and investment for the agricultural and rural development sector as regulated by the law. In charge of public budget planning and allocation for different sectors.
	Finance Department	Responsible for the integrated management of financing, accounting, and pricing.
	Science, Technology and Environment Department	Oversees the integrated management of agriculture and rural sector science and technology, including research, standards, measurement and results, including the planning and allocation of funding for research and extension.
	International Cooperation Department	Oversees the integrated management of international co-operation and international economic integration. It takes the leading role in co-ordinating with donors and other relevant agencies and NGOs in preparing, appraising and negotiating ODA and FDI projects and programmes in Vietnam's ARD Sector. It also contains the SPS Office.
	Legislation/Legal Department	Responsible for the integrated management of MARD's governance of law-related activities and tasks. Ensures the legality of policies developed by other MARD departments before issuing.
	Ministry Inspectorate	Inspects, verifies and recommends solutions to complaints and denunciations. Steers and guides on organisation and professional processes of administrative and specialised inspections.
	Division of Agricultural Enterprise, Renovation and Management	Assists the Minister in setting up, steering, instructing, monitoring and checking the implementation of re-structuring, renovation and development of state-owned enterprises within the domains of MARD.
General Offices	General Forestry Office (Directorate of Forestry)	Policy for public investment and management of forestry sector.
	General Fisheries Office (Fisheries Directorate)	Policy for public investment and management of aquaculture sector.
	General Irrigation and Water Management Office (Water Resource Directorate)	Policy for public investment and management of irrigation system.
Professional Departments	Department of Crop Production (Cultivation Department)	Responsible for policy and management of crop production including inputs such as seeds and fertilisers, and setting quality standards. Manage cultivation techniques, quality and utilisation of fertilisers. Steer the cultivation and production plans Set up the strategies, schemes, plans, procedures, norms, techniques and technologies for species of plants and fertilisers. Manage the attestation, corroboration of quality, field-testing, recognition and trademark protection of new species of plants and new fertilisers. Grant and revoke licenses and certificates. Set up the export/import lists of plant species and fertilisers. Incorporate the management over the plant gene stock.
	Department of Plant Protection	Works on legislation review, plant protection, plant quarantine, and pesticide/chemical control. It plays an advisory role for MARD and can propose policies related to its sphere of responsibilities. It has provincial branches known as the Plant Protection Department (PPD)
	Department of Livestock Husbandry	Performs professional functions related to the governance of the animal husbandry and livestock sector including draft laws, strategies and plans, quality certification and licensing. Responsible for inputs such as breeding and animal feed, and quality standards.
	Department of Animal Health	Performs professional functions related to the governance and inspection of veterinary activities nationwide.
	Department of Processing and Trade for Agro-forestry-fisheries Products and Salt Production	Performs the professional functions related to the governance of preservation and processing of agricultural, fishery and forest products and salt production, including managing the mechanisation and industrialisation of sectoral production.
	National Agro-Forestry-Fisheries Quality Assurance Department	Responsible for food safety administration for the products under MARD jurisdiction and the development of food safety policies in general in co-ordination with MOIT and MOH.
	Department of Collective Economics and Rural Development	Performs governance functions over co-operatives, farmer organisations and other agricultural production entities, including integration of policies and rural development programmes. Has responsibility for poverty reduction.
	Department of Construction Management	Submits to the Ministry proposals for capital construction programmes and projects, and investment decisions. Appraises the technical designs and cost estimates for projected building items. Approves construction designs and detailed cost estimates. Appraises biddings and selection of contractors, and perform the consultant role for investment and bidding for construction and assembly work. Certifies the quality of constructed works. Monitors and accelerates the progress of investment activities. Holds the position of standing member in the Council of Hand-and-Take-over of Ministerial and State-Level Works and Projects of the Sector.

Table 2.A1.2. Main tasks of units under the Ministry of Agriculture and Rural Development (*cont.*)

Category	Units	Activities
Non-productive units	Centre for Informatics and Statistics	Responsible for implementing all statistical activities (collecting, analysing and reporting) in the agricultural sector (within MARD, under MARD, and local state agricultural authorities) and provision of market information systems for agricultural products.
	National Agricultural Extension Centre	Follows MARD guidelines and strategies with demonstration models, information dissemination, training, service delivery and international co-operation in the fields of agricultural, forestry and fishery.
	National Centre for Rural Water Supply and Sanitation	Responsible for implementing the National Target Program on Rural Water Supply and Sanitation. The program involves installing community and household water systems, public latrines and water supply in schools and clinics, and training and capacity building.
	Vietnam Agriculture Newspaper	
	Vietnam Journal of Agriculture and Rural Development	
	Institute of Policy and Strategy for Agriculture and Rural Development (IPSARD)	The policy think-tank, providing analysis and results supporting strategy and policy formulation process in agriculture and rural development. Established by Decision 9/2006/WQD-TTg dated 9 September 2006 – initially as a separate, independent agency but became a line department in 2013.

Source: Ministry of Agriculture and Rural Development, 2014.

Table 2.A1.3. Selected final bound, MFN applied and preferential tariffs for MPS commodities, 2013

Product	HS tariff line	Final bound	MFN applied	ATIGA	ACFTA	AJCEP	AKFTA	AANZFTA
Beef and veal								
Carcasses and half-carcasses of bovine animals, fresh or chilled	0201.10	30	30	5	0	13	7	10
Fresh or chilled bovine cuts, with bone in (excl. carcasses and 1/2 carcasses)	0201.20	20	20	5	0	13	7	10
Fresh or chilled bovine meat, boneless	0201.30	14	14	5	0	13	7	10
Carcasses and half-carcasses of bovine animals, frozen	0202.10	20	20	5	0	13	7	10
Frozen bovine cuts, with bone in (excl. carcasses and half-carcasses)	0202.20	20	20	5	0	13	7	10
Frozen, boneless meat of bovine animals	0202.30	14	14	5	0	13	7	10
Pigmeat								
Fresh or chilled carcasses and half-carcasses of swine	0203.11	25	25	5	0	19	10	20
Fresh or chilled hams, shoulders and cuts thereof of swine, with bone in	0203.12	25	25	5	0	19	10	20
Swine, carcasses and half-carcasses, frozen	0203.21	15	15	5	0	n.a.	10	n.a.
Poultry								
Poultry, not cut in pieces, fresh or chilled	0207.11	40	40	5	5	13	7	20
Poultry, not cut in pieces, frozen	0207.12	40	40	5	5	13	7	20
Poultry, cuts and offal, frozen	0207.14	20	20	5	5	11	7	20
Eggs								
Fresh eggs of domestic fowls, in shell (excluding fertilised for incubation)	0407.21	40	30	5	n.a.	25	n.a.	20
Birds' eggs, in shell, preserved or cooked	0407.90	40	30	5	n.a.	25	n.a.	20
Cashew nuts								
Fresh or dried cashew nuts, in shell	0801.31	30	3	n.a.	0	n.a.	n.a.	n.a.
Fresh or dried cashew nuts, shelled	0801.32	25	25	5	0	n.a.	15	20
Coffee								
Coffee (excl. roasted and decaffeinated)	0901.11	15	15	5	5	13	7	10
Roasted coffee (excl. decaffeinated)	0901.21	30	30	5	10	25	15	20

Table 2.A1.3. **Selected final bound, MFN applied and preferential tariffs for MPS commodities, 2013** *(cont.)*

Product	HS tariff line	Final bound	MFN applied	ATIGA	ACFTA	AJCEP	AKFTA	AANZFTA
Tea								
Green tea in immediate packings of <= 3 kg	0902.10	40	40	5	10	25	15	20
Green tea in immediate packings of > 3 kg	0902.20	40	40	5	10	25	15	20
Black fermented tea and partly fermented tea, whether or not flavoured, in immediate packings of <= 3 kg	0902.30	40	40	5	10	25	15	20
Black fermented tea and partly fermented tea, whether or not flavoured, in immediate packings of > 3 kg	0902.40	40	40	5	10	25	15	20
Pepper								
Pepper of the genus Piper, neither crushed nor ground	0904.11	20	20	0	5	19	10	n.a.
Pepper of the genus Piper, crushed or ground	0904.12	20	20	0	5	19	10	n.a.
Maize								
Maize (excl. seed for sowing)	1005.90	30	30	0	7.5	12	10	12.5
Rice								
Rice in the husk, "paddy" or rough	1006.10	0-40	0-40	2.5	0	0	7.5	0
Husked or brown rice	1006.20	40	40	5	n.a.	25	15	20
Semi-milled or wholly milled rice, whether or not polished or glazed	1006.30	40-50	40	5	5-10	25-31	15	20
Broken rice	1006.40	40	40	5	5	25	15	20
Sugar								
Raw beet sugar (excl. added flavouring or colouring)	1701.12	100	25	5	n.a.	n.a.	n.a.	20
Raw cane sugar (excl. added flavouring or colouring)	1701.13 1701.14	85	25	5	n.a.	n.a.	n.a.	20
Refined cane or beet sugar, containing added flavouring or colouring, in solid form	1701.91	100	40	5	n.a.	n.a.	n.a.	20
Cane or beet sugar and chemically pure sucrose, in solid form (excl. cane and beet sugar containing added flavouring or colouring and raw sugar)	1701.99	85	40	5	n.a.	n.a.	n.a.	20
Rubber								
Natural rubber latex	4001.10	5	3	0	0	n.a.	n.a.	n.a.
Smoked sheets	4001.21	5	3	0	0	n.a.	n.a.	n.a.
Technically specified natural rubber (TSNR)	4001.22	5	3	0	0	n.a.	n.a.	n.a.

n.a. not applicable, i.e. no preferential tariff rate is provided.
MFN applied tariff is applicable.
ATIGA: ASEAN Trade In Goods Agreement; ACFTA: ASEAN-China Free Trade Agreement; AJCEP: ASEAN-Japan Comprehensive Economic Partnership Agreement; AKFTA: ASEAN-Korea Free Trade Agreement; AANZFTA: ASEAN-Australia New Zealand Free Trade Agreement.
Source: WTO Tariff Download Facility, *http://tariffdata.wto.org/Default.aspx.*

Table 2.A1.4. **Imported goods subject to line management licensing by MARD**

HS tariff lines	Description	Form of management
ex 3004; 30062000	Veterinary medicines and materials for making veterinary medicines	TBT/SPS licence
N.A.	Biological products for veterinary use registered for first-time use in Viet Nam	TBT/SPS licence
ex 3808	Pesticides and materials for making pesticides excluded in the List of pesticides allowed to be used in Viet Nam	Automatic licence
ex 3808	Pesticides and materials for making pesticides included in the List of pesticides subject to restricted use in Viet Nam	Automatic licence
ex 0106; 06; 07; 08; 09; 12	Plants and animal strains, and various types of insects not available in Viet Nam	TBT/SPS licence
ex 23	Animal feeds and materials for producing animal feeds newly used in Viet Nam	TBT/SPS licence
3101; 3102; 3103; 3104; 3105	Fertilisers newly used in Viet Nam	TBT/SPS licence
3001; 3002	Genetic sources of plants, animals and micro-organs used for scientific purposes	TBT/SPS licence
ex 01	Wild animals and plants subject to import control according to CITES Convention	Automatic licence

Source: WTO (2013), *Trade Policy Review of Vietnam: Report by the Secretariat*, WT/TPR/S/287/Rev.1, 4 November.

References

Abbott, P., J. Bentzen and F. Tarp (2006), "Vietnam's Accession to the WTO: Lessons from Past Trade Agreements", *Discussion Paper* No. PGR2.06.02, Central Institute of Economic Management (CIEM), Hanoi, *www.ciem.org.vn/Portals/1/CIEM/SurveyData/1245397077318.pdf.*

Arita, S. and J. Dyck (2014), "Vietnam's Agri-Food Sector and the Trans-Pacific Partnership", *Economic Information Bulletin* Number 130, Economic Research Service, USDA, Washington, DC, *www.ers.usda.gov/publications/eib-economic-information-bulletin/eib-130.aspx.*

Athukorala, P.-C., L.H. Pham and T.T. Vo, (2009), "Vietnam", in K. Anderson and W. Martin (eds.), *Distortions to Agricultural Incentives in Asia*, World Bank, Washington, DC, *http://siteresources.worldbank.org/INTTRADERESEARCH/Resources/544824-1146153362267/Asia_e-book_0209.pdf.*

Bao, D.H. (2012), *Total Factor Productivity in Vietnamese Agriculture and its Determinants*, Thesis submitted for the degree of Doctor of Philosophy in Economics of the University of Canberra.

Barker, R., C. Ringler, M.T. Nguyen and M. Rosegrant (2004), "Macro Policies and Investment Priorities for Irrigated Agriculture in Vietnam", *Comprehensive Assessment Research Report* 6, Comprehensive Assessment of Water Management in Agriculture, Colombo, Sri Lanka, *www.iwmi.cgiar.org/assessment/files_new/publications/CA%20Research%20Reports/CARR6.pdf.*

Benjamin, D. and L. Brandt (2002), "Agriculture and Income Distribution in Rural Vietnam Under Economic Reforms: A Tale of Two Regions", *William Davidson Working Paper* No. 519, University of Michigan Business School, *http://wdi.umich.edu/files/publications/workingpapers/wp519.pdf.*

Bjornstad, L. (2009), "Fiscal Decentralization, Fiscal Incentives, and Pro-Poor Outcomes: Evidence from Viet Nam", *ADB Economic Working Paper Series* No. 168, Asian Development Bank, Manila, *www.adb.org/sites/default/files/pub/2009/Economics-WP168.pdf.*

Briones, R.M, A. Durand-Morat, E.J. Wailes and E.C. Chaves (2012), "Climate Change and Price Volatility: Can we Count on the ASEAN Plus Three Emergency Rice Reserve?", *ADB Sustainable Development Working Paper Series*, No. 24, August, ADB, Manila, *www.10.iadb.org/intal/intalcdi/PE/2012/12264.pdf.*

Business Monitor International [BMI] (2011), *Vietnam Agribusiness Report Q4 2011*, BMI, London, *www.vietnam-report.com/wp-content/uploads/downloads/2012/02/B-VN-Agri-Q4-2011.pdf.*

Centre for International Economics [CIE] (1998), *Vietnam's Trade Policies 1998*, Centre for International Economics, Canberra and Sydney, *www.thecie.com.au/content/publications/CIE-vietnam_trade_policies.pdf.*

Cervantes-Godoy, D. and J. Dewbre (2010), "Economic Importance of Agriculture for Sustainable Development and Poverty Reduction: The Case Study of Vietnam", paper presented at Global Forum on Agriculture, 29-30 November, OECD, Paris *www.oecd.org/tad/agricultural-policies/46378758.pdf.*

Cook, R.C., T.T. Dao, D. Ellingson, J.J. van Gijn and T.E. McGrath (2013), "The Irrigation Service Fee Waiver in Viet Nam", *ADB Briefs* No. 12, ADB, Manila, *www.adb.org/sites/default/files/pub/2013/irrigation-service-fee-waiver-in-viet-nam-rev.pdf.*

Coxhead, I., N.B.N. Kim, T.T. Vu and T.P.H. Nguyen (2010), *A Robust Harvest: Strategic Choices for Agriculture and Rural Development in Vietnam*, United Nations Development Programme, Hanoi, *www.vn.undp.org/content/dam/vietnam/docs/Publications/26491_20459_SEDS.31032010__RP7_E_.pdf.*

Dang, K.S., N.Q. Nguyen, Q.D. Pham, T.T.T. Truong (2006), "Policy Reform and the Transformation of Vietnamese Agriculture", in *Rapid Growth of Selected Asian Economies-Lessons and Implications for Agriculture and Food Security: Republic of Korea, Thailand and Viet Nam*, FAO, Regional Office for Asia and the Pacific, *http://www.fao.org/docrep/009/ag089e/AG089E00.htm#TOC.*

Dao, T.A. (2014), "Institutional Structure of Food Safety Control in Food Value Chain", Report prepared for the Ministry of Agriculture and Rural Development.

Dao, T.A., D.T. Le and T.B. Vu (2005), "Enhancing Sustainable Development of Diverse Agriculture in Viet Nam", *CAPSA Working Paper* No. 86, Centre for Alleviation of Poverty through Secondary Crops' Development in Asia and the Pacific, United National Economic and Social Commission for Asia and the Pacific, *www.uncapsa.org/publication/wp86.pdf.*

Development Economics Research Group (DERG) (2012), *The Availability and Effectiveness of Credit in Rural Vietnam: Evidence from the Vietnamese Access to Resources Household Survey 2006-2008-2010*, prepared under the Agriculture and Rural Development (ARD) Programme, Royal Embassy of Denmark of Vietnam, *www.ciem.org.vn/Portals/1/CIEM/IndepthStudy/2012/13388684552500.pdf.*

Ellis, K., R. Singh, S. Khemani, H. Vu and N.D. Tran (2010), *Assessing the Economic Impact of Competition: Findings from Vietnam*, Overseas Development Institute, London, *www.odi.org.uk/sites/odi.org.uk/files/odi-assets/publications-opinion-files/6060.pdf.*

FAO (Food and Agricultural Organization of the United Nations) (2013), *Viet Nam Country Programming Framework 2012-2016*, FAO Representation in Viet Nam, *www.fao.org/fileadmin/templates/faovn/files/Administration/Vietnam_CPF_31_July_2013_Final-_Aug_printed.pdf.*

FAO (2011), Viet Nam and FAO Achievements and Success Stories, FAO Representation in Viet Nam, *www.fao.org/fileadmin/templates/rap/files/epublications/Viet_NamedocFINAL.pdf.*

Hong-Loan, T. and W.J. McCluskey (2012), "Property Tax Reform in Vietnam: A Work in Progress", *IMFG Papers on Municipal Finance and Governance*, No. 8, Institute on Municipal Finance & Governance, University of Toronto, *www.utoronto.ca/mcis/imfg/.*

Japan International Cooperation Agency (JICA) (2012), *Vietnam Country Study*, *www.centennial-group.com/downloads/VN%20country%20report3.pdf.*

Kirk, M. and D.A.T. Nguyen (2009), "Land-Tenure Policy Reform: Decollectivization and the *Doi Moi* System in Vietnam", *IFPRI Discussion Paper* No. 927, IFPRI.

Le, H.A. (2006), "Tax Policies and Agricultural Land Use", in *Agricultural Development and Land Policy in Vietnam*, by S.P. Marsh, T.G. MacAulay and V.H. Pham (eds.), ACIAR Monograph No. 123, Australian Centre for International Agricultural Research. Canberra, *http://ageconsearch.umn.edu/bitstream/114071/2/123.pdf.*

MARD (Ministry of Agriculture and Rural Development) (2014a), "Background Paper on Agricultural Policy Framework in Viet Nam During Period 1990-2013 Prepared for OECD *Review of Agricultural Policies in Viet Nam*".

MARD (2014b), "Background Paper on Vietnamese Domestic Agricultural Policies Prepared for OECD Review of Agricultural Policies in Viet Nam".

Marsh, S.P., H.A. Le and T.G. MacAulay (2206), "Credit Use in Farm Households in Vietnam: Implications for Rural Credit Policy", in *Agricultural Development and Land Policy in Vietnam*, by S.P. Marsh, T.G. MacAulay and V.H. Pham (eds.), ACIAR Monograph No. 123, Australian Centre for International Agricultural Research. Canberra, *http://ageconsearch.umn.edu/bitstream/114071/2/123.pdf.*

Martini, R. (2011), "Long Term Trends in Agricultural Policy Impacts", *OECD Food, Agriculture and Fisheries Papers*, No. 45, OECD Publishing, Paris, *http://dx.doi.org/10.1787/5kgdp5zw179q-en.*

McCaig, B. and N. Pavcnik (2013), "Moving out of Agriculture: Structural Change in Vietnam", *NBER Working Paper*, No. 19616, National Bureau of Economic Research, Cambridge, MA, *http://www.nber.org/papers/w19616.*

Newman, C., F. Wainwright, L.H. Nguyen, T.U. Bui, V.N.K Le, D.K. Luu (2012), *Income Shocks and Household Risk-Coping Strategies: The Role of Formal Insurance in Rural Vietnam*, *www.ciem.org.vn/Portals/1/CIEM/IndepthStudy/2012/13388683221090.pdf.*

Nguyen, D.A.T. (2010), "Vietnam's Agrarian Reform, Rural Livelihood and Policy Issue", *www.rimisp.org/wp-content/uploads/2010/05/Paper_Nguyen_Do_Anh_Tuan.pdf.*

Nguyen, H.C. (2006), "Input and Output Price Policy and its Impacts on Agricultural Production", in *Agricultural Development and Land Policy in Vietnam*, by S.P. Marsh, T.G. MacAulay and V.H. Pham (eds.), ACIAR Monograph No. 123, Australian Centre for International Agricultural Research, Canberra, *http://ageconsearch.umn.edu/bitstream/114071/2/123.pdf*.

Nguyen, H. and Grote, U. (2004), "Agricultural Policies in Vietnam: Producer Support Estimates 1986-2002", *MTID Discussion Paper*, No. 79, IFPRI, *www.ifpri.org/sites/default/files/publications/mtidp79.pdf*.

Nguyen, M.H. and T. Talbot (2013), "The Political Economy of Food Price Policy: The Case of Rice Prices in Vietnam", *WIDER Working Paper* No. 2013/035, World Institute for Development Economics Research, United Nations University, Helsinki, *www.wider.unu.edu/stc/repec/pdfs/wp2013/WP2013-035.pdf*.

Nguyen, V.B. (2012), "Agricultural Extension in Vietnam: Its Roles, Problems and Opportunities", paper presented at Roundtable Consultation on Agricultural Extension, Beijing, March 15-17, *www.syngentafoundation.org/__temp/BO_AG_EXTN_VIETNAM.pdf*.

Nielsen, C.P. (2002), "Vietnam's Rice Policy: Recent Reforms and Future Opportunities", *Working Paper 8/02*, Danish Research Institute of Food Economics, *www.gtap.agecon.purdue.edu/resources/download/1080.pdf*.

OECD (2014), *Agricultural Policy Monitoring and Evaluation 2014: OECD Countries*, OECD Publishing, Paris, *http://dx.doi.org/10.1787/agr_pol-2014-en*.

OECD (2012), "Viet Nam", in *Livestock Diseases: Prevention, Control and Compensation Schemes*, OECD Publishing, Paris, *http://dx.doi.org/10.1787/9789264178762-14-en*.

OECD (2009), *OECD Investment Policy Reviews: Viet Nam 2009: Policy Framework for Investment Assessment*, OECD Publishing, Paris, *http://dx.doi.org/10.1787/9789264050921-en*.

Orden, D., F. Cheng, H. Nguyen, U. Grote, M. Thomas, K. Mullen and D. Sun (2007), "Agricultural Producer Support Estimates for Developing Countries: Measurement Issues and Evidence from India, Indonesia, China, and Vietnam", *Research Report*, No. 152, IFPRI, *www.ifpri.org/sites/default/files/publications/ rr152.pdf*.

Pham, B.D. (2013), "Reviewing the Development of Rural Finance in Vietnam", *Journal of Economics and Development*, Vol. 15, No. 1, April, *www.vjol.info/index.php/KTQD/article/viewFile/11365/10348*.

Pham, D.M., D. Mishra, K.C. Cheong, J. Arnold, A.M. Trinh, N.H.T. Ngo, P.H.T. Nguyen (2013), *Trade Facilitation, Value Creation and Competitiveness: Policy Implications for Vietnam's Economic Growth*, Volume 2, The World Bank, Washington, DC, *http://documents.worldbank.org/curated/en/2013/07/18307918/trade-facilitation-value-creation-competiveness-policy-implications-vietnams-economic-growth-vol-2-3*.

Pham, H.N. (2010), "The Vietnamese Rice Industry During the Global Food Crisis", in D. Dawe (ed.), *The Rice Crisis: Markets, Policies and Food Security*, FAO and Earthscan, *www.fao.org/docrep/015/an794e/an 794e00.pdf*.

Pham, Q.D. (2006), "Agriculture Sector of Vietnam: Polices and Performance", Paper delivered to International Congress on Human Development, Madrid, *www.reduniversitaria.es/ficheros/Pham%20 Quang%20Dieu.pdf*.

Phan, S.H. (2014), "A Qualitative Review of Vietnam's 2006-2010 Economic Plan and the Performance of the Agriculture Sector", *Asian Journal of Agriculture and Development*, Vol. 11, No. 1, pp. 39-63.

Phan, S.H. and V.T. Trinh (2014), *Review of Agricultural Policies: Vietnam 1976-2013*.

Que, N.N. and N.T. Manh (nda), *Cashew Sector Development in Vietnam*.

Rama, M. (2008), "Making Difficult Choices: Vietnam in Transition", *Working Paper* No. 40, Commission on Growth and Development, The World Bank, Washington, DC.

Riedel, J. and W.S. Turley (1999), "The Politics and Economics of Transition to an Open Market Economy in Viet Nam", *Working Paper* No. 152, OECD Development Centre, OECD Publishing, Paris.

Sikor, T. (2011), "Financing Household Tree Plantations in Vietnam: Current Programmes and Future Options", *Working Paper*, No. 69, Center for International Forestry Research, Bogor, Indonesia, *www. cifor.org/publications/pdf_files/WPapers/WP69CIFOR.pdf*.

Tantraporn, A. and V. Tuchinda (2012), "AFTA and its Implications for Agricultural Trade and Food Security in ASEAN", in *Regional Trade Agreements and Food Security in Asia*, RAP Publication 2012/15, FAO Regional Office for Asia and the Pacific, Bangkok, *www.fao.org/docrep/018/i3026e/i3026e.pdf*.

Tisdell, C. (2010), "Economic Growth and Transition in Vietnam and China and its Consequences for their Agricultural Sectors: Policy and Agricultural Adjustment Issues", *Economics, Ecology and the Environment Working Paper*, No. 171, University of Queensland, Brisbane, *http://ageconsearch.umn.edu/bitstream/ 94305/2/WP%20171.pdf*.

Tran, C.T. (2014a), *Food Security Policies of Vietnam*, FFTC Agricultural Policy Database, Food and Fertilizer Technology Center for the Asia and Pacific Region, *http://ap.fftc.agnet.org/ap_db.php?id=212*.

Tran, C.T. (2014b), *Overview of Agricultural Policies in Vietnam*, Agricultural Policy Database, FFTC Agricultural Policy Database, Food and Fertilizer Technology Center for the Asia and Pacific Region, *http://ap.fftc.agnet.org/ap_db.php?id=195*.

Tran, C.T. (2014c), *Agricultural Insurance Policies in Vietnam*, Agricultural Policy Database, FFTC Agricultural Policy Database, Food and Fertilizer Technology Center for the Asia and Pacific Region, *http://ap.fftc.agnet.org/ap_db.php?id=276*.

Tran, C.T., B.L. Dinh and H.P. Hu (2013), *Vietnam: Rice Sector Policy Analysis*, SNV Netherlands Development Organisation, Vietnam Office, Hanoi.

Tran, C.T. and B.L. Dinh, (2014a), *The Frame of Agricultural Policy and Recent Major Agricultural Policies in Vietnam*, FFTC Agricultural Policy Database, Food and Fertilizer Technology Center for the Asia and Pacific Region, *http://ap.fftc.agnet.org/ap_db.php?id=261*.

Tran, C.T. and B.L, Dinh (2014b), *Trade Liberalization Countermeasures in Agricultural Sector of Vietnam*, FFTC Agricultural Policy Database, Food and Fertilizer Technology Center for the Asia and Pacific Region, *http://ap.fftc.agnet.org/ap_db.php?id=252#_ftn1*.

Tran, T.T. (2014), "Fertilizers in Vietnam", *ReSAKSS Asia Policy Note* 11, IFPRI, *www.ifpri.org/sites/default/files/publications/resakssasiapn11.pdf*.

Tsukaga, K. (2011), "Vietnam: Food Security in a Rice-Exporting Country", in S. Shigetomi, K. Kubo and K. Tsukada (eds), *The World Food Crisis and the Strategies of Asian Rice Exporters*, Spot Survey 32, IDE-JETRO, Chibu, *http://d-arch.ide.go.jp/idedp/SPT/SPT003200_005.pdf*.

van Tongeren, F. (2008), "Agricultural Policy Design and Implementation: A Synthesis", *OECD Food, Agriculture and Fisheries Papers*, No. 7, OECD Publishing, Paris, *http://dx.doi.org/10.1787/243786286663*.

VietCapital Securities (2011), *Vietnam Natural Rubber Sector*, VietCapital Securities, HCMC, *www.scec.com.vn*.

Vo, T.T. (2008), "Challenges, Prospects and Strategies for CLMV Development: The Case of Vietnam", in C. Sotharith (ed.), *Development Strategy for CLMV in the Age of Economic Integration*, ERIA Research Project Report 2007-4, Chiba: IDE-JETRO, pp. 497-515, *www.eria.org/publications/research_project_reports/images/pdf/PDF%20No.4/No.4-part3-13-Vietnam.pdf*.

Vo, T.T. (2005), "Vietnam's Trade Liberalization and International Economic Integration", *ASEAN Economic Bulletin*, Vol. 22, No. 1, April, pp. 75-91.

Vu, N.K. (2014), "Viet Nam's Participation in FTAs: History, Commitments and Challenges", *OREI Policy Brief*, No. 11, March, ADB, Manila, *http://aric.adb.org/pdf/policy_brief/vietnam-participation-fta-history-commitments-and-challenges.pdf*.

WB (World Bank) (2014), "Restructuring Corporate Income Tax and Value Added Tax in Vietnam: An Analysis of Current Changes and Agenda for the Future", Report No. AUS1671, East Asia and Pacific, World Bank, Washington, DC.

WB (2012), *Vietnam Development Report 2012: Market Economy for a Middle-Income Vietnam*, World Bank, Hanoi.

WB (2011), *Vietnam Development Report 2011: Natural Resources Management*, World Bank, Hanoi.

Wolz, A. and B.D. Pham (2010), "The Transformation of Agricultural Producer Cooperatives: The Case of Vietnam", *Journal of Rural Cooperation*, Vol. 38, No. 2, pp. 117-133, *http://ageconsearch.umn.edu/bitstream/163895/2/Wolz%20jrc38-2%282010%29.pdf*.

WTO (World Trade Organisation) (2013), "Trade Policy Review of Vietnam: Report by the Secretariat", WT/TPR/S/287/Rev.1, 4 November.

WTO (2006), "Report of the Working Party on the Accession of Viet Nam", WT/ACC/VNM/48, 27 October.

WTO (2004), "Accession of Viet Nam: Additional Questions and Replies", WT/ACC/VNM/33, 13 October.

Chapter 3

Viet Nam's policy environment for investment in agriculture

This chapter highlights key challenges to be addressed to improve the policy framework for sustainable investment in agriculture, drawing from the OECD Policy Framework for Investment in Agriculture. First, it examines the key trends of domestic and foreign investment in agriculture since the Doi Moi reforms in 1986, and provides an overview of Viet Nam's investment policy, focusing on efforts to promote public-private partnerships, reduce the role of state-owned enterprises and enhance food safety. Then, it examines policies and measures for investment promotion, including investment incentives and licensing procedures, and describes land tenure policy as secure land tenure is a key condition for sustainable investment in agriculture. It also analyses existing policies intending to facilitate access to finance by agricultural investors, and examines the constraints faced by investors arising from infrastructure development, trade policy, human resource development and research. Finally, it reviews policies aiming to promote responsible business conduct in agriculture. The last section summarises the key findings of the chapter.

3.1. Introduction

As detailed in the OECD Policy Framework for Investment in Agriculture (OECD, 2014a), improving the policy environment for sustainable private investment in agriculture is crucial to enhance agricultural productivity growth, stimulate value addition, maximise the development benefits of investments, and achieve food security. Investment in agriculture relies on an integrated policy environment where a wide range of sectoral and economy-wide policies contribute to a sound investment climate.

Over the last two decades, Viet Nam has been **successful at attracting investment** due to a favourable macroeconomic context and an enabling environment investment supported by the de-collectivisation of farms and the issuance of land rights. As a result, agricultural production almost doubled in volume terms between 1990 and 2012, outperforming all Viet Nam's major competitors in Asia. By 2012, Viet Nam had become the world's largest exporter of cashews and black pepper and one of the largest exporters of rice, coffee, cassava and natural rubber (Section 1.6 and Annex 1.A1). In a context where many growth rates are already declining and where most natural resources are already well exploited, and even sometimes over-exploited, Viet Nam would now need to promote sustainable intensification to tap into the potential offered by domestic and international markets for Vietnamese agro-food products.

Existing constraints that hinder private investment along agricultural supply chains should thus be addressed. High land fragmentation limits scale economies and the lack of transparency in land management constitutes a significant impediment to investment. Large investors have difficulties accessing long-term financing while small-scale producers continue to rely mostly on informal credit. Basic rural infrastructure has significantly improved over the past decade but has not been matching rapid economic growth, which resulted in serious infrastructure bottlenecks. Finally, the weak role played by farmers' organisations obliges investors to interact with numerous small-scale producers which increase transaction costs and uncertainty in a context of weak contract enforcement.

As per the World Bank **doing business** survey, Viet Nam ranks 99 out of 189 countries in 2014 in terms of business climate. While it compares favourably with the Philippines, Indonesia, Cambodia, Lao PDR, and Myanmar, it falls behind Singapore, Malaysia, Thailand and China. According to the Provincial Competitiveness Index (PCI) survey, Foreign Invested Enterprises (FIEs) noted that Viet Nam fared well on: expropriation risk; policy stability; and the influence of FIEs over policies that affect their business. It fared reasonably well on the burden of tax rates relative to competitors. In turn, Viet Nam was significantly less attractive when it comes to corruption, regulatory burdens, the quality of public services (such as education and health care), and the quality and reliability of infrastructure (PCI, 2013). This is in line with the World Economic Forum report (Figure 3.1). Vietnamese respondents cited political instability, unskilled labour, inflation, inadequate infrastructure, tax regulations and corruption as major constraints faced by investors. They considered that access to finance was the greatest problem for doing business.

Figure 3.1. **The most problematic factors for doing business, 2014**

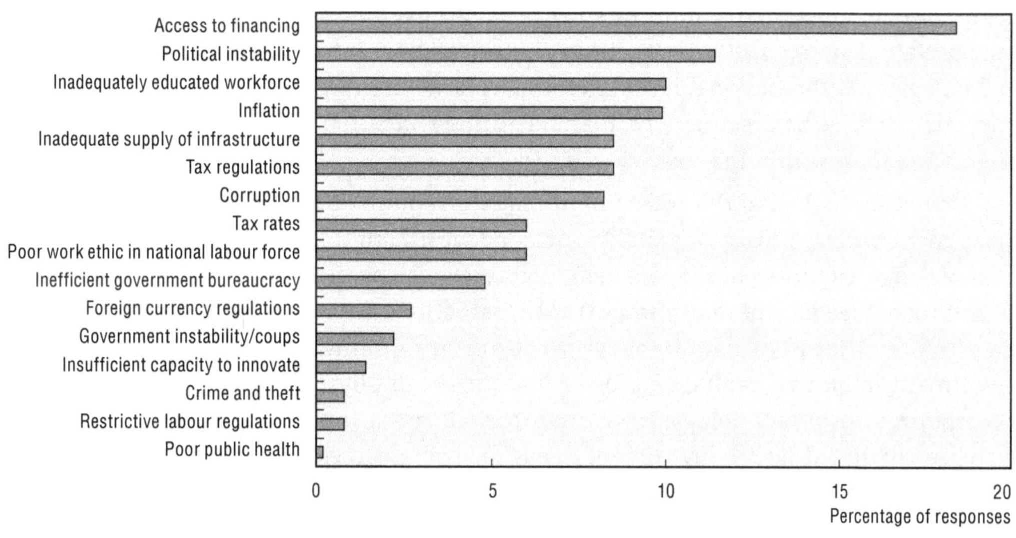

Source: WEF (2013), *The Global Competitiveness Report 2013-2014*, World Economic Forum.

StatLink http://dx.doi.org/10.1787/888933223942

The government has developed **several policies** and legal instruments to respond to these challenges and allow domestic and foreign investors to seize growing opportunities in the agro-food sector. The domestic market offers an important potential for high-quality food products as a result of population and income growth. Viet Nam has the fastest-growing middle and affluent class in Southeast Asia. Between 2012 and 2020, this population of consumers will rise from 12 million to 33 million (BCG, 2014). Investment policy has increasingly supported private sector development since 1986. A new Investment Law has been adopted in November 2014, reducing the number of sectors in which investment is conditional and streamlining investment procedures. Roadmaps have been developed to equitise state-owned enterprises. Significant efforts have recently been made to promote public-private partnerships for various commodities. Due to an impressive land titling programme, land use right certificates have been granted on approximately 85% of the agricultural land.

Some policies may be revised and **further efforts** could be made to attract further private investment. Cost-benefit analyses could be undertaken to evaluate the opportunity cost and the impact of generous investment incentives. Clarifying which incentives are granted to investors would help reduce uncertainty as such incentives are currently granted on a case-by-case basis. Restrictions on land size and on the transfer of land use rights could be revised to facilitate land consolidation. Administrative hurdles to set up a business and to pay taxes could be streamlined further. Extension services and research and development could be strengthened and re-oriented to better address the needs of the private sector. Finally, the division of responsibilities between various government levels could be clarified to ensure consistent policy design and implementation.

3.2. Trends in investment in agriculture

Definition of investment

In the **narrow sense**, investment refers to the change in the physical capital stock, i.e. in physical inputs used in the production process for one year or more. Such investment includes land acquisition, the purchase of equipment, machinery or livestock, plantation development, the construction of storage facilities, dams and dikes for irrigation or transportation systems, or investment in electrification or information and communication technologies. At the national level, investment flows are measured by gross fixed capital formation defined as procurement, manufacturing and purchasing of new physical capital originating from within the country and new or used physical capital from abroad.

However, analysing the investment climate requires **broadening the definition** of investment beyond physical capital to include investment in human capital, i.e. the stock of knowledge, expertise or management ability, as well as investment in technology, such as improved seeds. There are important interactions between these various types of investments that play complementary roles in the production process. For instance, investment in human capital can foster investment in physical capital and facilitate the absorption of new technology. Investments in all these inputs are therefore needed to increase growth, although investment needs may vary across countries depending on the stage of agricultural development.

Investment can be financed by both **public and private sources**, including domestic savings of households and private companies, government savings, external borrowing, and foreign investment. Public investments are commonly associated with public expenditures. However, only certain types of public expenditures lead to physical and human capital formation. While Chapter 2 of this report focuses on public expenditures, this chapter analyses how public policy and investment in public goods and human capital help create an attractive business climate for private investment, both domestic and foreign.

Total investment in agriculture

Investment in agriculture has **increased** over the last two decades, as demonstrated by the increase in the agricultural capital stock per worker. Viet Nam has experienced one of the fastest growth rates of agricultural capital stock in the East Asia and Pacific region (Table 3.1).

Table 3.1. **Agricultural capital stock**

	Agricultural capital stock per worker *in million constant 2005 USD*			Annual growth rate of total agricultural capital stock	
	1990	2000	2007	% 1990-2000	% 2000-07
East Asia and the Pacific	1 050	1 186	1 294	2.2	2.0
Cambodia	1 351	1 227	1 149	1.5	1.4
Indonesia	1 737	1 770	1 944	1.5	1.6
Malaysia	9 620	11 174	12 453	1.1	0.3
Thailand	1 339	1 431	1 601	0.1	1.4
Viet Nam	1 279	1 936	2 251	5.9	3.5

Note: Agricultural capital stock includes land development, livestock, machinery and equipment, plantation crops (trees, vines and shrubs yielding repeated products), and structures for livestock.
Source: FAO (2012), *The State of Food and Agriculture – Investing in Agriculture for a Better Future.*

Prior to 1990, most investment in Viet Nam originated from **the public sector**. Over the period 1995-2010, public expenditures in agriculture multiplied tenfold, of which around 75% were allocated to irrigation (Nguyen, 2012). However, according to the SPEED database of the International Food Policy Research Institute, the share of agriculture in public expenditures amounted to only 3.9% in 2010, which was lower than in other neighbouring countries, such as Thailand (5.8%), the Philippines (5.9%), Malaysia (6.7%), Myanmar (8%), and China (9%).

Over the last decades, **private investment** in agriculture (including fisheries) has been increasing, accounting for 56% of agricultural investment in 2008, with the rest coming from state-owned enterprises (34%) and foreign investors (10%). Indeed, the number of private enterprises and their capital stock grew sharply over the period 2000-08 (Table 3.2). Over this period, the number of state-owned enterprises (SOEs) shrunk by nearly 50% through closures, mergers and privatisation (Dao and Nguyen, 2013).

Table 3.2. **Number of enterprises and capital stock by economic activity and ownership**

	2000		2008	
	No. of enterprises	Capital stock (in USD billion)	No. of enterprises	Capital stock (in USD billion)
By economic activity				
Agriculture	3 378	1.9	8 619	4.7
Non-agriculture	3 890	75.8	197 070	380.6
By ownership				
SOEs	5 759	52.7	3 287	153.6
Non-state	35 004	8.0	196 776	165.6
Foreign investment	1 525	17.0	5 626	66.1
Total	**42 288**	**77.7**	**205 689**	**385.3**

Note: Including SOEs, collective and private enterprises, private limited companies, joint stock companies having capital of state, joint stock companies without capital of state enterprises, foreign and joint venture enterprises.
Source: Nguyen, T.D.N. (2012), *Private Sector Investments in Viet Nam: Agriculture Investment Trends – The Role of Public and Private Sector in Vietnam*, Ms. Nguyen Thi Duong Nga, Hanoi Agriculture University.

While the **number of SOEs** economy-wide fell, their share in other indicators – capital, fixed assets, bank credit – shows a still strong, although declining, position in the economy. SOEs still represented 40% of total capital stock in 2008. They have a dominant position in terms of output and revenue in certain sectors. In 2009, they produced over 90% of the total output of fertiliser, coal, electricity and gas, and water supply (WB, 2011a).

Foreign direct investment in agriculture

Since the *Doi Moi* policy of renovation and economic reform started in 1986, Viet Nam has been **increasingly successful** in attracting foreign direct investment (FDI). The Law on Foreign Investment was implemented from 1988. The first half of the 1990s is generally referred to as the "investment boom": from a complete ban before 1987, FDI inflows reached USD 2.6 billion in 1997. The East Asian financial crisis resulted in a sharp decline in FDI inflows that fell for five consecutive years after 1997. A second wave of FDI began in 2003 as countries in the region recovered from the crisis, a bilateral trade agreement was signed with the US, and Viet Nam acceded to the WTO in 2007. FDI inflows peaked at USD 9.6 billion in 2008 although the global economic crisis caused a fall thereafter. FDI stocks have been growing steadily since 1986 (Figure 3.2). However, according to the Provincial Competitiveness Index (PCI) survey, 54% of the Foreign Invested Enterprises (FIEs) currently in Viet Nam considered other countries before investing in Viet Nam – most commonly China (11.1%), Thailand (10.6%), and Cambodia (7.7%), Indonesia (7.3%) and Malaysia (6.5%) (PCI, 2013).

Figure 3.2. **FDI inflows and stocks in selected ASEAN countries, 1986-2013**

USD billion at current prices and current exchange rates

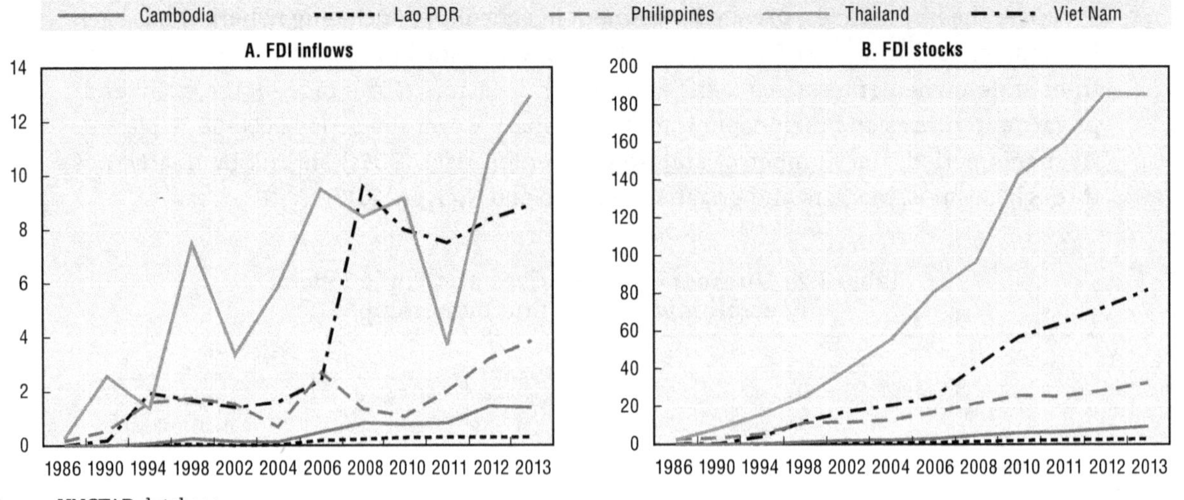

Source: UNCTAD database.

StatLink http://dx.doi.org/10.1787/888933223951

FDI inflows in agriculture have represented an average of 1.47% of total FDI flows over the period 2000-13. They increased significantly over the last two years (Figure 3.3). Over 1991-95, it peaked at about USD 1.4 billion but dropped to USD 463 million in 2007. As of 20 August 2014, the number of accumulated FDI projects in agriculture, forestry and fisheries was 512 (3.1% of the total number of FDI projects) with a total registered capital of USD 3.43 billion (1.4% of the total registered capital of FDI projects) (MARD, 2014d).

Figure 3.3. **Foreign direct investment inflows by sector, 2000-13**

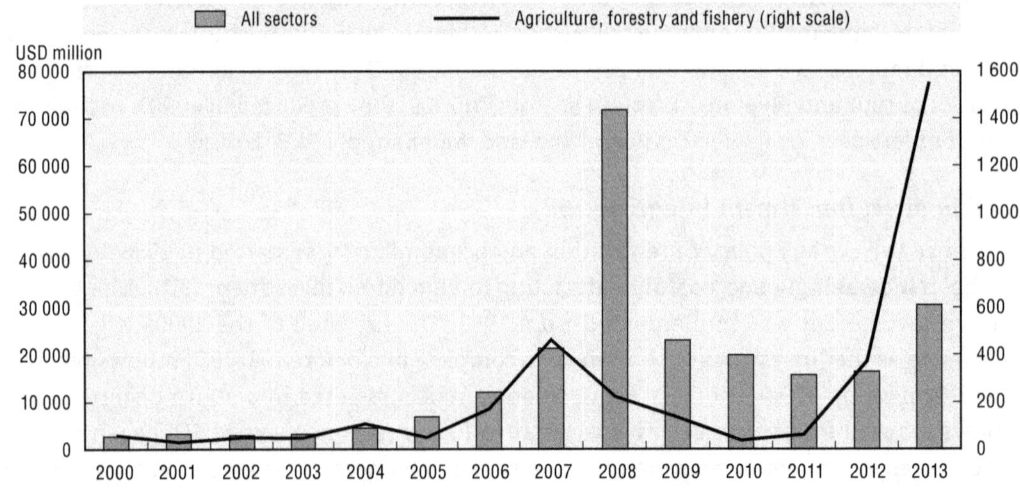

Source: Foreign Investment Agency, Ministry of Planning and Investment.

StatLink http://dx.doi.org/10.1787/888933223965

FDI in agriculture (including agro-processing, forestry and fisheries) is concentrated in **selected sub-sectors**. In 1988-2007, agro-processing projects accounted for 54% of the capital registered for FDI in agriculture, followed by forest plantation and forestry product processing (25%), livestock and animal feed (13%), and crop production (less than 10%) (FIA,

2012). From 1998 to 2012, FDI projects in agro-processing represented USD 1.2 billion, followed by crop production (USD 276 million), livestock production (USD 190 million), fisheries (USD 128 million), and forestry (USD 79.9 million). The remaining investments in agriculture amounted to USD 501 million (MARD, 2014d).

FDI in agriculture, which has been mostly greenfield,[1] comes **from over 50 countries**, of which Chinese Taipei, Japan, China, and Thailand are top investors with capital registered accounting for about 60% of FDI in agriculture (Quang and Ngoc, 2011). As of 20 August 2014, Chinese Taipei emerged as the largest partner with 183 projects, accounting for 35.7% of the number of FDI projects in agriculture and 20% of the value of investment. Top investors in agriculture also include Thailand (11.2% of investment value), British Virgin Islands (9.9%), Singapore (9.8%), Hong Kong (8.2%), France (6.4%), Japan (4.2%), Malaysia (3.6%), Australia (3.4%) and Switzerland (2.9%) (MARD, 2014d). These statistics tend to underestimate the level of investment from European and North American investors, many of which (e.g. Coca-Cola, Procter and Gamble, Unocal and Conoco Phillips) licence their investments through third countries, predominantly Hong Kong, Singapore and the British Virgin Islands (UNCTAD, 2008).

European investors, such as Nestlé and Unilever, and Asian investors invest substantially in **agro-processing**. Although FDI targeting coffee export is common (Louis Dreyfus, Armajaro, and ED&F Man), FDI is limited to sourcing and primary processing and no coffee estate is foreign owned. Soluble coffee has received significant FDI, including from Olam, Nestlé, and Viz Branz. The beverage manufacturing industry has also received FDI, with the involvement of major players such as Heineken, SAB Miller, and Carlsberg (WB, 2014a). The Charoen Pokphand group from Thailand focuses on the production and trade of seeds, animal feed, piglets, chicks, fishery seedlings, and food processing. It ranked 28th in Viet Nam Record 500, the list of the 500 largest companies, in 2011. The Singaporean company, Olam Viet Nam, is the largest exporter of cashews, pepper and instant coffee, with eight large factories. According to the Singapore Business Group, Viet Nam proves to be an exciting destination for retail due to its large consumer market and its rapidly growing middle class and young population (FIA, 2014).

3.3. Investment policy

Although investment policy has improved over the last three decades by levelling the playing field between domestic and foreign investors and simplifying administrative procedures, reforms could be pushed further. Further efforts could also be made to promote public-private co-operation, strengthen the inclusion of producers into the supply chain, level the playing field between state-owned and private enterprises, and increase food safety.

Strategies, policies and legal instruments regulating agricultural investment

Investment policy has **evolved significantly** since the *Doi Moi* reforms. In 1986, the Communist Party opened the way to transform a centrally planned economy into an open, state-regulated market economy. The development of the private sector was encouraged, firstly in the agricultural sector. Farm households were recognised as the main unit in agricultural production and entitled to manage land with 10-15 year terms. Input and output markets were liberalised, and price subsidies eliminated. Restrictions to internal trade were gradually removed. The participation of the private sector in external trade was encouraged by simplifying export and import licensing. The integration of Viet Nam into regional and global political and economic communities was marked by its entry into ASEAN (1995), APEC (1998), and WTO (2007) (see Chapter 2 for further detail). Viet Nam signed the ASEAN Comprehensive Investment Agreement (ACIA) that took effect on 29 March 2012 and that aims to create an open, transparent and competitive investment environment among ASEAN countries.

Several laws regulate investment. The **Investment Law** approved in November 2014 replaces the Investment Law of 2005. It came into effect on the first of July 2015. It regulates investment incentives and guarantees offered to investors, investment procedures, and offshore investment. It clarifies the definition of foreign investment,[2] simplifies licensing procedures and reduces the number of sectors where investment is prohibited or conditional.[3] Based on Viet Nam's socio-economic development planning and strategy issued every five years, the MPI issues a list of: i) investment projects that are particularly encouraged; ii) investment projects that are encouraged; iii) geographical regions where investment is encouraged; iv) sectors in which investment licensing is conditional; and v) restricted sectors in which no licences are granted (PWC, 2008). The new law provides equal treatment to all investors (domestic and foreign, public and private). A foreign investor is permitted to own an unlimited ratio of the capital of an economic organisation, except for specific cases.[4]

In addition to the Investment Law, the **Law on Enterprises** was approved on the same day as the Investment Law. It replaces the law of 2005. It defines the types of enterprises and establishment procedures, and regulates all forms of private enterprises, their organisation and operations. The Law on Tax Management of 2006 provides equality among economic entities as well as between domestic and foreign investors (Nguyen, 2012). The Competition Law was introduced in 2004.

These laws first enacted in 2005 unify rules for domestic and foreign investors, draw a clearer boundary between the role of the state and business, and better protect investors' legitimate interests (OECD, 2009). They also led to the gradual decentralisation of policy design and implementation.[5] They induced a boom in private sector activity and rapid growth in the share of private firms. By 2012, there were 663 800 established enterprises nationwide, of which 468 000 were operating, with SMEs constituting approximately 97% of the total number (MPI, 2012). Their numbers grew on average by 28% a year between 2000 and 2009. The formal private sector now employs 2.9 times the number of workers employed by SOEs (Anh and Duc, 2010; GSO, 2010).

The government actively supports private investment in the agricultural sector through **various plans and programmes**. The five-year economic plan 2006-10 aimed to encourage investment in agriculture through the National Target Programme of New Rural Development running from 2010 to 2020 – for which however no budget has been publicly released.[6] This programme has been controversial as performance indicators are designed at the central level while each commune has its own characteristics (Phan et al., 2014). The master plan to develop agricultural production by 2020 with a vision to 2030 was approved in May 2012 and the plan for restructuring the agricultural sector towards improving value-added and sustainable development in June 2013. These strategies aim to: increase agricultural output and export turnover; encourage land concentration; increase value-added; and promote sustainable development (MARD, 2014a). MARD also designed a plan to develop agricultural enterprises in 2011-15. It sets the target of having 10% newly established enterprises, 10 000 new jobs, and 30% of SMEs that benefited from trainings by 2015 (Nguyen, 2012).

MARD developed a **strategy to attract FDI** in the agriculture, forestry and fishery sector up to 2030. It has already been endorsed by the Minister and should be adopted soon by the Prime Minister. A large number of priority sectors have been identified.[7] The strategy aims to attract joint ventures to improve, in particular, technology, marketing and branding. Based on this strategy, MARD plans to design policies to attract FDI in agriculture and establish co-ordination mechanisms. It would design an information system on investment, including on business climate policies, and reform the administrative procedures related to the investment process. In 2021-30, MARD would continue to improve institutions and policies to attract FDI in agriculture, and periodically monitor and evaluate their effectiveness. It will support infrastructure development, provide further training and improve the quality of training systems and technical manpower to meet the requirements of investors, especially in priority sectors and areas (MARD, 2014d). An action plan drawing from this strategy will be developed and a Decree may also be issued in 2015 to help implement it – although working groups involving various Ministries are already in place to implement it.

Despite great progress in improving the legal and regulatory framework for investment, **reforms are far from complete**. Viet Nam still faces a number of challenges in putting in place the legal and regulatory mechanisms required to establish and manage a market economy. The legal framework still contains elements of a planned-economy approach, such as for investment entry and establishment. It tends to be excessively burdensome and complex. As laws and regulations grow more complex, they also become less readily understood by those who must administer them and are susceptible to uneven implementation across provinces (UNCTAD, 2008). The government management apparatus and procedures can be cumbersome and ineffective. The division of responsibility and co-ordination among various ministries and agencies remains relatively weak, which leads to inconsistent and confusing policies at various government levels. While many policies are sound, they are not fully enforced due to the weak institutional capacity and the lack of financial resources, leading to poor monitoring and oversight (JICA, 2013).

Another concern lies in the emphasis put by local governments on attracting FDI versus supporting local existing enterprises. As per the PCI survey, 32% of respondents believe that some provincial leaders cater to the needs of foreign investors at the expense of domestic enterprises. This rate is much higher in some provinces, especially Tuyen Quang (49%), Nam Dinh (46%), and Ha Nam (44%) (PCI, 2013).

Public-private partnerships

Public-Private Partnerships (PPPs)[8] can enhance the co-operation between public and private actors, thereby increasing returns from public funds through cost and risk sharing and securing contributions that are more adapted to both public and private demand. For both the public and private sectors, the benefits from PPPs come from the pooling of resources and the complementarity of capacities (OECD, 2013b). PPPs can aid private sector investment by sharing risks, providing strategic input and helping to minimise bottlenecks. The decision for the government to adopt a PPP approach to pursue a given objective should be guided by the balance of costs and benefits, compared with other alternatives. If properly implemented, PPPs in Viet Nam could support the development of efficient and competitive supply chains by enhancing rural infrastructure (Section 3.7), increasing access to credit, providing market-oriented R&D, and improving product quality.

Some PPPs are **already operating** in the agricultural sector. For instance, the Netherlands and Rabobank Foundation funded a project to strengthen sustainable cocoa development with Cargill and Mars providing in-kind contribution, such as technical training, demonstration models and assets, and MARD contributing in the form of equipment, office and staff salary. GENTRACO, an animal feed joint stock company, offers a model of public-private co-operation in rice aiming to improve rice quality, increase farmers' incomes and develop farmers' co-operatives. The project is fully funded by the private sector. It involves the local government, provincial Departments of Agriculture and Rural Development (DARD), state-owned commercial banks, international donors on the public side and local private enterprises, farmers and farmers' co-operatives on the private side (MARD, 2014c). In June 2014, the Vietnamese and Japanese governments agreed to collaborate to strengthen food value chains through PPPs. In September 2014, Cool Japan Fund, a public-private fund founded in November 2013, announced its decision to invest USD 7.35 million to form a joint venture. This joint venture would build and operate high quality cold storage in Viet Nam, with Japan Logistic Systems Corp. and Kawasaki Kisen Kaisha, Ltd., to expand the distribution network for high quality, fresh, chilled and frozen food products imported from Japan.

The main **conditions for forming a successful PPP** are: common objective, mutual benefits, complementarity of human and financial resources, and clear institutional arrangements. Good governance, transparency and public leadership are essential to ensure success. Consultation with stakeholders and the establishment of dispute settlement and exit strategies are also important (OECD, 2013b). However, existing agricultural PPPs in Viet Nam currently operate out of a formal institutional framework, which undermines their progress and effectiveness. In the case of the cocoa project, partners face difficulties related to public procurement procedures, which delays activities. The feedback from the Netherlands and businesses suggests that the ODA's legal framework[9] is not suitable to attract and maintain private investment due to bureaucratic procedures and the lack of an appropriate mechanism through which state agencies can operate as real partners. Projects of the Research Institute of Southern Fruit (SOFRI) and GENTRACO are based on the Commercial Law which does not provide suitable frameworks for co-managing activities and lead to low contract enforcement (MARD, 2014c).

To address this issue, **Decree** 15/2015/ND-CP has been approved on 14 February 2015 and been implemented starting on 10 April 2015.[10] It covers not only infrastructure as was the case before but also other sectors, including "agricultural infrastructure and rural development services associated with agro-processing and the consumption of agricultural products" (MARD, 2014c). MARD will issue circulars on PPPs in value chains and PPPs in irrigation infrastructure to complement this decree. When drafting these circulars, best practices could be drawn from the successful examples of various PPPs in Indonesia and Thailand (Box 3.1).

In addition to drafting the decree, the government has actively supported the development of PPPs over the last few years. MARD has established **various PPP task forces** (Box 3.2), officially launched in 2010 in co-operation with the World Economic Forum (WEF). Building on the work of the coffee PPP task force, the Vietnam Coffee Coordination Board was established in 2013 as a joint public-private commodity association chaired by the Deputy Minister of MARD and comprising representatives of the government, Dak Lak and Lam Dong peoples committees and producers, Viet Nam Coffee and Cocoa Association (VICOFA), as well as domestic and foreign enterprises.[11] To strengthen PPP co-ordination, MARD has recently established the Secretariat of the Partnership for Sustainable Agriculture in Viet Nam (PSAV Secretariat) and developed a website to provide updated information on its various efforts to promote PPPs.[12]

Box 3.1. **Lessons learned from Public-Private Partnerships in Indonesia and Thailand**

Drawing from five agribusiness PPP cases in Indonesia, the success of PPPs requires to: involve all partners in the partnership process from the beginning; establish resource-sharing among partners based on their capacity; share benefits fairly among partners; support small farmers to build co-operatives and have viable businesses to work with private companies; and establish trust and commitment among partners. The partnership design should be tailored to the specific context. While subsidies or grants can provide a suitable incentive at the start of a partnership, they should be discontinued once the partnership has matured. Finally, farmers' experience should be recognised and they should not be forced to adopt a certain type of commodity without market certainty.

In Thailand, the main lessons learned from analysing five agribusiness PPPs are complementary to those from Indonesia. PPPs require a clear identification of roles and responsibilities. Well-defined timelines increase the likelihood of success – generally short time frames from the project proposal stage to implementation help to retain enthusiasm and commitment among partners. Stakeholders should know their markets and set realistic targets – in each of the PPPs analysed, the private sector partner was already operating in the business area. The private sector should clearly benefit from the partnership to increase the likelihood that it fulfils its commitment to make the project succeed.

Source: FAO (2013a) and FAO (2013b).

Box 3.2. **Agricultural Public-Private Partnerships in Viet Nam**

Seven PPP task forces have been set up, bringing together over 30 organisations, and 16 models of co-operation between farmers and multinational enterprises have been developed to help improve agricultural productivity, quality, and value-added by promoting new varieties, cultivation methods and quality standards and stronger links between producers and processors. Ultimately, the task forces aim to increase farmers' income and promote sustainable agriculture.

MARD and about 15 international groups, including ADM, Bunge, DuPont, Monsanto, and Swiss Re, participate in these seven PPP task forces:

- Coffee PPP task force, including Nestlé, Yara, Syngenta, Bayer, BASF, Cisco, EDE consulting, Dakman, Sara Lee, and Vinacafe, and operating in Lam Dong and Dak Lak provinces.
- Tea PPP task force to enhance tea exports and improve tea quality. It operates in Phu Tho province with the co-operation of Unilever which collects approximately 30 000 tonnes of tea leaves per year.
- Vegetable and fruits PPP task force, including PepsiCo, Syngenta and Yara, to apply Good Agricultural Practices (GAPs), focusing initially on potato growing in Lam Dong province (South), and in Hai Duong, Bac Giang, Vinh Phuc, and Hanoi (North).
- Fishery PPP task force, including Metro Cash & Carry, Cargill, and Fresh studio. Metro established a centre for fishery product collection, processing and packaging in partnership with MARD.
- Common commodity PPP task force which aims to strengthen the sustainable production of maize and soybeans. Monsanto conducted a study on maize GMOs in partnership with the Agricultural Genetic Institute.

Box 3.2. **Agricultural Public-Private Partnerships in Viet Nam** *(cont.)*

- Pepper and spices PPP task force focusing on sustainability, and more particularly the responsible use of agro-chemicals.
- A PPP task force specialised in finance and microfinance, called the agri-finance PPP task force, which comprises MARD, state-owned banks, the Ministry of Finance, the World Bank, ADB, FAO, IFAD, and bilateral co-operation agencies such as JICA and the Dutch co-operation. HSBC, Standard Chartered Bank, BIDV, Agribank and Bindibank participate. It aims to connect banks with agricultural producers.

Source: Nguyen (2012); IPSARD (2014); MARD (2015).

The PPP task forces established in Viet Nam are supported by the "**New Vision for Agriculture**" which serves as a platform to build collaboration among stakeholders to help agriculture become a driver of food security, environmental sustainability and economic opportunity. Drawing from the successful experience of Viet Nam, the initiative has spread to other countries, and the WEF and the ASEAN Secretariat launched Grow Asia in May 2014. Grow Asia will explore avenues for investment in the agricultural sector and implement market-based solutions by encouraging partnerships between ASEAN governments, the private sector, international organisations, civil society, research institutions, academia, and farmers' associations.

Trading contracts and farmers' organisations

Strong contract enforcement and dynamic farmers' organisations can help reduce transaction costs and uncertainty, ensure that large investors gain from co-operating with smallholder farmers, and increase the efficiency of supply chains, thereby incentivising investment. **Co-operation and linkages** between the production and the consumption of agricultural products are encouraged through the development of large fields, with large field being defined as a form of production based on the co-operation between farmers and enterprises on a large scale to centralise farm products and increase quality and competitiveness (Decision No. 62/2013-QD-TTg). Sanctions can be applied in case the contract is breached and various types of support are provided to enterprises,[13] farmers' organisations[14] and farmers[15] involved in large fields. The programme "Development of Rural Business in the period 2010-15 and Orientation to 2020" aims to increase the proportion of agricultural products traded through contracts to 25-30% by 2015 and 45-50% by 2020.[16] Although foreign enterprises are allowed to enter into contract farming agreements, they are not allowed to buy raw materials directly from producers.[17]

While Decision No. 62/2013-QD-TTg strengthens institutional and spatial linkages to reduce transaction costs and improve the power balance between contract parties, it does not improve contract enforcement. In practice, contract farming has been successful in aquaculture and horticulture. The types of pricing mechanisms included in contractual clauses have become more flexible to take into consideration the fluctuations of markets prices in order to address the lack of contract enforcement. Productivity increases due to a better access to improved seed varieties may offset lower market prices and discourage side-selling. Social pressure associated with the contracting of co-operative groups also helps improve compliance.

Co-operatives that could effectively help connect farmers with traders often do not function well. Farmers lack trust in the large co-operatives that existed prior to 1986 even though they have been transformed and restructured. Following the policy efforts easing the creation of voluntary informal co-operatives (*to hop tac*) that can be registered at commune level through relatively simple procedures, smaller co-operatives created around specific commodities such as milk, vegetables and horticulture, function better and provide input supply and marketing services. The Co-operative Law of 2012 indicates that MARD at the provincial level should provide the following support to co-operatives: training, credit, infrastructure, technology transfer, access to land, marketing and trade promotion. However, the limited access of co-operatives to credit due to a lack of collateral and poor management remain a major impediment to their development (MARD, 2014b; MARD, 2015).

As a result, local traders, millers, or transport operators function mostly through informal commercial networks. The functions of the rice association comprising most leading exporters consist mainly of monitoring and distributing export quota and applying certain administrative rules, such as buying paddy with floor prices for buffer stock (ADB, 2013b).

State-owned enterprises

Although SOEs do not play a major role in agricultural production anymore (with the exception of rubber and sugar), they continue to enjoy **near-monopoly status** in the production of several agricultural inputs and have a significant stake in rice exports (JICA, 2013; Sections 1.9 and 2.2 of this report). They tend to be among the less dynamic firms as they operate inefficiently and/or are under financial pressure. Anh and Duc (2010) estimate the Incremental Capital Output Ratio (ICOR) of SOEs – the investment required to generate an additional unit of output – in 2007 as 8.28, of FIEs as 4.99 and of private firms engaged in labour-intensive activities as 3.74.

The privileges granted to SOEs hinder the **participation of the private sector** in certain key areas of the economy, thus inhibiting associated productivity and efficiency gains. SOEs benefit from an easier access to land, raw materials, finance, procurement contracts, and R&D compared to their peers in the private sector (MARD, 2014a). As per the PCI survey, about one third of respondents continue to cite bias toward centrally-managed SOEs as an obstacle to their business, and that figure has increased slightly since 2012. The bias is highest in public procurement (35% agree), but is also visible in land access (27%), credit access (27%), and ease of administrative procedures (26%). The level of policy bias varies significantly across the country. In some provinces, over half the respondents agree that provincial officials favour SOEs in land and credit access (PCI, 2013).

There are signs however that the government is moving towards a **more open and competitive market**. In mid-2012, two Prime Ministerial Decisions announced plans to restructure SOEs to increase competitiveness. Roadmaps have been designed for privatising various SOEs. For instance, VINATEA should be fully privatised by the end of 2014 (Sections 1.9 and 2.2).

Food safety

A weak and complex food safety management system can undermine investment by preventing investors from securing a supply of high quality and safe food products, thereby increasing their risks, on the one hand, and by hindering the participation of small enterprises that are not able to go through the complex process of food safety procedures, on the other hand.

Viet Nam's rapid expansion in the production of agro-food products has not been associated with high quality. Food safety is a critical issue. Fishery products and various packaged foods have experienced a relatively high level of consignment rejections by trading partners due to the presence of harmful substances, product deterioration or mislabelling. Vietnamese consumers are increasingly concerned about the food safety of various locally produced perishable foods (MARD, 2012).

Despite efforts to increase food safety,[18] the regulatory food safety regime could be improved further (Section 2.4). The involvement of various government agencies, including MARD and the Ministry of Health, that lack the necessary skills for monitoring food quality, as well as the loose relationship between the central and local government agencies, create confusion for enterprises. While large enterprises have a relatively easy access to new documents and policy guidelines, small and micro enterprises, especially in remote areas, report that they cannot access such information. Besides, terms used in some legal documents can be difficult to understand, and some documents contradict each other and are revised frequently. Many standards are still not harmonised with international regulations such as the Codex Alimentarius, and technical regulations on food production are lacking – typically for vegetables, tea, fruits, meat and salt (Nguyen, 2012; MARD, 2014e).

3.4. Investment promotion and facilitation

This section examines first the various investment incentives granted to agricultural investors before looking at the administrative hurdles for registering their business.

Investment incentives

Numerous investment incentives are offered to small and large investors. They can take various forms: subsidised interest rates, reduction or exemption of land rent and land use tax, reduced corporate income tax and value added tax, or reduced import and export tax. National laws, numerous national decrees and provincial regulations lead to a complex web of investment incentives. The absence of a strong independent Investment Promotion Agency accentuates this complexity. Indeed, promotion activities are performed by a mix of agencies, including the Foreign Investment Agency in MPI, Viet-Trade in the Ministry of Trade and Industry, the International Cooperation Department in MARD, and the promotion departments of individual provinces.

As the legislation is **complex and quite unclear**, it is difficult to have a good understanding of which investment incentives should be granted and of how the incentives included in the legislation are applied in practice. This provides a lot of leeway for Provincial People's Committees (PPCs) to grant incentives on a case-by-case basis. It also makes the system difficult to understand for investors who cannot know prior to their investment which incentives they will be granted. Furthermore, procedures to be granted incentives can be difficult.[19] Finally, general policies that would not target any specific product or region may be more effective than the current investment incentives aiming to promote specific products or regions. Box 3.3 summarises good practices on investment incentives, drawing from OECD analysis.

According to the new **Investment Law**, domains that can benefit from investment incentives include the production and processing of agricultural, forestry and aquaculture products as well as the production of plant varieties, animal breeds, and biotechnology products. The categories of investments benefiting from investment incentives[20] are quite

Box 3.3. Good practices in using investment incentives

Prudent use of tax incentives

Tax systems may impose a non-uniform effective tax rate on different businesses, dependant on their size, ownership structure, business activity or location. Where tax relief is targeted, policy makers should ensure that the different treatment can be properly justified. The standard justification is that tax incentives can correct for market imperfections, based on the assumption that private investors do not take into account the benefits to the larger society of certain types of investment, such as for example renewable energy, which leads to under-investment. Asymmetric information on markets or products or monopoly power of large firms may also make entry difficult for SMEs. It is often easier to administer a tax incentives programme than to deliver a similarly-targeted expenditure programme.

Evaluating the costs and benefits of tax incentives

Thorough analysis of the effectiveness and cost-efficiency of proposed tax incentives should be conducted prior to their introduction as well as ex-post, to assess the extent to which they meet their intended objectives. Such evaluation should take into account:

- Benefits, including: a) direct impact of the incentives on investment flows; b) indirect and induced impact due to inter-industry transactions and changes in income and consumption; c) positive externalities, such as technology and know-how transfers; and d) social and environmental benefits;
- Costs, including: a) primary revenue foregone; b) revenue leakages due to unintended and unforeseen tax-planning opportunities; c) costs incurred by taxpayers to comply with tax incentives; and d) administrative costs from running tax incentives.

Transparency and good governance of tax incentives

Tax incentives in which the government has a great deal of discretion increase an investor's uncertainty about the tax system and may inadvertently discourage, rather than encourage, investment. A poorly designed tax system, where the rules and their application lack transparency, are overly complex or unpredictable, may add to project costs and uncertainty. Excessive administrative discretion in the hands of tax officials can seriously increase the risk of corruption and undermine good governance objectives fundamental to securing an attractive investment environment. Additionally, any provisions over which tax authorities have discretion as to their application create opportunities for rent-seeking. As such, general tax incentives and those that involve little or no discretion in their application are preferred.

The granting of tax incentives for investment can often be done outside of a country's tax laws and administration, sometimes under multiple pieces of legislation. Where various Ministries are involved in designing and administering tax incentives, they may not co-ordinate their incentive measures with each other or the national revenue authority, with the result that incentives may overlap, be inconsistent, or even work at cross-purposes. Consolidating all tax incentives, along with their eligibility criteria, into the main body of the tax law, increases transparency and may remove any doubt that the tax administration is empowered to administer them.

Source: OECD *Policy Framework for Investment*, 2015 edition.

broad which gives PPCs the possibility to issue better targeted regulations – although even provincial regulations often leave space for providing investment incentives on a case-by-case basis.

Incentives offered include: exemptions or reductions of land use tax; exemption of land rent during up to 15 years; funding of 20% of the land or water surface rent during five years; funding of 50-70% of the costs related to vocational training, advertising and product promotion, and technology development; and financial support for various investments, such as facilities for drying rice, corn, sweet potato, or cassava, processing coffee, or storing and processing agricultural products.[21] Priority projects are exempted from land rent during 3-15 years and projects on agricultural land are exempted from land use tax until 2020[22] – although in practice, no small-scale farmer pays a land use tax (Section 2.3 for further detail on the land use tax).

Individuals, households, and co-operatives engaged in agricultural production are not liable to **Corporate Income Tax** (CIT), except those with high incomes and producing on a large scale as defined by the government[23] (Nguyen, 2012). However, the legislation on CIT remains vague and procedures for accessing corporate incentives are complex and inconsistent across various documents (IPSARD, 2014). Table 3.3, which tries to synthesise the various CIT preferential rates offered to investors, demonstrates that categorising various investments to determine appropriate CIT rates remains difficult.

Table 3.3. **Corporate Income Tax incentives, 2013**

Project type	CIT rate[1] %	Incentive period[2]	CIT exemption holiday[3]	50% CIT reduction when CIT exemption period has expired
Breeding, rearing, growing and processing of agricultural, forestry and aquaculture products, salt production, creation of new plant and animal varieties	20	10 years	2 years	3 years
Areas with difficult socioeconomic conditions, industrial zones, export processing zones, high technology zones and economic zones; manufacture of machinery and equipment serving for agriculture, forestry, fish farming; manufacture of irrigation equipment; production and refining of feed for cattle, poultry and aquatic resources.	20	10 years	2 years	4 years
Agricultural service co-operatives and people's credit fund	20	No limit		
Encouraged investments and areas	15	12 years	3 years	7 years
Special encouraged investments and/or areas	10	15 years	4 years	9 years
Forestation, tending of forests; breeding, rearing and growing agricultural, forest and aquaculture products in areas with difficult socioeconomic conditions; production of artificial strains, new plant varieties, livestock breeding; production, mining and refining of salt; preservation of agricultural and aquaculture products and foodstuffs.	10	No limit		

1. The standard CIT rate, effective as of January 2012, is 22% but should be reduced to 20% by 2016. This tax can go as low as 10% for encouraged geographic areas or sectors.
2. The incentive period runs generally from the first year of generating revenue, but the tax exemption and reduction period may not begin until taxable income is generated or from the fourth year of generating revenue in the event of no taxable income within the first three years. Thus, the tax exemption and tax reduction periods cover fewer years than the incentive period.
3. CIT exemption begins once a company starts generating taxable revenue.

Source: Canadian Trade Commission (2011); Law No. 32/2013/QH13 amending and supplementing a number of articles of the Law on Corporate Income Tax; and FIA website.

Most exported agricultural products are exempted from **value added tax** (VAT). Since July 2013, the Department of Tax has applied the scheme "check first, reimburse later": it examines the bill from the first seller to check if export companies meet all conditions and can be refunded VAT. VAT reimbursement procedure for exported commodities is complicated as exporters purchase products through many intermediaries. Furthermore, VAT fraud is widespread and disadvantages producers who need invoices to receive financial support while buyers want to evade invoices to avoid paying VAT (IPSARD, 2014).

In addition to the general incentives described above, some incentives are granted for specific commodities (Box 3.4) or in special locations (Box 3.5).

Box 3.4. Investment incentives for specific agricultural commodities

Some incentives are granted to specific commodities:

- Farmers and enterprises investing in large-scale **pork** production benefit from a 50% reduction of the land use tax, and enterprises investing in pork production in remote areas are exempted from land use tax and benefit from a reduced income and VAT tax. Banks can also offer preferential loans to such enterprises. But in practice, these preferential loans are difficult to access as banks require numerous documents.
- For provinces with favourable natural conditions for **coffee** production, coffee enterprises can access loans equal to the value of their coffee output at 0% interest rate within the first two years. From the third year onwards, these enterprises benefit from interest rates that are 50% lower than official rates to buy domestically produced machines and equipment. Moreover, if coffee prices are much lower than production costs, farmers can borrow money at 0% interest rates to purchase inputs and fertilisers. They also benefit from a 50% reduction of the land use tax to plant coffee. Enterprises are exempted from land use tax and CIT during five and three years respectively to build coffee storages. They benefit from a 50% reduction of CIT in the 4th and 5th years.
- **Tea** producers can benefit from an interest rate of 9% for a loan of USD 250-1 400 per hectare to replant or restore tea plantations. Each province issued additional policies to encourage tea production. For instance, 35% of transport costs of tea export enterprises operating in Lam Dong province are covered. Provinces having advantages in tea production, such as Son La and Tuyen Quang, reduce land use taxes for households growing and expanding tea plantations within 6 to 13 years. In 2010, Nghe An province invested in transport and irrigation for tea production. The provinces of Lam Dong, Nghe An, Lao Cai, Phu Tho, Lang Son, Son La, Tuyen Quang and Thai Nguyen, encourage farmers to use high yield tea varieties by subsidising around 20-25% of the new variety cost.
- **Rubber** farmers benefit from an interest rate of 9% per year and are exempted from land use tax in remote areas.
- **Pepper** farmers benefit from 15% discount on the interest rate for investment in mountainous areas, islands, and new economic zones. They also benefit from low land use tax if they invest on new land, barren hills, and wild land. Enterprises benefit from preferential interest rates of 3-9% per year for export products. They also benefit from lower land use tax if they have direct contracts with farmers.
- **Sugar** producers can borrow up to USD 500 without mortgage. Enterprises benefit from low interest rates and large loans from 5-15 years. They are exempted or benefit from a 50% reduction of land use tax during seven years or more depending on regional conditions. They can access a support fund to import machine and equipment if the exchange rate fluctuates significantly.

Source: MARD (2014a); Phan (2014).

Box 3.5. Investment incentives in selected regions

As per the Investment Law of 2005, geographical areas that benefit from investment preferences comprise areas with difficult or especially difficult socio-economic conditions as well as industrial, export processing, high-tech and economic zones.

The Saigon High-Tech Park (SHTP) for instance aims to attract in particular biotechnology applied to agriculture. Investment incentives include: CIT exemption for the first four years, followed by nine years at 5% and ten years at 10%; VAT and import duty exemption for certain equipment and machinery; VAT and export duty exemption for high-technology products; similar personal income tax for local and foreign workers; one-stop investment application service; and on-site and electronic customs clearance.

The city of Can Tho in the Mekong delta, a special economic zone, is set to attract USD 26 million of investment to build the high-tech Thoi Hung agricultural zone covering 500 ha. The project is expected to produce crops and livestock meeting international standards and using cutting-edge technology to process agro-food products. It should help expand the application of advanced agricultural technology through a network of satellite farms and businesses linked within the zone. Incentives offered to foreign investors in the zone include 50% land use tax and the possibility of acquiring free land for constructing workers' residential buildings and public facilities.

Provincial governments also offer specific incentives beyond national incentives. Here are a few examples:

- Hai Phong City: CIT rates of 10% during 15 years to enterprises established in particularly difficult socio-economic areas and industrial parks; CIT reduction to enterprises whose employees are from an ethnic minority under the condition that the enterprises invest their savings in supporting such employees through vocational training, housing, social and labour insurance; land rent exemption during up to 15 years; possible financial assistance to prepare project documents up to USD 1 124 per project; 50-100% of the costs of compensation, relocation, site clearance, and completion of the procedure for leasing land; up to 25% of the costs of site levelling; training of the labour employed in FDI projects at local vocational training centres, and up to 30% of the training costs of labour employed in FDI projects.
- Quang Nam province grants incentives to specific activities, including: plantation and processing of agricultural products; production of artificial seeds, plant seeds and animal breeds; and creation of new plant varieties. These incentives include: one-stop shop for investment activities; 20-30% of the cost of workers' training; land rent exemption during 50 years and possibly an additional 20 years; and higher CIT reduction or exemption than those offered at central level.

Source: MARD (2009); Viet Nam Briefing (2014); SHTP (2014).

Investment licensing

Although efforts have been made to lessen bureaucratic hurdles, registering businesses and products can still be a **complex and time-consuming process** (Box 3.6). To reduce bureaucratic corruption, relevant ministries are required to simplify administrative processes and procedures relating to foreign investment and establish one-stop shops[24] (IPSARD, 2014). Administrative procedures were tackled under Project 30 implemented in particular by the Ministry of Science and Technology. A total of 5 700 administrative procedures were under inventory and reviewed in 2007-10, of which 258 were identified as

Box 3.6. **Licensing and registration procedures**

The **licensing process** is both bureaucratic and complicated, facilitating corruption. Unofficial payments are frequently made to speed up the decision making process, usually at the expense of other companies seeking to abide by anti-corruption legislation. The overlap and duplication of officials' responsibilities, ineffective leadership and supervision by senior staff, and inefficient working practices facilitate corruption. Recent reforms have reduced some of the complexity but their impact on corruption levels has been insignificant. While launching a business takes ten procedures on average, completing licensing requirements still takes more than 100 days. However, most provinces have invested great effort into facilitating business entry through reforms of business registration offices, reduced licensing requirements, and one-stop shops. As a result, according to PCI, the median province now scores close to a nine on the ten-point index. While initial drafts of the new Investment Law proposed to harmonise licensing procedures for domestic and foreign investors, the final law reintroduced specific requirements for foreign investors.

Procedures for **registering plant protection products and seeds** are particularly unpredictable. Enterprises face procedural bottlenecks to introduce new technologies. A temporary ban on registering plant protection products has been introduced, officially to regulate the number of generic products available on the market. Fertilisers' registration has shifted from MARD to the Ministry of Industry and Trade which is confusing for investors as procedures have thus changed. Registering seeds is also a complex and long process that can take up to three years. In contrast, seeds do not need to be registered in Thailand and the Philippines and registration takes only up to 1.5 years in Indonesia. The regulation for seeds is being revised, but its direction remains unclear. Local governments have significant power which increases uncertainty for investors: if a seed has been cleared at central level, it can be rejected at local level. A moratorium on registering feed products had also been put in place in September 2013 and was lifted in July 2014. A major issue faced by input companies lies in the presence of counterfeit inputs, with for instance up to 50% of the market of chemicals being counterfeit.

Source: PCI (2013); VBF (2013); MARD (2014a); WB (2014b).

hindrances to residents and enterprises that should be given priority.[25] In addition, the government created the Administrative Procedure Control Agency, a permanent unit that reviews the flow of new regulations and manages a new national database of administrative procedures (OECD, 2013c; WB, 2014a).

3.5. Land tenure policy

Secure land rights are a necessary condition of any investment in agricultural production. They are critical to ease the process of land acquisition, incentivise long-term investment in land and sustainable land management, and facilitate access to credit by allowing land to be used as collateral.

Viet Nam has very **low agricultural land endowment per capita** and is dominated by small-scale farming, which hinders the integration of producers into supply chains. Each farm holds less than 0.5 ha on average (Section 1.8 for further detail). Processing factories have difficulties securing a regular supply from small scale and fragmented farm production. For example, PACIFIC Hoa Binh currently contracts with about 600 farmers growing cucumber on 350 ha and spread over several provinces (Nguyen, 2012). Limited access to land is a major constraint for the establishment of large commercial farms (WB, 2013a).

The state owns all land. According to the Constitution, land is the property of the "entire people" and is allocated or leased by the state to organisations, households or individuals. The state issues **land use right certificates** (LURCs) that can be sold, rented, exchanged, mortgaged and bequeathed. Land users may legally acquire land use rights through purchase, lease, inheritance, or grant from a family member. LURCs are necessary for state recognition of a user's rights, formal land transactions and access to formal credit (USAID, 2013).

Land registration

Land titling has proceeded unevenly. Beginning in the late 1980s, the state started allocating land use rights to farmers. Viet Nam's land titling process was one of the most ambitious ever attempted in the developing world both in scale (nearly 11 million land titles had been issued to rural households by 2000) and the speed with which it was implemented (Cervantes-Godoy et al., 2010). According to the Ministry of Natural Resources and Environment (MONRE), in 2012 alone, 1.8 million LURCs were issued and 2.6 million LURCs renewed (CIEM, 2013). By 2012, 75% of all land had been mapped and LURCs had been issued to cover 85% of agricultural land (Section 1.8).

Farmers who receive agricultural and production forest land are granted **red books** to document their land use rights. Those contracted to protect natural forest are granted green books. All changes in the land laws are reflected in such books (Phan et al., 2014). Red books include all land parcels allocated to one individual. If a parcel belongs to individuals that do not belong to the same household, it appears in several red books. Members of a household do not have individual land use rights (CNRS, 2010).

Land titling can help enhance **land tenure security** and thus foster investment. Indeed, as per CIEM's 2012 household survey, 82% of the land plots with a LURC had irrigation infrastructure against only 55% of those without a LURC. The largest difference was found in Lai Chau province where 66% of the plots with a LURC had irrigation infrastructure against only 25% of those without a LURC (CIEM, 2013). Farmers with a formal land title have significantly higher rice yields than those with no defined rights (Van den Broek et al., 2007).

However, acquiring agricultural land is often a **time-consuming and complex process**. Foreign investors cannot be granted red books and can only lease land from the government, and not directly from an individual. Bribes are often paid for access to land-related information and, given the lack of supply and potential profitability of the resource, the size of these bribes can be considerable. As with business licensing above, navigating the land administration process is time-consuming, which provides fertile ground for corruption. Businesses also find that the same procedures are applied differently across provinces, which causes confusion and further delay (VBF, 2014). Provinces are poorly prepared to deal with the significant powers with which they are endowed (CNRS, 2010).

Land legislation

Land laws have been adjusted five times since 1986 (1988, 1993, 1998, 2003 and 2013) and gradually supported the development of a land market. In November 2013, the National Assembly adopted a new **Land Law**. While this law strengthens the development of a land market, by for instance allowing foreign entities to be allocated land to build housing for sale and lease for the first time, it keeps several restrictions on the duration of land use rights, land areas per household, the choice of crops and land transfers and exchanges. Such regulations intend to guarantee equal access to land among the rural population, but they also limit land consolidation, thus the creation of viable, market-oriented farming, and can force rural households to leave on small farms and use their land for agricultural production, thereby hindering an optimal use of the land. The exemption of land use tax discourages land use change or land sale when changing activities.

As per the new Land Law, the **duration of land use rights** is extended from 20 to 50 years for all agricultural land and 70 years for forest and other categories of non-agricultural land and for large projects or projects located in poor rural areas. The process for renewing land use rights at the end of the 50-year period is simple as the land use rights holder should only indicate his willingness to continue using the land. The term for leasing land which is part of the agricultural land fund used for public purposes should not exceed five years.

Land area per household or individual is limited when land is allocated by the state. Land area for annual crops per household or individual directly involved in agricultural production should not exceed 3 ha for land in centrally-managed provinces and cities in the Southeast region and Mekong Delta region, and 2 ha for land in other centrally-managed provinces and cities. The land area for perennial crops should not exceed 10 ha in the delta area and 30 ha in the midland and mountainous areas. Protection or production forest should not exceed 30 ha per household or individual. When a household or individual is allocated different categories of land including land for annual crops, aquatic farming, and salt production, the total area should not exceed 5 ha. When it is also allocated land for perennial crops, the area of such land should not exceed 5 ha in the delta areas and 25 ha in midland and mountainous areas. When it is also allocated production forest, the area of such land should not exceed 25 ha. Agricultural land obtained through the transfer, lease, sublease, inheritance or donation of land use rights or through capital contribution in the form of land use rights, is not covered by allocation limits.

In practice, such limits can be applied on an ad-hoc basis as the law offers **some flexibility**. For instance, the law allows increasing the allocation norm by ten times under unspecified conditions, and states that norms can be adjusted "in accordance with the specific conditions of each locality and in each period". This leaves the application of these rules vague and uncertain, and at the discretion of the local People's Committee. One has to expect erratic application of these larger land size limits, and an added potential for demands by officials for side payments. To encourage land consolidation, removal of these size limits entirely would have been preferable.

Land use changes require the permission of competent state agencies for the following conversions: from wet rice to perennial crops, forests, aquaculture or salt production;[26] from other annual crops to aquaculture or salt production; from special, protection or production forests to agriculture; and from agriculture to non-agriculture (Tran, 2014a). In particular, the law protects rice land by limiting the change from rice land to non-agricultural land. As per the Land Law, when such land change is essential, the state should take measures to increase the area or improve the productivity of rice land. It should support and invest in the construction of infrastructure and the application of modern technologies in the areas selected for high-productivity and high-quality rice cultivation.

People who are allocated or lease wet rice land from the state for non-agricultural purposes should pay a lump sum to compensate for lost wet rice land or to improve rice land productivity. Users of rice land are responsible for increasing soil fertility.

Restrictions on land use vary across provinces. In terms of the choice of crops, the North faces more restrictions than the South for choosing crops, building fixed structures or converting agricultural land to non-agricultural land – although more households are required to grow rice in the South, with for instance all households required to grow rice in all seasons in Lam Dong province. In Lai Chau and Phu Tho provinces, up to 85% of the plots face restrictions on building structures and converting agricultural land (CIEM, 2013).

These restrictions have **negative effects** as documented by Markussen et al. (2009). Farmers are denied the opportunity to use lands to their most profitable use, thus limiting the opportunity to diversify their crop mix and hence smooth their income sources, a significant issue given the volatility of today's commodity markets. At the household level this lowers farm incomes or increases poverty, and at the country level it reduces aggregate efficiency, lowers national income, and adds little or nothing to food security, given the very large rice surpluses (exports) that Viet Nam produces.

Land transfers are also strictly regulated. The maximum land area for transferring land use rights is as follows: 15 ha in provinces and centrally-run cities in the South East and the Mekong Delta, and 10 ha in the remaining provinces and centrally-run cities, for annual crops; 50 ha in communes, wards, and towns in flat land, and 150 ha in communes, wards, and towns in mid-land and mountainous areas, for perennial crops. The lack of cadastral surveys, local capacity constraints, and various administrative and regulatory barriers make the transfer of land a difficult and potentially costly process, resulting in a thin land rental market. The 2008 VASS Participatory Poverty Assessment found that farms without LURCs, or with LURCs close to expiry (as most currently were, given the 20-year leases initiated in 1993), were not able to transfer land use rights. According to the law, agricultural land recovered from households and individuals should be allocated or leased to households and individuals without productive land.

According to Article 190 of the 2013 Land Law, households and individuals using agricultural land allocated by the state or obtained through the exchange, acquisition, inheritance or donation of land use rights from other land users, are entitled to **exchange** their land use rights only with households and individuals in the same commune, ward or township. Although some legislation aims to facilitate land accumulation and promote specialised, large-scale and mechanised cultivation areas,[27] land exchanges still occur slowly due to difficult procedures (Tran et al., 2013).

Land is not administered with the same efficiency everywhere and significant differences exist between land categories and regions (Box 3.7 provides further detail on the division of responsibilities). **Land administration** requires important staff and financial needs, particularly at the lowest levels (communes, districts) where staff is usually not or poorly trained (CNRS, 2010). In addition, the decentralisation of land management and licensing to local governments remains unclear, resulting in several conflicting land use licenses granted by local governments (MARD, 2014a).

Land use planning

Regular and participatory land use planning is critical to ensure a sustainable, efficient and equitable use of the land, particularly in a context where pressure on land resources is

Box 3.7. **Responsibilities of various land administration bodies**

The National Assembly promulgates land laws and approves decisions related to nationwide land use planning. A hierarchy of authorities at the central, provincial, district and communal levels administers land policies. MONRE is the primary central-level administrative body for land, water and mineral resources. MARD manages forest land through provincial DARD and agricultural offices, state forestry enterprises, forest management boards, or forest communities.

With at least one land staff per commune, People's Committees at all levels (provincial, city, district, commune, ward and township) implement the land policy defined at central level, as follows:

- PPCs are responsible for: developing and approving land use plans; allocating or leasing land in line with these plans; and recovering the land that is unused, used for improper purpose, or contracted, leased, lent, or occupied illegally.
- District-level People's Committees decide on the allocation and lease of land and the changes in land uses. They must get the written approval from PPCs before making any decision for agricultural land exceeding 0.5 ha.
- Commune-level People's Committees lease agricultural land for public purposes in their respective commune, ward or township. They are responsible for managing and protecting unused land and registering land in local records. They receive applications for land use rights and transfer them to district-level People's Committees who review them and should grant LURCs within 50 working days.

Source: MARD (2014a).

high due to low available land per capita combined with rapid economic growth and urbanisation. The development of land use plans can help regulate the conversion of agricultural land into urban land and ensure it does not have adverse social impacts.

Agricultural land use plans are prepared every five years, with the necessary budget provided to districts by the central government (MARD, 2014a). The Land Law of 2013 recentralises the decision-making power on land use planning from the commune to the district, from the district to the province, and from the provinces to the Prime Minister's Office (Oxfam, 2013). However, this planning, whose role is to maintain central government control over land allocation and use, is described as being much more theoretical than real. Land use plans should be approved by the next level up, which is MONRE for provinces. MONRE does not have the capacity to make a detailed assessment of all plans, which leads to long delays in getting them approved. Many districts and half of the communes have no plans. Districts and provinces can thus easily take decisions outside any planned or centralised framework which facilitates the conversion of agricultural land into industrial or public land (CNRS, 2010).

The Land Law of 2013 requires People' Committees to organise **public consultation** on land use planning. At national and provincial levels, information on land use planning should be published on the websites of MONRE and PPCs. At district level, meetings and direct consultations should be organised and information published on the websites of district-level People's Committees. However, community participation in land use planning is still limited (Oxfam 2012d). According to public administration surveys, only 22% of respondents said that they had been given an opportunity to make comments about local land use plans, and of these only two out of five said their comments had been taken into consideration (Oxfam, 2013).

Land expropriation and compensation

As per the Land Law of 2013, land can be recovered only for the **purposes** of national defence or security and socio-economic development for national and public benefits. These projects of national importance should be approved by the Prime Minister and aim for instance to build urban and residential areas in rural areas, to develop industrial clusters or concentrated zones for producing and processing agricultural products, or conserve special or protection forests (MARD, 2014a).

The **conversion** of agricultural land, especially rice land, into industrial zones, remains a controversial issue at the National Assembly (MARD, 2014a). Indeed, about 700 000 land-related complaints have been lodged in the 2009-11 period according to the National Assembly data (Bloomberg, 2013). In the 2001-10 decade, it is estimated that one million ha of farmland, 10% of all agricultural lands across Viet Nam, were appropriated from farmers for conversion to non-farm uses, and that these were predominantly private, not public "takings" (Wells-Dang, 2013).

Land users are **compensated with land** only in certain cases. As per the Land Law, the state should compensate land users that have LURCs with similar land. When it recovers over-limit agricultural land, households and individuals should receive land-based compensation if their LURCs have been granted as inheritance, gift, grant, or transfer. For households and individuals without LURCs, compensation should be given only for land within allocation limits but other types of support could be provided for land beyond such limits as decided by PPCs.[28]

The law states that the **value of compensation** should be based not only on the price of land but also the value of crops. If no land is available for compensation, land users should be compensated based on the market price of recovered land as decided by the PPC. Land users should also be compensated for the loss of the crops cultivated on the land at the time of the expropriation.[29] Households and individuals should not only be compensated but also provided with job and vocational training, occupation transition, and job search by the PPCs as approved when negotiating compensation, support, and resettlement plans. Affected people must be informed and consulted during the preparation of training and occupation transition plans.

Although the law requires the state to pay compensation based on the market price, the state lacks procedures for assessing market price and often takes into account annual agricultural incomes rather than the value of adjacent land, ongoing or planned investment projects or plans to change land status. The value of compensation is often based on the market price for that particular use of land (i.e. agricultural use value). This leads to **expropriation at low cost** and enables speculators to acquire agricultural land cheaply, change its status and rent it out at a much higher price. Indeed, state agencies setting agricultural land prices have kept these as low as 30% of the estimated market price in order to promote land conversion (Oxfam, 2013). Fewer than 10% of those who had lost land considered that the compensation received was close to market value (Wells-Dang, 2013). In fact, prevailing market prices on the open market are much higher than the market price for land restricted to agriculture use. It is partly an issue of imbalances in information and power between the investor taking the land and the farmer, plus the corruption that can occur between the investor taking the land and the state officials that administer the transfers and allocations of LURCs. It is also a matter of the rules for compensation.

Corruption and numerous administrative payments on compensation are critical issues. Farmers are particularly adversely affected because most of them know little about the land legislation, and local officials are not well trained (CNRS, 2010). Observers report that current compensation procedures are slow, unpredictable and lacking in transparency. Inadequate compensation is a source of widespread grievance that sometimes leads to violence. Approximately 70% of all complaints directed at the government each year are administrative complaints regarding land, and 70% of the land complaints relate to compensation (USAID, 2013).

Although the legislation provides strong provisions on **resettlement plans**, in line with the FAO Voluntary Guidelines on Responsible Governance of Tenure of Land, Fisheries and Forests in the Context of National Food Security,[30] such provisions are rarely enforced. As per the Land Law, PPCs should develop resettlement plans before land expropriation. They should co-ordinate with the commune-level People's Committee to conduct consultations on the compensation, support and resettlement plan by organising meetings with land users and disseminating the plan at the offices of the commune-level People's Committee and at common public places. Consultations must be recorded in minutes certified by representatives of the commune-level People's Committee and by affected land users. These minutes should indicate the number of land users who are for or against the compensation, support and resettlement plans.

There is a widely shared view that the Land Law of 2013 made a number of **modest improvements**, but that the essential **points of controversy** in the many land disputes were left largely unaddressed (Wells-Dang, 2013; USAID, 2013). Thus the level of land conflicts may not be significantly reduced. Some general critiques of the law are that state acquisition of land ("recovery") should be restricted to public uses, not for "socio-economic development", and that investors should purchase land use rights from farmers in a voluntary transaction without coercion, that public disclosure and open hearings should be required before land appropriations, that there should be rules requiring state agencies to respond to complaints, that farmers should have equal rights to investors and urban residents generally, that land use restrictions on agricultural crop type should be removed, and compensation should be based upon open market land prices, not just agricultural land values (Wells-Dang, 2013; USAID, 2013).

3.6. Financial sector development

Efficient financial markets can allocate capital to innovative and high return investment projects of both large and small agricultural investors, thus increasing revenues and generating economic activities. This section provides an overview of the challenges faced by agricultural investors to access credit by briefly describing existing financial institutions in the agricultural sector, highlighting the constraints faced by investors to access formal loans, and examining policies aiming to address these constraints.

Existing institutions

Since 1986, Viet Nam's **banking system** has experienced a major transformation in terms of organisational structure, evolving from a single-tier system with a state-owned monopoly to a two-tier system comprising a growing number of banks. Five state-owned commercial banks still account for the dominant share of the market. 28 foreign banks, 37 joint stock banks, 5 joint venture banks, 9 finance companies, 12 leasing companies, and 960 public credit unions are also operating (OECD, 2009). The establishment of a stock market in 2000 helped improve access to capital markets (MARD, 2014a). In 2013, as per the State Bank of Viet Nam, USD 17.4 billion of bank loans were extended to the agriculture, forestry and fishery sectors, which amounts to half of the loans extended to the industry and to commercial activities.

Three major formal sources provide financial services to the agricultural sector: two state-owned banks, namely Viet Nam Bank for Social Policies (VBSP) and Viet Nam Bank for Agriculture and Rural Development (VBARD), as well as co-operative banks renamed as such by the Co-operative Law of 2012 and formerly known as People's Credit Funds. VBSP, established in 2003, is the major formal provider of financial services in rural areas. It targets micro and small entrepreneurs in marginalised areas by providing subsidised credit on a social policy basis and financial services accessible without formal collateral. Established in 1988, VBARD is the largest commercial bank with an extensive network of 2 340 branches and transaction offices and 40 000 staff located in 61 provinces. Co-operative banks were established in 1993 as savings and credit co-operatives at commune level (ARCM, 2014). As per the Co-operative Law in 2012, they should now extend credit to co-operatives.

Since 2006, VBSP has been able to gain a larger share of the rural credit market with a proportion of loans on the rural market increasing from 26% in 2006 to 42% in 2010 while VBARD loans decreased from 40% to 24% over the same period. Informal sources of credit recorded relatively insignificant changes over this period (Figure 3.4). The rising prominence of VBSP is in line with government efforts aiming to provide credit at zero or low interest rate. VBARD's decreasing prominence can be attributed to the ongoing commercialisation of the bank and the increasing scrutiny of loan applicants (DERG, 2012).

Figure 3.4. **Sources of credit in rural areas, 2006-10**

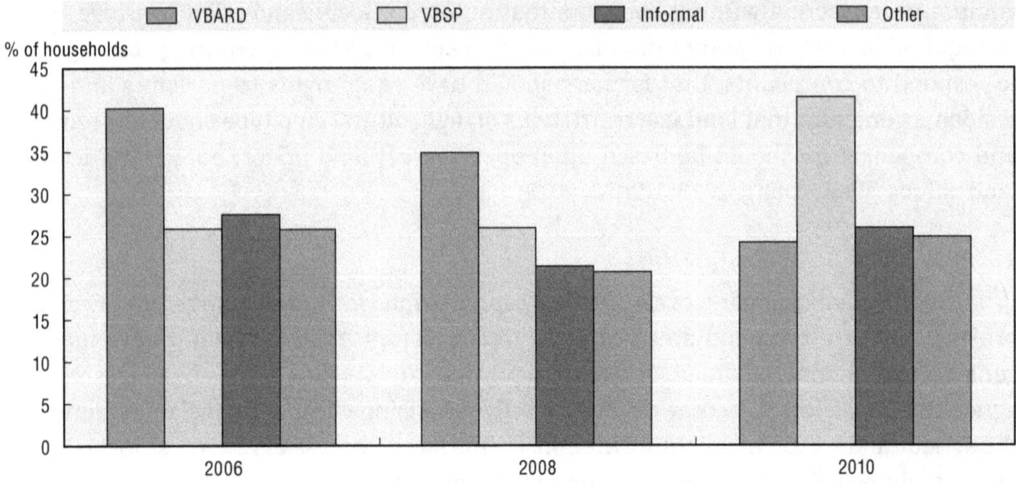

Note: Since households may hold loans from more than one source, the proportion of loans by source adds up to more than 100%. Data is taken from the Vietnam Access to Resources Household Survey implemented in 2006, 2008 and 2010 in 12 provinces. The 2 200 households for which a full panel is available are spread over 437 communes and 130 districts. VBSP: Viet Nam Bank for Social Policies; VBARD: Viet Nam Bank for Agriculture and Rural Development.

Source: DERG [Development Economics Research Group] (2012), The Availability and Effectiveness of Credit in Rural Vietnam: Evidence from the Vietnamese Access to Resources Household Survey 2006-2008-2010.

StatLink http://dx.doi.org/10.1787/888933223977

The approach of VBARD and VBSP is quite different. VBARD lends mostly on a commercial basis for productive investments while VBSP provides subsidised credit primarily for consumption purposes. The establishment of VBSP is largely seen as improving financial services in rural areas since it allowed for a full separation between preferential and commercial credit (Section 2.3). The case study of Phu Thuong Commune indicates that VBSP charges an average monthly interest rate of 0.5-0.65% for loans of usually 12 to 72 months, while VBARD charges an average monthly interest rate of 1-1.2% for loans of 12 to 60 months (Nguyen, T.T.H., 2010). By providing subsidised credit, VBSP faces difficulties in ensuring that its loans are put to their stated use and repaid. Through commercial loans, VBARD can better monitor its clients (DERG, 2012).

Mass organisations facilitate access to credit. Households that are members of Farmer's Unions are more likely to access loans for agricultural purposes while households that are members of Women's Unions are more likely to access loans for non-farm and investment purposes. Indeed, VBARD and VBSP use local organisations, including political and administrative organisations and indigenous village groups such as Viet Nam Farmers' Union and Viet Nam Women's Union, to facilitate loan repayment and ensure loans are put to their intended use (IPSARD, 2010a). Loans are extended either through joint liability groups in which all members should repay each other's debts, or through joint borrowing groups in which lending is extended directly to groups (Pham, 2013). The semi-formal sector is relatively small since it represents only 5-10% of the rural credit market. It comprises: state-controlled mass organisations providing various credit services to their members at the grassroots level; specialised microfinance funds; and several non-governmental organisations, such as Action Aid and Viet Nam Plus (ARCM, 2014).

Despite the presence of the above-mentioned institutions, 49% of rural households were still unable to access banking services in 2010 (DERG, 2012). As a result, the **informal sector** remains a major source of rural credit. While large enterprises can borrow from formal sources, SMEs usually borrow from informal sources such as families and relatives (Kemper and Klump, 2014). The informal sector comprises financial assistance from family, friends, moneylenders, and traditional rotating savings and credit associations (ROSCAs), such as Ho (in the North) and Hui (in the South) groups. Small-scale producers also rely heavily on value chain financing from traders and exporters who often pre-finance inputs, provide small cash advances to meet immediate needs and sometimes pre-purchase entire crops upfront by providing cash or part-payment at the start of the season and are then responsible for harvesting also. Loans from relatives typically carry zero or low interest rates, while other informal lenders (i.e. moneylenders and ROSCAs) charge about 7-10% per month, about two or three times the rate charged by formal financial institutions that charge 1.7-2.8% per month (ARCM, 2014).

Challenges of access to credit

Difficult access to credit from banks and high interest rates are among the two most serious constraints for the survival and development of enterprises, as reported by enterprise leaders in the latest GSO survey in 2012 (Nguyen, 2012). Several reasons can explain this.

Financial markets in rural areas remain **very concentrated**. Although co-operative banks and other private financial institutions have been established, so far they have not achieved significant coverage in rural areas and the major players remain VBARD and VBSP. Such concentration can explain several constraints limiting access to credit. First, Viet Nam performs poorly as regards access to commercial bank branches compared to its peers in Southeast Asia. It is also well above the average in the percentage of outstanding loans and deposits in GDP, which could signal lax contract enforcement and real estate speculation (Table 3.4). Second, the average loan size remains small at around USD 200 for formal and semi-formal lenders and at USD 110 for informal lenders. Third, lending is overwhelmingly extended to groups rather than individuals due to the lower transaction costs associated with group lending (Quach et al., 2005). Finally, high interest rates constrain access to finance, as reported by 27% of enterprise leaders in the latest GSO survey in 2012. The survey that covered 9 331 enterprises showed that nearly one third of the respondents had to pay an annual interest rate higher than 19% (Nguyen, 2012).

Table 3.4. **Access to and use of financial services, 2012**

Indicator	Unit	Viet Nam	Indonesia	Thailand	Philippines	Cambodia
Commercial bank branches	Per 1 000 km^2	6.91	9.24	12.55	17.26	2.54
	Per 100 000 adults	3.18	9.59	11.57	8.13	4.38
Outstanding loans from commercial banks	% of GDP	111.88	32.85	74.08	20.94	38.34

Source: IMF (2013), *Financial Access Survey: Viet Nam, http://fas.imf.org/.*

The lack of sufficient **collateral** also limits access to credit by small-scale farmers. Only a small number of households have the collateral required by formal financial institutions which accept only legally registered assets as collateral – the primary asset being the LURC. Each household can have only one LURC, which provides eligibility for only one loan at a time. However, given the weak development of the land market, using LURCs as collateral might not provide banks with meaningful security. There are only few cases where land has been liquidated in the event of a farmer's collapse. Movable assets, such as televisions, bicycles, and animals, often do not qualify as collateral, except in theory for VBARD. When they do qualify, administrative procedures require that local Peoples' Committees certify the list of assets and their total value (Quach et al., 2005; Pham, 2013). Recently, the government has made efforts to promote value chain financing in order to improve access to credit despite the lack of collateral by adopting Resolution 14 (MARD, 2015).

Despite several policies aiming to address such constraints (Box 3.8), in practice, smallholder farmers cannot access subsidised credit nor credit guarantees due to high transaction costs and a lack of collateral.

Box 3.8. **Key policies for improving access to credit**

The following policies intend to improve access to credit by agricultural investors:

- Some loans can be accessed **without collateral**: up to USD 2 600 for individuals and households engaged in agricultural production; up to USD 10 460 for households carrying out business or production activities, or providing services to agriculture in rural areas; and up to USD 26 140 for co-operatives and farm owners.[1] Furthermore, SMEs can use assets purchased thanks to the loan extended by Agribank as collateral for this very same loan.
- Farm households, individuals, co-operatives and enterprises enjoy subsidised credit when purchasing **assets for agricultural production** (fertilisers, pesticides, machinery,

Box 3.8. **Key policies for improving access to credit** *(cont.)*

or equipment) with a loan that can cover the entire value of goods, but that cannot exceed USD 393 per ha, at an annual interest rate of 4%. A local content requirement rate of at least 40% of the purchasing assets is required to benefit from the subsidy.[2] Such requirement can hinder access to the subsidy as much of the equipment (e.g. dryers and combine harvesters) produced domestically is quite expensive and thus not affordable for SMEs.

- Enterprises, co-operatives, households and individuals investing in or manufacturing machines equipment and facilities aiming to reduce **agricultural losses** can benefit from subsidised interest rates for long and medium-term loans covering up to 70% of the investment value and during up to 12 years. These loans can cover the full value of the cost of such goods and interest rates can be fully subsidised in the first two years.[3]
- **Export credit** is provided for tea, pepper, cashew nuts, processed vegetables, sugar, meat, poultry, and coffee. The maximum lending rate is 85% of the export contract value. The loan term should be of 12 years, and interest rates aligned with VBARD rates.[4]
- VBARD provides credit to farmers living in **mountainous areas** or on islands or belonging to the Khmer minorities at an interest rate that is 30% lower than the market rate.[5]
- Enterprises producing **coffee** and operating in provinces with favourable natural conditions may borrow an amount equal to the total value of their output at 0% interest rate during the first two years. In the third year, interest rates are only half of market rates provided that enterprises buy domestically produced machines and equipment. Moreover, if coffee prices are much lower than production costs, farmers can access loans at 0% interest rates to purchase materials and fertilisers.
- Domestic firms engaged in the rice, coffee, fruit and vegetables industries can benefit from **credit guarantees** for 23 commodities.[6] Since 2011, the Viet Nam Development Bank (VDB) has been responsible for providing credit guarantees to SMEs that lend from commercial banks to facilitate the development of certain sectors, including agro-forestry. In order to qualify for a loan guarantee, a SME must have an effective investment project approved by the VDB and at least 15% of its equity invested into such projects. VDB may guarantee either a portion of or the entire loan but the guarantee can account for a maximum of 85% of the project capital.

1. Decree No. 41/2010/ND-CP on credit for agriculture and rural development.
2. Decisions No. 497/2009/QD-TTg and No. 2213/2009/QD-TTg and Circulars No. 09/2009/TT-NHNN and No. 02/2010/TT-NHNN.
3. Decision No. 68/2013/QD-TTg.
4. Decrees No. 133/2013/ND-CP, No. 54/2013/ND-CP, No. 75/2011/ND-CP, as well as Circulars No. 77/2013/TT-BTC and No. 52/2008/QD-BTC.
5. Decision No. 189/1999/QD-NHNN1.
6. Decision No. 2011/QD-TTg.

Source: Nguyen (2012); MOJ (2010); Viet Nam Briefing (2010); *Business Times* (2011); IPSARD (2014); MARD (2014a). See also Section 2.3.

A **credit bureau** can reduce the dependence on conventional collateral as the reputation of borrowers can replace collateral. It can also lower operational costs and speed up the time required to obtain loan approval. By law, credit organisations are responsible for reporting high value credit loans to the public registry, the Credit Information Centre (CIC). CIC collects information from credit institutions operating in accordance with the Law on Credit Institutions as well as from institutions that join voluntarily. CIC data covers over 24 million borrowers, including over 400 000 corporate borrowers, the remaining borrowers comprising individuals and credit card holders. In 2012, CIC provided

2 million credit information reports, an increase of over 18% compared to 2010 (SBV, 2012). Furthermore, the International Finance Corporation (IFC) has been supporting Viet Nam in establishing a credit reporting system over the last ten years (Box 3.9).

Box 3.9. **Activities of the International Finance Corporation in Viet Nam**

IFC has supported Viet Nam in developing a credit reporting system since 2005. In the first phase of the project up to 2010, IFC supported the development of an enabling regulatory framework for establishing a private credit bureau. It also supported the banking community to set up the first private credit bureau. Decree No. 10 that regulates credit information activities was enacted in 2010. The first private credit bureau (PCB) obtained its operations license from SBV in 2013. As of September 2014, it comprised a total of 25 members, including 11 bank shareholders as well as other commercial banks and a financial institution. CRIF, a global company specialised in the development and management of credit reporting systems, is its international technical and strategic shareholder. Data has been collected from about 15 members and only those 15 members have received pilot services from PCB. The second phase of the project has been launched mid-2014 and focuses on: expanding the coverage of credit information to help Credit Reporting Service Providers reach out microfinance institutions, rural and SMEs lenders, retailers and other public sources; strengthening the supervisory capacity of credit reporting service providers; supporting public education on credit reporting; and improving financial consumer protection.

The establishment of PCB is particularly important to improve access to credit by SMEs. As reported by a World Bank survey covering over 5 000 businesses worldwide and conducted in 2003, establishing a credit bureau resulted in a reduction of the percentage of small firms reporting financial difficulties from 49% to 27%, and in an increase of the probability of SMEs accessing credit from 28% to 40%.

IFC has also been working with the Ministry of Justice and the National Registration Agency for Secured Transactions (NRAST) to secure transactions since 2005. It helped improve the legal framework and develop a web-based registration system for secured transactions. Improved regulations on secured transactions have been enacted since 2010. They facilitate the use of broader types of movable collateral and on-line registration of security interests over movable collateral. The online registration system for secured transactions was launched in March 2012. It allows creditors, borrowers, and the public to search, register, amend and terminate security interests over movable collateral online. The system has recorded 289 163 initial registrations since it was launched on 20 June 2014. IFC continues supporting NRAST to further improve the enabling regulatory for secured transactions that should comply with international best practices.

Source: Interview with IFC in September 2014.

Although they have been actively encouraged by the government, agricultural **insurance markets** are virtually non-existent as the agricultural sector is a risky sector with unaffordable premiums for smallholder farmers and the capital market remains underdeveloped. Approximately 1% of farmers are insured against crop damage, 0.24% for cattle, 0.1% for swine and 0.04% for poultry. A plan piloted in 21 provinces establishes a progressive premium where, in the event of a disaster, poor farming households receive 100% reimbursement, nearly poor farming households receive 80%, other farming households 60%, and agricultural organisations 20% (Section 2.3).[31]

3.7. Infrastructure development

Well-developed rural infrastructure, including good irrigation networks and transportation and storage systems as well as a reliable access to energy and information and communication technologies, can effectively attract private investors in the agricultural sector and increase competitiveness.

Most indicators have pointed to continued advances in terms of basic rural infrastructure over the past decade, with now over 90% of the rural population having access to electricity and over 98.5% having access to roads (MARD, 2012). However, Viet Nam's recent rapid economic growth has resulted in serious **infrastructure bottlenecks**. New infrastructure is generally located in urban areas to connect major cities, airports, sea ports, industrial parks, tourist resorts, or areas related to security and defence, while rural infrastructure is usually in poor conditions and not properly maintained (MARD, 2014a). WEF's 2013-14 Global Competitiveness Report rated Viet Nam 110 out of 148 countries for the availability and quality of infrastructure behind most of its ASEAN peers. It ranks relatively well for mobile phone subscriptions but relatively less well for roads, railroads and electricity access.

Policy context

In 2008, annual public **investment in infrastructure** accounted for 9-10% of GDP (WB, 2008). The World Bank and the Asian Development Bank advised that such investment should be increased to 11-12% of GDP to maintain the current growth rate. Currently, total investment in infrastructure development is funded by the state budget (45%), international sources (40%) and the private sector (15%) (Lovells, 2009). MPI has estimated that infrastructure development requires approximately USD 500 billion in the next ten years. The state budget, ODA and other public financial sources can provide only USD 200 billion, the rest should be funded by the private sector (Tuoi Tre News, 2014).

The national target **programme of new rural development** approved in 2010 aims, among other objectives, to improve socio-economic infrastructure.[32] While, after three years of implementation, it has not achieved expected results due to limited staff capacity and budget, it has been relatively successful at supporting infrastructure development in rural areas by requiring farmers to contribute in labour with the government contributing in-kind with equipment and materials (WB, 2014d). Approximately 38 000 kilometres of roads and 15 000 kilometres of canals have been upgraded or built.

Despite this plan, infrastructure development remains constrained by a **weak management of public funds**, including: an inefficient allocation of public resources; project implementation by decentralised local governments and SOEs which delays implementation; lack of links between infrastructure investment and national strategic priorities; decentralisation that led to competition between localities which may provide incentives for improvement and disseminating good practices but also hindered holistic development and resulted in fragmented, suboptimal infrastructure projects with low utilisation rates (WB, 2011a).

While several measures have been taken to increase private sector participation in infrastructure development, **private investment** in improving rural infrastructure remains low. As highlighted in the previous section on investment policy, the government is actively supporting PPPs, including for infrastructure development. A PPP decree has been approved in February 2015. The MPI supports provinces in implementing PPPs in areas such as roads and inland ports. A PPP office and an inter-ministerial steering committee were recently created. A Viability Gap Fund is slated to be implemented by the end of 2015 to invest USD 1 billion as a state contribution to PPP projects (USAID, 2014).

Enterprises supporting rural infrastructure can be granted **investment incentives**. They benefit from a reduction of 20% of the land and water rent in the first five years if they contribute to improving infrastructure for agricultural production[33] (IPSARD, 2014). They can also benefit from a CIT rate of 20% if they invest in infrastructure either for one of the encouraged investment sectors, including agriculture, or in a socio-economic under-developed area, and of 10% if they satisfy both conditions or invest in "especially encouraged investments or areas".

Private investors continue to face **several challenges** when investing in infrastructure. Commercial banks have been lending to local governments for infrastructure development, but at a limited scale, as they do not have any recourse mechanism for lending to local governments and they face a maturity mismatch between short-term deposits and the long-term financing needed for infrastructure development. Weak assessments of demand by local governments, high construction costs relative to peer countries, the persistent participation of SOEs and weak contract enforcement, also hinder private investment in infrastructure (WB, 2014c). For instance, Hoa Binh agro-processing company upgraded irrigation networks and secondary roads in 2006. However, farmers broke agreements and did not supply maize to the company. The lack of efficient court enforcement means that such agreements can fail unless investors pay an additional cost for a private enforcement mechanism, such as setting up check-up points in producing areas to prevent farmers from selling products to traders (Nguyen, 2012).

Transport

Viet Nam's **road system** consists of a network of over 250 000 km. The longest and most important route is the Hanoi – Ho Chi Minh City line, which stretches for about 1 730 km. The government has recently mobilised significant capital to improve the highway system with the financial support from international lending agencies. It aims to reduce travel time along this line from 30 to 10 hours with a new high-speed connection. The rail network has over 2 600 km of single-track line. The rail lines connecting Viet Nam to China were re-opened a few years ago while new rail lines connecting Viet Nam with Lao PDR and Cambodia should be developed. Waterways are a particularly important mode of transport, covering around 17 000 km. Viet Nam has also 11 major seaports and over 100 smaller seaports. Ho Chi Minh City serves most of the south and boasts modern container loading facilities, while Hai Phong serves most of the north (PWC, 2008).

The **planning process** of the Ministry of Transport is divided between different sectoral departments, and projects that may require interdisciplinary planning are split into sub-projects assigned to different agencies. This inefficient planning process results in fragmented transport networks. The significant influence of local governments on the issuance of new port development licenses elevates the risk of fostering demand-supply mismatches, contributes to a highly fragmented port system and leads to overcapacity and severe price competition among marine terminal operators that undermine the financial sustainability of facilities.

Inadequate transport infrastructure has led to **several inefficiencies**, including a high reliance on trucking as a mode of transport compared with other cheaper options such as rail for long distance shipments, congestion related to inadequate highway infrastructure, an old national truck fleet that increases maintenance expenses, and frequent truck breakdowns that contribute to unpredictable transit times (WB, 2014c). Infrastructure is inadequate to respond to weather risks. In the rainy season, many secondary roads are inaccessible, and agricultural products are stuck in producing areas while enterprises face serious shortages of raw materials (Malesky, 2011). As per the WEF's infrastructure quality rankings of 2013, Viet Nam ranks lowest in terms of road quality when compared to major regional economies and one of the lowest in ports and air transport quality.

Storage

In the Mekong Delta, **rice storage** constitutes a major constraint. In the past when there was generally a single rice crop, farmers regularly stored and dried paddy in their homes after harvest. Now, with two or three crops per year, they do not have sufficient capacity for in-home storage or drying. Furthermore, co-operatives or private mills rarely provide high-quality storage capacity. Semi-dried (low season harvest) or wet paddy (high season harvest) is transported by small barges of about 50 tonnes per day by local traders. Traders bring wet paddy to drying service providers and then sell the paddy to millers who husk it. Before milling, paddy tends to be stored outside or in the barges, usually under some kind of shading or roofing. It is estimated that between the farmers' field and the first stage of processing, approximately one million tonne of paddy is damaged or physically lost every year (ADB, 2013b).

Several **policies and projects** intend to improve storage infrastructure, particularly for rice. Rice export enterprises need to have more than 5 000 tonnes of specialised storage and at least one milling factory with a capacity of over ten tonnes per hour to benefit from government storage projects.[34] In 2009, the government planned the construction of a storage system of about 2.5 million tonnes of rice in the Mekong River Delta. However, project implementation has been slow and only 40% of the project has been completed (Tran, 2014b).

Electricity

Rural electrification has been a **remarkable achievement**. According to Viet Nam Electricity Group, 99.1% of communes and 97.6% of rural households had electricity by the end of 2013, thus exceeding the goals stated by Decision No. 800/2010/QD-TTg governing the programme 'New Rural Communes' for the period 2010-20 (MARD, 2014a). In terms of overall electrification rate, Viet Nam ranks only behind Thailand (99.3%) and ahead of the Philippines (89.7%) and Indonesia (64.5%) (IEA, 2011). However, it ranks at the bottom among ASEAN countries when it comes to the number of procedures, the time, and the cost required by an enterprise to access electricity when compared to its regional peers (Table 3.5).

Table 3.5. **Access to electricity, 2014**

	Rank	Procedures[1]	Time[2]	Cost[3]
Indonesia	121	6	101	370.6
Malaysia	21	5	32	49.1
Philippines	33	5	42	118.2
Thailand	12	4	35	67.3
Viet Nam	156	6	115	1 726.4

1. Procedures refer to interactions between company employees and steps related to the internal electrical wiring.
2. Time is recorded in calendar days and captures the median duration that the electricity utility and experts indicate is necessary in practice, rather than required by law, to complete a procedure with minimum follow-up and no extra payments.
3. Cost is recorded as a percentage of the economy's income per capita. All the fees and costs associated with completing the procedures to connect to electricity are recorded.

Source: WB (2014), *Doing Business Report in Viet Nam*, a co-publication of the World Bank and the International Finance Corporation.

Electricity generation is dominated by hydropower (39%) and gas-fired plants (38%) with the rest supplied by coal-fired (14%), oil-fired (5%), and diesel generators (6%) (WB, 2013b). The government plans to: rehabilitate electricity networks in about 3 000 communes; determine how to achieve the target of electrifying all households; ensure sustainability of rural electricity networks; and ensure that electricity is affordable to the poor[35] (WB, 2011b).

Information and communication technologies

In 2013, Viet Nam ranked 81 among 152 countries in the **ICT Development Index** of the International Telecommunication Union (ITU), ahead of Thailand (92) and Indonesia (95), but behind Malaysia (58) and China (78) and the world leader Korea (1). Access to ICTs is relatively good compared to other ASEAN economies in terms of fixed, mobile and internet penetration rates (Table 3.6). The high level of adoption of mobile phones can facilitate the use of mobile banking and payment systems, drawing from the successful experience of M-Pesa in Kenya for instance. Foreign enterprises have already showed interest in investing in mobile banking, such as the Singaporean company Tagit that partners with the domestic enterprise Smartlink.[36]

Table 3.6. **Telecommunications subscribers/users per 100 inhabitants, 2013**

	Mobile	Fixed	Internet
Cambodia	134	3	6
Indonesia	122	16	16
Malaysia	145	15	67
Philippines	104	3	37
Thailand	138	9	29
Viet Nam	131	10	44

Note: Figures are based on ranking of 152 countries.
Source: ITU (2013), Statistics on ICTs, International Telecommunication Union, *www.itu.int/en/ITU-D/Statistics/Pages/stat/default.aspx*.

The government has supported ICT development and made **major investments** in ICT network modernisation and capacity upgrading. It started to license private companies in 1995, breaking the monopole of the Viet Nam Post and Telecommunications Corporation (VNPT). The pricing of telecommunication services has been liberalised, resulting in a more competitive ICT sector (OECD, 2013c; WB, 2014a). Viet Nam aims to complete the coverage of broadband connections in all communes and wards and expand broadband mobile coverage to 85% of the population by 2015 and to 95% of the population by 2020. By 2020, 50-60% of households would have computers and internet connections (VBF, 2010).

3.8. Trade policy

Open, transparent and predictable agricultural trade policies both domestically and across borders can improve the efficiency of resource allocation, thus facilitating scale economies, reducing transaction costs and boosting productivity and rates of return on investment. They should be implemented in a consistent manner.

As detailed in Chapter 2, significant steps have been taken to liberalise agricultural trade since 1986. The integration of the domestic market into the global economy culminated with WTO membership in 2007. According to **OECD trade facilitation indicators**,

Viet Nam performs better than the average of Asian and lower middle income countries in the areas of involvement of trade community, appeal procedures, and governance and impartiality (Figure 3.5). Its performance for advance rulings, fees and charges, automation, streamlining of procedures and internal border agency co-operation is below the averages of Asian and lower middle income countries (OECD, 2014b). Cumbersome and complicated procedures can indeed hinder trade. Inconsistent interpretation, implementation, and enforcement of government regulations across provinces leads to import and export clearance processing that are longer and less predictable than in peer countries (WB, 2014b).

Figure 3.5. **Trade facilitation performance: OECD indicators, 2014**

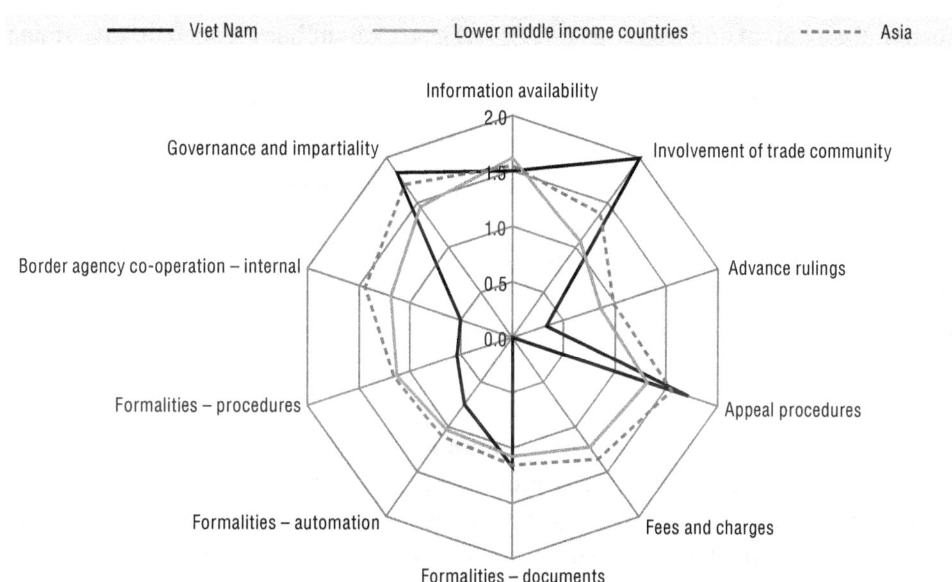

Note: Best performance = 2. Analysis is based on the latest available data as of January 2013 and on the set of indicators constructed in the OECD *Policy Paper* No. 144 of 2013 for 107 countries outside the OECD area.
Source: OECD (2014b), *OECD Trade Facilitation Indicators – Viet Nam.*

StatLink http://dx.doi.org/10.1787/888933223981

Viet Nam could draw considerable benefits in terms of trade volumes and trade costs from significant improvements in the areas of advance rulings, automation, and streamlining of procedures. Continued efforts in the areas of information availability and appeal procedures would bring further benefits (OECD, 2014b).

3.9. Human resources, research and innovation

Strong human capital and dynamic agricultural innovation systems are critical to increase investment in agriculture. Policies should support high-quality education and well-functioning extension and advisory services to enhance human capital. They should promote partnerships between national and international research and connect research with demand to build effective innovation systems.

Viet Nam has obtained **impressive results in the education sector** compared to countries with similar economic development: over 90% of the working-age population is literate and more than 98% of children of primary school age attend schools. However, labour productivity remains low, amounting to 23.3% of Malaysia's and 37% of Thailand's in

2010 (OECD, 2013c). Viet Nam ranks eleventh out of twelve East Asian countries in terms of human resource capacity with a score of 3.79 out of 10 (OECD/WB, 2014).

The results of the STEP Employers Survey conducted in 2012 by Viet Nam's Central Institute for Economic Management (CIEM) with the support of the World Bank point to **mismatches between supply and demand of skills**. Around 47% of firms claimed that the education system failed to meet the skill needs of the workplace. Of international firms, 66% stated there was a mismatch, while 36% of local firms made the same claim. Indeed, the rapid change in demand for skilled labour is not being met by the market supply. The evidence of bottlenecks is confirmed by the fact that firms often spend up to two months recruiting for vacant positions (OECD/WB, 2014).

In order to rectify this mismatch, Viet Nam needs to improve schooling and technical and vocational education and training (TVET). The government has stepped up **investment in public education** and training institutions. It also offers generous incentives such as low cost land, credit subsidies, and tax holidays to encourage domestic and foreign private investment in TVET, while subsidising firms that provide in-house training (OECD, 2013c).

The government would also need to strengthen **extension services** to better address the needs of farmers. Extension services face several challenges, including limited human resources (one public extension worker per 330 farming households and expenditures on extension services amounting to about USD 3.30 per farming household), weak experience and limited services for SME development, dominance of a top-down approach, a lack of services tailored to different types of farms, weak participation of the private sector, and poor monitoring system (Section 2.3 for further detail).

Dynamic **agricultural research and development** has played a major role in driving the impressive increases in agricultural production since the mid-1980s by supporting plant breeding programmes, introducing new breeds and crops and improving pest and disease management. In 2014, 15 research institutes and four universities operated under MARD (Section 2.3).

Viet Nam's research and innovation capacity is **limited by various factors**, including: the relatively small proportion of university lecturers and researchers qualified at PhD level (around 16%); the bureaucratically fragmented and cumbersome mechanisms used to allocate research funds; the fragmented provision of research services (1 600 science and technology research institutes and centres in 2011); a shortage of world-class scientists; a lack of co-operation between leading scientists in research institutes and universities; and continued separation between research and teaching (OECD, 2013a).

In the agricultural sector, the rigid application of policies and technologies without considering local conditions leads to a waste of financial and human resources. Policies to facilitate the autonomy of R&D institutes remain weak. R&D programmes tend to focus on agricultural production rather than on a broader approach including marketing and profitability elements and focusing on productivity growth. The pace of productivity gain for important crops has slowed, while the incidence of disease for livestock and aquaculture seems to have overshadowed any advances in technology to result in more volatile productivity swings. The agricultural innovation system is perceived to involve an ageing cadre of researchers, not responsive to farmers' demand and collaborating weakly with other institutions and the private sector (MARD, 2012).

Over the last ten years, **significant efforts** have been made to address these challenges, particularly to increase the autonomy of R&D institutes and better link research with demand. Decree No. 115/2005 ND-CP aims to build a more competitive and market oriented research system and provides the framework for a blend of basic, applied and near market research to be implemented through more autonomous, competitive and commercially-oriented research institutions. The technical assistance offered by the Asian Development Bank also helped strengthen the agricultural science and technology system. In May 2008, MARD approved Decision No. 1613/QD-BNNKN to strengthen grassroots extension services. In June 2008, it approved Decision No. 1874/QD-BNN-KHCN to make agriculture research programmes more client-oriented. As of October 2007, 40 Provincial Extension Advisory Councils (PEACs) had been established to improve the co-ordination of agricultural research and extension services at local level. These PEACs allowed formalising the consultation process for extension planning and monitoring and evaluation. About half of the seats in the PEACs are occupied by representatives from the public sector (ADB, 2008).

However, Viet Nam lags behind other countries in the region in terms of **research funding**. In 2012, investment in science and technology development reached USD 650 million, or 0.27% of GDP. Though this proportion was higher than for Indonesia (0.05% of GDP) or the Philippines (0.12% of GDP), it was lower than for Thailand (0.3% of GDP), Malaysia (0.5% of GDP) or Singapore (2.2% of GDP). Private expenditure on science and technology is much more limited, representing around 30% of public expenditure (OECD, 2013a). While government expenditure on agricultural research increased from USD 10 million in 2000 to USD 40 million in 2012, funding as a percentage of GDP remains relatively low at around 0.03% (Section 2.3). As a result, most research is carried out by the state research agencies with limited funding and cannot meet the practical requirements of farmers and enterprises.

According to the **Intellectual Property Rights** Index (IPRI) of the World Economic Forum 2014,[37] Viet Nam is ranked 66 out of 97 countries. In the Asia-Pacific region, it falls behind Malaysia, Thailand and Indonesia. Viet Nam could strive to better protect patents and plant varieties to improve the performance of its agricultural innovation system. Indeed, through adequate IPR protection, rights-holders can exclude competitors from use of an innovation for a limited period of time, thereby encouraging private investment in innovation. The strengthening of IPR protection in recent decades worldwide has been associated with an increase in private sector investment in agriculture-related research and development and a surge in innovation leading to improved plant varieties, agricultural chemicals, and production technologies. At the same time, concerns have emerged with respect to some aspects of the present approaches to IPR protection in agriculture, particularly with respect to patents and breeder's rights (OECD, 2012).

3.10. Responsible business conduct

Large-scale enterprises operating in the agricultural sector can bring the necessary expertise, financing capacities and marketing networks to enhance the competitiveness of agricultural production and supply chains. They can foster employment creation, including through backward and forward linkages and multiplier effects. However, their operations can also have adverse human rights, social and environmental impacts. Policies, laws, and regulations must be well-designed and effectively implemented to ensure that enterprises behave responsibly and bring both economic and social benefits at the national and local levels, while guaranteeing a sustainable use of natural resources.

While States have to comply with their international obligations, enterprises are also expected to observe certain standards derived from international obligations. For instance, the OECD Guidelines for Multinational Enterprises provide for voluntary principles and standards of responsible business conduct (RBC) for enterprises consistent with applicable laws and internationally recognised standards.[38]

This section examines existing laws and regulations related to two major RBC issues in the agricultural sector, namely environmental protection and transparency. The observance of existing land tenure rights is also a critical element of RBC but is addressed in the section above on land tenure that describes policies related to public consultations, compensation and resettlement.

Environmental protection

Policy makers face the challenge of creating the right incentives to protect the environment and optimise resource use from an economic, environmental and social perspective. **Environmental policy** should promote sustainable farming practices, such as agro-forestry and agro-ecology, as well as resource use efficiency to increase production relative to inputs used, and ensure that prices reflect the scarcity value of natural resources and the cost of environmental impacts. Sustainable resource management can allow investors to maximise returns on their investment by harnessing long-term economic benefits. While the legislation on environmental protection is quite advanced in Viet Nam, it is not well enforced which increases the risks related to environmental degradation and climate change.

Pollution

Viet Nam's agricultural growth has been primarily based on the **extensive use of natural resources** and the intensive use of fertilisers and pesticides that led to toxic residues in agricultural products and increased resistance of diseases (Section 1.7). Fertiliser use per hectare is currently almost twice as high as the average in Southeast Asia. Rural industries – both craft and larger scale – have an uneven record with regard to environmental management, with growing concerns about their contribution to groundwater and surface pollution (MARD, 2012; ADB, 2013a; Tran, 2014a). These developments have adverse impacts not only on the health of rural communities and on the livelihoods of farmers whose land and/or water has been polluted by industrial effluents, but also threaten access to international markets due to the environmental concerns of consumers and regulators (MARD, 2012).

Viet Nam has **state-of-the-art environmental legislation** but environmental damage continues in many areas, often unchecked, due to weak monitoring, compliance and enforcement. The current legislation[39] sets environmental standards, including on pollutants of surface and groundwater and on effluents flowing into water resources, and defines environmental protection requirements, dealing for instance with waste collection and treatment, water and air pollution and toxic substances. All projects that may have a significant environmental impact are required to undertake Environmental Impact Assessments (EIAs) covering potential impacts during project design and operations, as well as mitigation measures and contingency plans[40] (MARD, 2014a).

The legislation establishes the "polluter pays" principle and promotes environmentally friendly technologies. An environmental police responsible for auditing enterprises has been created. An environmental tax for projects exerting a "long-term adverse impact" and an environmental fee for projects having an "adverse impact" have been introduced (UNCTAD, 2008). However, the inefficiency of the tax administration constitutes a major challenge in collecting such taxes.

Penalties for breaching the environmental legislation can be applied.[41] Organisations and individuals may also need to: suspend their operations, relocate, restore the original state of the environment, compensate damages to the concerned parties and pay back the illicit profits gained from violating the law. However, law enforcement remains an issue. For instance, while the existing legislation states that a firm failing to pay fines and wastewater charges and/or to establish a treatment system should be temporarily closed, provincial Departments of Natural Resources and Environment have been unable to close such firms (MARD, 2014a).

Water management

Sustainable water use in agriculture represents a **major challenge**, particularly as agriculture represents an estimated 95% of freshwater withdrawals in Viet Nam, with approximately 45% of agricultural land being irrigated (USAID, 2013). While water demand will increase due to population growth and urbanisation, the availability of freshwater resources may be reduced due to climate change (Tran, 2014a). The largely unregulated use of pesticides, herbicides, and antibiotics has increased water pollution in the Mekong and Red River Deltas. As a result of excessive pumping of ground water in the Mekong Delta, the land is subsiding adding to the problem of saltwater intrusion (IPSARD, 2010b).

Legislation[42] provides for a strong licensing system for water use and wastewater discharge, encourages water conservation, and decentralises water management. Investors exploiting water resources or discharging wastewater into water sources should: consult representatives of local communities in affected geographical areas on their investment plans; attach these opinions to their projects' dossiers for submission to government agencies; and publicise information on their projects before proceeding implementation. Organisations and individuals should constantly protect water sources they are exploiting and supervise water pollution, deterioration or depletion committed by others. Those whose activities cause a decrease in water resources, land subsidence or water pollution or salinisation should remedy consequences and, if causing damage, pay compensation. Despite this strong legislation, monitoring water use and regulating water pollution is difficult. As highlighted in Chapter 2, the exemption of irrigation fee is a major concern as it may lead to unsustainable water use.

Climate change

Over recent years, much of the discussion about environment has focused on the **consequences of climate change**. Projections indicate that large sections of the two Deltas and coastal areas will be flooded and watershed and forest zones will experience more extreme weather. Under some scenarios, rising temperatures will extend the dry season and increase the frequency of droughts, result in rising sea levels in the Mekong Delta and the coastal Middle Region, and increase salt penetration as well as the risks related to pests and diseases, thereby forcing changes in patterns of grain cropping and plantation production (IPSARD, 2010b; ADB, 2013a).

Reducing greenhouse gas (GHG) emissions in the agricultural sector is particularly important not only to mitigate climate change but also because the sector is the fourth largest sectoral contributor to such emissions worldwide after energy, industry and forestry. MARD has approved a programme to reduce GHG emissions by 2020 which aims to: reduce GHG emissions in agriculture and rural areas by 20%; ensure that 3.2 million ha of rice apply advanced methods, such as the System of Rice Intensification and Alternative Wetting and Drying; and promote a more efficient use of agricultural inputs.[43] The legislation also supports: research on, selection and production of plant varieties and animal breeds able to minimise GHG emissions and adapt to climate change; minimum tillage and techniques for reducing the use of water and fertilisers to minimise methane gas emissions in rice fields; the reduction of plants contributing to GHG emissions; and the increase in bioenergy crops[44] (Tran et al., 2013).

Reducing GHG emissions also relies on sustainable forest management. The expansion of agricultural land resulted in considerable deforestation which is only partly remedied by reforestation efforts undertaken over the last decade (Section 1.7). While the overall forested area increased, primary forests continue to disappear. Clearing of forest land without crop rotation and the conversion of poor quality forest into agricultural land resulted in land degradation, especially in the Northern mountains and mangrove forests (MARD, 2012). The ban on large-scale logging, widespread public support (including incentives) for reforestation,[45] efforts to protect coastal marshlands and mangrove areas, and the expansion of parks and protected forests, have been impressive and valuable achievements (IPSARD, 2010b). Since 2012, forest management has been decentralised, financial support provided to communes and the competence and effectiveness of rangers improved.[46] REDD+ pilots have been expanded nationwide[47] (Tran, 2014a).

Transparency

The **anti-corruption** legal framework has improved significantly over the past few years with the adoption of the Anti-Corruption Law in 2005 and the National Strategy on Anti-Corruption to 2020. The Anti-Corruption Law revised in 2007 criminalises attempted corruption, passive and active bribery, extortion, bribing of foreign officials, abuse of office, and money laundering.

However, **Viet Nam scores relatively poorly** in international rankings. According to Transparency International, it scored below Malaysia, Thailand, Indonesia, and the Philippines in 2013. The Global Integrity Report of 2011 ranked Viet Nam as very weak on all categories (non-governmental organisations, public information and media; elections; government conflicts of interest safeguards and checks and balances; public administration and professionalism; government oversight and controls), except anti-corruption legal framework, judicial impartiality, and law enforcement professionalism, for which it was ranked as weak. With 55% of the respondents to the survey believing that corruption was increasing in Viet Nam, the country may lose out on foreign investment to Cambodia, Lao PDR and Myanmar instead of competing with Singapore, Malaysia and Thailand (VBF, 2014).

The overwhelming perception is that corruption is considerably more prevalent within the **public sector** than in the private sector. Respondents were asked to identify the three sectors that the government needs to prioritise in its anti-corruption efforts. Customs (55.2%), taxation (46.2%) and land administration (39.8%) were the areas chosen (VBF, 2014).

Indeed, corruption in **land management** is problematic, as stated in the National Anti-Corruption Strategy. The complexity of delivering LURCs can encourage corruption, with investors paying bribes to land officials in exchange of information privileges and expedited procedures. According to the World Bank Enterprises Survey of 2009,[48] one out of three enterprises answered that informal payments or gifts were expected or requested when applying for land use rights (WB, 2011c; U4, 2014). Wells-Dang (2013) argues that the process of converting farmland into land used for real estate and industrial development for private investment is the activity most susceptible to local-level corruption, including in the issuance of the LURCs, the acquisition of land, particularly its re-allocation or "conversion" to development interests, and in the compensation levels given to farmers. Land has been evaluated as the sector most vulnerable to local-level corruption, with 86% of respondents agreeing that corrupt practices were widespread in their provinces (Oxfam, 2013). The decision to give power to the provinces and districts was identified as one of the main causes of corruption, especially the power to decide on land attributions, rentals and changes of use (CNRS, 2010). Such bribery is facilitated by complicated processes and limited information about applicable fees (TI, 2011).

3.11. Summary

- Since the *Doi Moi* reforms began in 1986, the **capital** of private enterprises has risen sharply and now exceeds the capital of SOEs. Viet Nam has attracted foreign investors, although FDI in agriculture remains low in relation to the size of the sector and its role in international trade, representing 5% of total FDI in 2013 and concentrating mostly in agro-processing.
- **Investment policy** has evolved significantly over the last three decades by levelling the playing field between domestic and foreign investors and simplifying administrative procedures. A new Investment Law adopted in December 2014 reduces the number of sectors in which investment is conditional and streamlines investment procedures further. Efforts are being made to promote public-private partnerships in agriculture but are still at an early stage.
- Several major **challenges** remain to be addressed to improve the policy framework for private investment along agricultural supply chains. They include: administrative hurdles to set up a business and to pay taxes, limited access to finance, poor infrastructure, unskilled labour, and corruption. The gradual decentralisation of policy design and implementation has led to sometimes confusing and inconsistent policies which creates uncertainty for investors.
- The weak role played by **farmers' organisations** hinders the emergence of efficient and reliable supply chains. While the new Law of Co-operatives of 2012 provides further state support to co-operatives, their limited access to credit undermines their development. Traders and large investors have thus to interact with numerous small-scale producers which increase transaction costs and uncertainty in a context of weak contract enforcement while processing factories cannot secure a regular supply of raw materials.
- Despite efforts to move towards a more open and competitive market, **SOEs** continue to play a major role in the production, processing and trade of some major agricultural commodities. They benefit from an easier access to land, raw materials, finance, procurement contracts, and research and development compared to their peers in the private sector, which undermines private investment and impedes associated productivity and efficiency gains.
- The instability and generally low and inconsistent quality of raw materials undermines the development of the processing industry. Several efforts have been undertaken to improve **food safety**, including by passing a Law on Food Safety and designing national quality standards, but the implementation remains weak.

- A complex web of legislative instruments composed of general laws and numerous decrees determines **investment incentives**, some of which are region or product-specific. As the legislation remains quite vague and often unclear, incentives are mostly granted on a case-by-case basis at provincial level. Incentives may thus not be effective at encouraging investment as investors cannot know which incentives they will be granted prior to investing. No cost-benefit analysis has been undertaken to evaluate the opportunity cost and the impact of these incentives.
- By 2012, **land use rights** certificates (LURCs) had been issued to farmers to cover 85% of agricultural land. Land use rights duration and the land area are limited and land use changes and transfers are strictly regulated and costly, which hinders land consolidation in a context where each farm holds on average less than 0.5 ha. Acquiring agricultural land can be a time-consuming and complex process which facilitates corruption. Due to weak capacity, land use plans are not developed or, when they exist, they have not been developed in a participatory manner. Expropriations associated with inadequate compensation for lost access to land and corruption has led to violent land conflicts.
- Limited access to long-term financing for large investors and the inability to **access credit** for small-scale producers constrain investment. The financial sector in rural areas is concentrated, with VBARD and VBSP being the major banks providing formal credit in the agricultural sector, including subsidised credit. However, costly and complex procedures, high borrowing costs, the lack of financial education, and above all, the lack of collateral, often prevent small-scale farmers from accessing loans. Indeed, 49% of rural households were unable to access banking services in 2010 and the informal sector remains the most important source of rural credit.
- **Rural infrastructure** has significantly improved over the past decade, with 98% of the rural population having access to electricity and a high penetration rate of mobile phones. Still, serious infrastructure bottlenecks hinder private investment in rural areas. New infrastructure is generally located in urban areas while rural infrastructure is usually in poor conditions and poorly maintained. The decentralisation coupled with an inefficient planning process led to fragmented infrastructure networks.
- As underlined by OECD **trade facilitation** indicators, cumbersome and complicated procedures hinder trade and thus investment. The inconsistent enforcement of regulations across provinces leads to export clearance processes that are longer and less predictable than in peer countries.
- Viet Nam has obtained impressive results in the **education** sector compared to countries with similar economic development, with over 90% of the working-age population being literate. However, around 66% of international firms state that there is a supply-demand mismatch of skills. Extension services face several challenges, including limited human resources, weak experience, dominance of top-down approach, a lack of services tailored to different types of farms, and low participation of the private sector.
- While dynamic agricultural **research and development** has played a major role in driving the impressive increases in agricultural production since the mid-1980s, the pace of productivity gain for important crops has slowed. This may be partly due to the focus on agricultural production rather than on productivity or marketing, the small proportion of researchers qualified at PhD level and the ageing cadre of researchers, the cumbersome mechanisms used to allocate research funds, and the continued separation between research and teaching.

- Agricultural growth has been primarily based on the extensive use of natural resources and the intensive use of inputs, which led to **natural resource** depletion, biodiversity deterioration and toxic residues in agricultural products. Recent efforts to enhance environmental protection, promote sustainable water use and forest management, reduce GHG emissions, and respond to climate change, should be sustained and strengthened, particularly by improving the implementation and enforcement of the existing legislation.
- Bribery and the **lack of transparency** constitute significant impediments to investment and have been identified as a major concern by foreign investors. Although the anti-corruption legal framework has improved significantly over the past few years, corruption remains a significant impediment to investment, particularly as regards customs, taxation and land administration.
- Addressing the above challenges is particularly important to **tap into the potential** offered by the domestic and international markets for Vietnamese agro-food products. Viet Nam's middle class has doubled over the last five years and will continue to expand quickly over the coming decades. Its rising disposable income leads to a growing demand for goods and services, including for better quality and premium products.

Notes

1. Greenfield investment is defined as investment directed at building or creating new infrastructure and not renovating existing infrastructure, the latter being referred to as "brownfield" investment.
2. 51% of charter capital or more is held by foreign investors, or the majority of the general partners are foreigners if the business organisation is a partnership.
3. However, investment remains conditional in a wide range of agricultural activities, including: trading and testing of pesticides and organic fertilisers; slaughtering, quarantine, processing, preservation of animals and animal products; trading of animal breeds and products and of plant varieties; trading of genetically modified food; breed production services; veterinary services; and trading of veterinary medicines, biological preparations, vaccines, and chemicals serving veterinary medicine.
4. Those include: investment in listed companies, public companies, securities-trading organisations, or securities investment funds that are subject to the law on securities; investment listed in a specific law that limits foreign ownership; investment in SOEs that conduct equitisation or convert the ownership into another form that follow the law on equitisation and conversion of SOEs.
5. Decree No. 108/2006/ND-CP provides that ministries, Provincial People's Committees (PPCs) and the management boards of industrial parks and economic zones are responsible for providing guidelines on investment in the localities under their authority. Directive No. 15/2007/CT-TTg of the Prime Minister requires all relevant ministries to decentralise the management of foreign investments, especially as regards investment approval and certificates, and to enhance co-ordination between central and local agencies and between relevant ministries and branches (IPSARD, 2014).
6. Each commune would need about USD 600 000 to meet the 19 criteria and 39 indicators set by the managing board of the programme.
7. These sectors include: biotechnology; technology and facilities required for rice and specialty food processing, environmental protection, storage, hygiene and food safety, and for stabilising agricultural markets; production of coffee Arabica and processing of cocoa, coffee and tea, development of competitive brands; rubber processing; seafood processing (such as smoked fish, fish oil or fish feed); fruit trees, focusing on intensive farming, nurseries, standards and quality seeds; plant protection; veterinary products for seafood; consulting services on cultivation techniques, animal husbandry, aquaculture, animal health, post-harvest technology or plant protection in the form of 100% foreign-owned or joint ventures associated with local research institutions.
8. OECD defines PPPs for the delivery of public services as "long term agreements between the government and a private partner whereby the latter delivers and funds public services using a capital asset and sharing the associated risks".

9. Decree No. 38/2013/ND-CP.
10. This Decree revises Decision No. 71/2010/QD-TTg on PPPs and Decree No. 108/2009/ND-CP.
11. The Coffee Co-ordination Board includes: co-ordinating coffee production, processing and trade programmes; recommending sector policy and strategy and supporting their implementation; providing and sharing information and research.
12. The website is accessible at *www.pppoffice-mard.org.vn*.
13. As per Decision No. 62/2013-QD-TTg, this support includes: land use fee exemption for land rented from the state to build processing, warehouses, housing for workers or public houses for large fields; funding of the activities around export contracts or storage programmes; partial funding for activities of soil improvement, transport, irrigation, and power systems serving agricultural production in large field projects; up to 50% of the cost of training and technical guidance to farmers, including the cost of materials, food, housing, transportation.
14. As per Decision No. 62/2013-QD-TTg, this support includes: same incentives as the first two incentives mentioned above for enterprises; up to 30% in the first year and 20% in the second year of the cost of plant protection activities (chemicals, labour, machine hire); up to 50% of the cost of training courses to the staff of co-operatives and co-operative unions on contracts and production techniques; up to 100% of the costs of training and technical guidance to farmers to produce agricultural products under contract.
15. As per Decision No. 62/2013-QD-TTg, this support includes: support for training and technical guidelines on the production and marketing information in large field projects; up to 30% of the cost of quality plant varieties used as inputs for the first purchase in large field projects; up to 100% of the storage cost during 3 months.
16. Prime Minister's Decision 23/QD-TTg dated 6 January 2010.
17. Circular No. 8/2013/ TT-BCT of the Ministry of Industry and Trade.
18. For instance, Good Agriculture Practices (VietGAPs) were developed in 2008, based on Hazard Analysis Critical Control Point principles. They specify voluntary technical requirements for agricultural products to guarantee their quality and safety. Producers should register their practices and assess their production and post-harvest activities according to several sanitary criteria. Then an external auditor evaluates producers' internal controls – this can cost up to USD 1 300 per ha which deters small producers. The government aims to have half of the land area of vegetables and tea certified by VietGAPs by 2015. While various types of support are provided to implement VietGaps (Decision No. 01/2012/QD-TTg on policies supporting the application of VietGAPs in agriculture, forestry and fisheries), implementation costs and the lack of staff have constrained its coverage. While VietGAPs are not recognised on international markets, they are based on GlobalGAP practices. Viet Nam has also participated in the development of ASEAN GAP standards which help producers upgrade to GlobalGAPs. The upgrading from VietGAPs to GlobalGAPs has already occurred to some extent in the horticulture sector in the South through group certification.
19. As per the new Investment Law, if the investment does not require a certificate of investment registration, the investor should determine the investment incentives that can be granted and follow procedures for investment incentives at the respective tax, finance or customs authorities.
20. Decree No. 210/2013/ND-CP on incentives for enterprises investing in agriculture and rural areas, which replaced Decree No. 61/2010/ND-CP. Various Ministries, including MARD, have issued circulars to implement this Decree. These circulars have not been implemented yet as all relevant ministries should issue circulars before any of them starts being implemented.
21. Decree No. 210/2013/ND-CP on incentives for enterprises investing in agriculture and rural areas.
22. Decree No. 20/2011/ND-CP, which guides the implementation of Resolution No. 55/2010/QH12 on the exemption and reduction of agricultural land use tax in 2011-20.
23. Law on Corporate Income Tax of 2003.
24. Directive No. 15/2007/CT-TTg of the Prime Minister dated 22 June 2007.
25. Resolution 25/ND-CP of 2010 of Government Cabinet dated 19 March 2010.
26. Restrictions apply to these specific conversions as converting land used for perennial crops, forests, aquaculture or salt production back to rice land is long and costly whereas converting land used for annual crop production back to rice land is easier.
27. Decision No. 150/2005/QD-TTg of June 2005 on land exchanges.

28. Decree on Compensation of 2013.

29. For annual crops, the compensation should be equal to the output value of the harvest period, i.e. the highest yield of harvest periods in the preceding three years of the local main crop and the average price at the time of the land recovery. For perennial crops, the compensation should be equal to the current value of the planting area calculated in local prices at the time of the land recovery. For the plants not harvested yet but that can be moved to other locations, the moving costs and the actual damage due to moving and replanting should be compensated.

30. These guidelines are the first global guidelines on the governance of tenure. They provide a reference framework to improve the governance of tenure of land, fisheries and forests that supports food security and contributes to the efforts towards the eradication of hunger and poverty. Recognising the central role of land in development, they promote secure tenure rights and equitable access to land, fisheries and forests. They set out principles and internationally accepted practices that may guide the preparation and implementation of policies and laws related to tenure governance. They were developed though intergovernmental negotiations led by the Committee on World Food Security that endorsed them on 11 May 2012.

31. Decision No. 315/2011/QD-TTg on implementing pilot agricultural insurance in 21 provinces in 2011-13.

32. The 797 communes located in poor districts should be fully funded by the state budget to build in particular village roads, in-field roads, canals, and the infrastructure necessary for concentrated production sites.

33. Decree No. 210/2013/ND-CP on policy for encouraging investment in agriculture and rural areas.

34. Decree No. 109/2010/ND-CP.

35. Decision No. 21 of 2009 on electricity tariffs dated 12 February 2009.

36. Mobile banking is supported by the Prime Minister's Decision No. 2453/QD-TTg dated 27 December 2011.

37. The IPRI includes the following three components: i) legal and political environment which provides an insight into the impact of political stability and rule of law and includes items that are broad in scope. This component has a significant impact on the development and protection of the two other components; ii) physical property rights; and iii) intellectual property rights which reflect two forms of property rights and include items that account for both de jure rights and de facto outcomes.

38. Other major standards in the agricultural sector include in particular: the Principles for Responsible Investment in Agriculture and Food Systems of the Committee on World Food Security; the FAO Voluntary Guidelines on the Responsible Governance of Tenure of Land, Fisheries and Forests in the Context of National Food Security; the Principles for Responsible Agricultural Investment that respect rights, livelihoods and resources developed by FAO, IFAD, UNCTAD and the World Bank; the Guiding Principles on Business and Human Rights; the International Labour Organization Tripartite Declaration of Principles concerning Multinational Enterprises and Social Policy; and the Convention on Biological Diversity.

39. Law on Environmental Protection of 2005.

40. An environmental protection undertaking must be registered with the local authorities for all projects not subject to EIAs – this is a simple registration requirement that does not require the authorisation from local authorities. Such undertaking should provide basic information on the raw materials and fuel used, the type of waste produced and the measures planned to minimise and treat waste to comply with the law.

41. Decree No. 179/2013/ND-CP.

42. Law on Water Resources of 2012.

43. Decision No. 3119/QD-BNN-KHCN of 2011.

44. Decision No. 543/QD-BNN-KHCN of 2011 on "the action plan to respond to climate change in the agricultural sector in 2011-15 and vision to 2050".

45. Because of the strong reduction of the forest area in 1975-92, the government has increased the area of planted production forest since 1993 through the Programme 327 in 1993-98 to reforest barren land and the Programme 661 in 1998-2010, also called the Five Million Hectares Reforestation Programme. These programmes provided farmers with low interest rates, low land taxes and free seedlings (Phan et al., 2014).

46. Decision No. 7/2012/QD-TTg on the intensification of forest protection practices.
47. Decision No. 799/QD-TTg of 2012 on the "national action programme on the reduction of GHG emissions by reducing deforestation and forest degradation, promoting sustainable forest management, and enhancing forest carbon stocks from 2011 to 2020". Reducing Emissions from Deforestation and Forest Degradation (REDD) is an effort to create a financial value for the carbon stored in forests, offering incentives for developing countries to reduce emissions from forested land and invest in low-carbon paths. REDD+ goes beyond deforestation and forest degradation, and includes sustainable forest management and the enhancement of carbon stocks.
48. The sample comprises 1 053 enterprises in 14 provinces. Land-related responses are based on the responses of 197 enterprises that applied for LURCs.

References

ADB (2013a), "Enhancing Agricultural Productivity in Viet Nam: Agenda for Reform", *CLMV Project Country Paper* No. 12, August 2013.

ADB (2013b), *Rice Value Chain Study in Mekong River Delta*, PhD. Dao The Anh, Thomas Reardon, Kevin Chen, Thai Van Tinh, Vu Nguyen, Nguyen Ngoc Vang, Nguyen Van Thang, Le Nguyen Doan Khoi, May 2013.

ADB (2008), "Socialist Republic of Viet Nam: Strengthening Agriculture Science and Technology Management", *Technical Assistance Consultant's Report*, Project number: TA 4619, November 2008.

Anh, N.T.T. and L.M. Duc (2010), "Efficiency and Competitiveness of the Private Sector in Vietnam", CIEM/DFID, Hanoi, *http://vnep.org.vn/Upload/Competitiveness%20of%20Vietnamprivate%20sector%20DFID-first%20draft.pdf.*

ARCM (2014), "Viet Nam Country Profile", Asia Resource Centre for Microfinance *www.bwtp.org/arcm/Vietnam/I_Country_Profile/Vietnam_country_profile.html* accessed on 31 July 2014.

BCG (2014), "Vietnam and Myanmar: Southeast Asia's New Growth Frontiers", *www.bcgperspectives.com/content/articles/consumer_insight_growth_vietnam_myanmar_southeast_asia_new_growth_frontier/.*

Bloomberg (2013), "Vietnam Tightens Land Seizure Law After Farmers Protest", *www.bloomberg.com/news/2013-12-08/vietnam-tightens-land-seizure-law-after-protests-southeast-asia.html.*

Business Times (2011), "New Credit Guarantee Regulations for SMEs", 19 January 2011, *http://businesstimes.com.vn/new-credit-guarantee-regulations-for-smes/.*

Canadian Trade Commission (2011), "An Investment Guide to Viet Nam", *www.tradecommissioner.gc.ca/eng/document.jsp?did=91398&cid=539&oid=595#II8.*

Cervantes-Godoy, D. and J. Dewbre (2010), "Economic Importance of Agriculture for Sustainable Development and Poverty Reduction: The Case Study of Vietnam", paper presented at Global Forum on Agriculture, 29-30 November, OECD, Paris, *www.oecd.org/tad/agricultural-policies/46378758.pdf.*

CIEM (Central Institute for Economic Management) (2013), *Characteristics of the Vietnamese Rural Economy: Evidence from a 2012 Rural Household Survey in 12 Provinces of Viet Nam*, Hanoi.

CNRS (2010), *Land Reform in Viet Nam, The Analysis of the Roles Played by Different Actors and Changes Within Central and Provincial Level Institutions*, Fortunel, F., M. Mellac and T.D. Dann, June 2010, ADESS – Aménagement, Développement, Environnement, Santé et Sociétés – UMR 5185.

Dao The Anh and Nguyen Van Son (2013), "Vietnam Agricultural Value Chain in the FTA of Asian Region", in *FFTC.NACF International Seminar on Threats and Opportunities of the Free Trade Agreements in the Asian Region*, Sept. 29-Oct. 3, 2013, Seoul, Korea, *http://ap.fftc.agnet.org/ap_db.php?id=106&print=1.*

DERG [Development Economics Research Group] (2012), *The Availability and Effectiveness of Credit in Rural Vietnam: Evidence from the Vietnamese Access to Resources Household Survey 2006-2008-2010*, Prepared under the Agriculture and Rural Development (ARD) Programme, Royal Embassy of Denmark of Vietnam, *www.ciem.org.vn/Portals/1/CIEM/IndepthStudy/2012/13388684552500.pdf.*

FAO (2014), "Public Sector Support for Inclusive Agribusiness Development – An appraisal of Institutional Models in Viet Nam", *Country Case Studies – Asia*, Rome.

FAO (2013a), "Agribusiness Public-Private Partnerships – A Country Report of Indonesia", *Country Case Studies – Asia*, Rome.

FAO (2013b), "Agribusiness Public-Private Partnerships – A Country Report of Thailand", *Country Case Studies – Asia*, Rome.

FIA (2014), Website of the Foreign Investment Agency, *http://fia.mpi.gov.vn/detail/931/agriculture-offers-food-for-thought*, 25 August 2014.

FIA (2012), "Report on FDI in Viet Nam", *http://fia.mpi.gov.vn/*.

GSO (2010), *General Statistics of Vietnam, www.gso.gov.vn/default_en.aspx?tabid=491*.

IEA (2011), *Database of the World Energy Outlook 2011*, International Energy Agency.

IMF (2013), *Financial Access Survey: Viet Nam, http://fas.imf.org/*.

IPSARD (2014), *Policy and Regulatory Review and Assessment*, Project ID 600060, January 2014.

IPSARD (2010a), *The Availability and Effectiveness of Credit in Rural Vietnam: Evidence from the Vietnamese Access to Resources Household Survey 2006-2008-2010*, Development Economics Research Group (DERG) of the University of Copenhagen (UoC), Central Institute for Economic Management (CIEM), Ministry of Planning and Investment (MPI) of Vietnam, Centre for Agricultural Policy (CAP) of IPSARD, MARD, Prepared under the Agriculture and Rural Development (ARD) Programme, Royal Embassy of Denmark of Vietnam.

IPSARD (2010b), *Land Policy for Socioeconomic Development in Viet Nam*, Dang Hoa Ho and Malcolm McPherson, Fulbright Economics Teaching Programme and Ash Center for Democratic Governance and Innovation, John F. Kennedy School, Harvard University, 27 May 2010.

ITU (2013), Statistics on ICTs, International Telecommunication Union, *www.itu.int/en/ITU-D/Statistics/Pages/stat/default.aspx*.

JICA (2013), *Viet Nam Country Study*, January 2013.

Kemper, N. and R. Klump (2014), *Land Reform and the Formalisation of Household Credit in Rural Viet Nam*, Goethe University, Frankfurt, Germany, *www.pegnet.ifw-kiel.de/members/kemper.pdf*.

Lovells (2009), *Investing in infrastructure in Viet Nam*, October 2009, *www.hoganlovells.com/files/Publication/9fcb228e-522c-4eb1-bcf0-f8e46fdd45f3/Presentation/PublicationAttachment/8970c342-5b36-4ef7-9446-fbcdc6014cfc/Investing_in_infrastructure_in_Vietnam_newsflash.pdf*.

Malesky, E. (2011), "The Vietnam Provincial Competitiveness Index: Measuring Economic Governance for Business Development 2011, Final Report", *USAID/VNCI Policy Paper #16*, Vietnam Chamber of Commerce and Industry and United States Agency for International Development's Vietnam Competitiveness Initiative, Hanoi.

MARD (2015), *Comments received from MARD on the Review*, May 2015.

MARD (2014a), *Responses to the PFIA Questionnaire*, Phan Sy Hieu and Ta Kim Cuc, March 2014.

MARD (2014c), *Public Private Partnership for the Sustainable Development of Agriculture in Viet Nam*, June 2014.

MARD (2014d), *FDI Strategy to Attract FDI in the Agriculture, Forestry and Fishery Sector Up to 2030*.

MARD (2014e), *Report: Institutional Structure of Food Safety Control in the Food Value Chain*, by Dao The Anh, Le Ba Anh, Nguyen Thi Mai Hien, Nguyen Thi Ha, and Bui Quang Duan.

MARD (2012), *Proposal on Restructuring the Agricultural Sector Towards Greater Value Added and Sustainable Development*, Hanoi, October.

MARD (2009), *Manual Guidelines for FDI to Agriculture, Forestry and Fisheries in Viet Nam*.

Markussen, T., F. Tarp and K. Van den Broek (2009), "The Forgotten Property Rights: Restrictions on Land Use in Vietnam", *Discussion Paper* 09-21, University of Copenhagen, Department of Economics.

MOJ (2010), Decree No. 41/2010/ND-CP on Credit Policies for Agricultural and Rural Development, *www.moj.gov.vn/vbpq/en/Lists/Vn%20bn%20php%20lut/View_Detail.aspx?ItemID=10730*.

MONDAQ (2012), "Viet Nam: Telecommunications in Viet Nam", 5 January 2012, *www.mondaq.com/x/118644/Telecommunications+Mobile+Cable+Communications/Telecommunications+in+Vietnam*.

MPI (2012), "Report on Socio-Economic Status for the First 7 months of 2012 and Implementation of Resolution 01/ND-CP", submitted to regular government meeting in July 2012, Ministry of Planning and Investment.

Nguyen, T.D.N. (2012), "Private Sector Investments in Viet Nam: Agriculture Investment Trends – The Role of Public and Private Sector in Vietnam", Ms. Nguyen Thi Duong Nga, Hanoi Agriculture University, *www.fao.org/fileadmin/templates/tci/pdf/CorporatePrivateSector/Vietnam_-_Private_Sector_Investments_in_Agriculture__Final_Report.pdf*).

Nguyen, T.T.H. (2010), *Credit Market Segmentation in Rural Areas: A Case Study in Phu Thuong Commune, Phu Vang District, Thua Thien Hue Province*, HUAF, Nguyen Thi Thanh Huong, Hue University, Viet Nam, *http://stud.epsilon.slu.se/2933/1/huong_nguyen_110627.pdf*.

Nguyen, X.T. (2009), *Infrastructure Constraints in Viet Nam*, Documentation for Political Dialogue prepared by Harvard and PNUD, Nguyen Xuan Thanh.

OECD (2015), *Policy Framework for Investment*. 2015 edition, OECD Publishing, Paris, *http://dx.doi.org/10.1787/9789264208667-en*.

OECD (2014a), *Policy Framework for Investment in Agriculture*, OECD Publishing, Paris, *http://dx.doi.org/10.1787/9789264212725-en*.

OECD (2014b), "OECD Trade Facilitation Indicators – Viet Nam".

OECD (2013a), *Effectiveness of Research and Innovation Management at Policy and Institutional Level – Cambodia, Malaysia, Thailand and Viet Nam*, Olsson, A. and L. Meek, Programme on Innovation, Higher Education and Research for Development (IHERD).

OECD (2013b), "Public-Private Partnerships: Overview", Working Party on Agricultural Policies and Markets, 19-21 November 2013.

OECD (2013c), *Economic Outlook for Southeast Asia, China and India 2014: Beyond the Middle-Income Trap*, OECD Publishing, Paris, *http://dx.doi.org/10.1787/saeo-2014-en*.

OECD (2012), *Sustainable Agricultural Productivity Growth and Bridging the Gap for Small-Family Farms, Interagency Report to the G20 Mexican Presidency*, contributions by Biodiversity, CGIAR Consortium, FAO, IFAD, IFPRI, IICA, OECD, UNCTAD, Coordination team of UN High Level Task Force on the Food Security Crisis, WFP, World Bank, and WTO, 12 June 2012.

OECD (2009), *OECD Investment Policy Reviews: Viet Nam 2009: Policy Framework for Investment Assessment*, OECD Publishing, Paris, *http://dx.doi.org/10.1787/9789264050921-en*.

OECD/The World Bank (2014), *Science, Technology and Innovation in Viet Nam*, OECD Publishing, Paris, *http://dx.doi.org/10.1787/9789264213500-en*.

Oxfam (2013), "Promoting Land Rights in Viet Nam: A Multi-Sector Advocacy Coalition Approach, Andrew Wells-Dang", Paper prepared for presentation at the Annual World Bank Conference on Land and Poverty 2013, Washington, DC, 8-11 April 2013.

PCI (2013), *The Viet Nam Provincial Competitiveness Index 2013, VCCI and USAID*, available at *www.pcivietnam.org*.

Pham, B.D (2013), "Reviewing the Development of Rural Finance in Vietnam", *Journal of Economics and Development*, Vol. 15, No. 1, April, *www.vjol.info/index.php/KTQD/article/viewFile/11365/10348*.

Phan, S.H. (2014), "A Qualitative Review of Vietnam's 2006-2010 Economic Plan and the Performance of the Agriculture Sector", *Asian Journal of Agriculture and Development*, Vol. 11, No. 1, June 2014.

Phan, S.H. and V.T. Trinh (2014), *Review of Agricultural Policies: Vietnam 1976-2013*.

PWC (2008), *Price Waterhouse Coopers, Vietnam: A Guide for Business and Investment: www.pwc.com/en_VN/vn/publications/assets/vietnam_guide.pdf*.

Quach, M.H., A.W. Mullineux and V. Murinde (2005), *Access to Credit and Household Poverty Reduction in Rural Viet Nam: A Cross-Sectional Study*, MH Quach The Birmingham Business School, The University of Birmingham, *www.grips.ac.jp/Vietnam/VDFTokyo/Doc/1stConf18Jun05/OPP01QuachPPR.pdf*.

Quang, Chu Tien and Ha Huy Ngoc (2011), "Đầu tư trực tiếp nước ngoài (FDI) vào lĩnhvực nông nghiệp: T hực trạng và chính sách", *Journal of Party Communist* Vol. 9 (225).

SBV (2012), *Annual Report 2012*, State Bank of Viet Nam.

SHTP (2014), Website of the Saigon High-Tech Park, accessed at *www.eng.shtp.hochiminhcity.gov.vn/Pages/default.aspx*.

TI (2011), "Corruption in the Land Sector", *Working Paper*, Transparency International, # 04/2011.

Tran, C.T. (2014a), *Paper on Vietnamese Domestic Agricultural Policies prepared for OECD Monitoring and Evaluation Report*.

Tran, C.T. (2014b), *Overview of Agricultural Policies in Vietnam*, Agricultural Policy Database, FFTC Agricultural Policy Database, Food and Fertilizer Technology Center for the Asia and Pacific Region, *http://ap.fftc.agnet.org/ap_db.php?id=195*.

Tran, C.T., B.L. Dinh and H.P. Hu (2013), *Vietnam: Rice Sector Policy Analysis*, SNV Netherlands Development Organisation, Vietnam Office, Hanoi.

Tuoi Tre News (2014), "Viet Nam Needs USD 500 Billion for Infrastructure Development in 10 Years", 8 May 2014, *http://tuoitrenews.vn/society/21484/Vietnam-needs-500bn-for-infrastructure-development-in-10-years-ministry*.

U4 (2014), "Overview of Corruption and Anti-Corruption in Viet Nam", U4 Expert Answer, available at *www.u4.no/*.

UNCTAD (2008), *Investment Policy Review of Viet Nam, www.unctad.org/en/docs/iteipc200710_en.pdf*.

USAID (2014), "Viet Nam Promotes Public-Private Partnerships in Infrastructure", *www.usaid.gov/results-data/success-stories/Vietnam-promotes-public-private-partnerships-infrastructure*.

USAID (2013), "Vietnam – Property Rights and Resource Governance Profile", *USAID Country Profile Series*.

Van den Broek, K., C. Newman and F. Tarp (2007), "Land Titles and Rice Production in Vietnam", *Trinity Economics Papers*, tep1207, Trinity College Dublin, Department of Economics.

VBF (2014), *Anti-Corruption and Corporate Governance: Specific Proposals for Change*, Corporate Governance and Transparency Working Group, Vietnam Business Forum.

VBF (2010), "Speeding up Viet Nam's ICT Engine", 17 November 2010, Viet Nam Business Forum, *www.vccinews.com/news_detail.asp?news_id=22023*.

Viet Nam Briefing (2014), "Foreign Investment Needed for Infrastructure and Agriculture Projects in Viet Nam", 9 May 2014, *www.Vietnam-briefing.com/news/us60-billion-foreign-investment-needed-infrastructure-agriculture-projects-Vietnam.html/#*.

Viet Nam Briefing (2010), "Viet Nam to Give Export Credit Guarantees to 23 New Commodities in 2011", 9 November 2010, *www.vietnam-briefing.com/news/vietnam-give-23-commodities-export-credit-guarantees-2011.html*.

WB (2014a), *Making Foreign Direct Investment Work for Sub-Saharan Africa, Local Spillovers and Competitiveness in Global Value Chains, Directions in Development – Trade*, Thomas Farole and Deborah Winkler.

WB (2014b), *Doing Business Report in Viet Nam*, a co-publication of the World Bank and the International Finance Corporation.

WB (2014c), *Efficient Logistics: A Key to Viet Nam's Competitiveness*, Luis C. Blancas, John Isbell, Monica Isbell, Hua Joo Tan, Wendy Tao, *www-wds.worldbank.org/external/default/WDSContentServer/WDSP/IB/2013/11/27/000461832_20131127115422/Rendered/PDF/830310PUB0978100Box379862B00PUBLIC0.pdf*.

WB (2014d), *Viet Nam: Rural Connectivity and Agriculture Logistics in Domestic Market Supply Chains – Synthesis Report*, World Bank and Australian Aid, March 2014.

WB (2013a), *Light Manufacturing in Viet Nam: Job Creation and Prosperity in a Middle-Income Economy*, Directions in Development, Washington, DC: World Bank, Dinh, Hinh T., *http://dx.doi.org/10.1596/978-1-4648-0034-4*, License: Creative Commons Attribution CC BY 3.0.

WB (2013b), "Viet Nam and Energy", *http://go.worldbank.org/J25FRKU830*.

WB (2011a), "Market Economy for a Middle-Income Vietnam, Vietnam Development Report 2012", *World Bank Report* No. 65980.

WB (2011b), *Viet Nam State and People, Central and Local, Working Together: The Rural Electrification Experience*, March 2011, *www-wds.worldbank.org/external/default/WDSContentServer/WDSP/IB/2011/03/25/000356161_20110325024020/Rendered/PDF/604380WP0P11051ification0Experience.pdf*.

WB (2011c), *Recognising and Reducing Corruption Risks in Land Management*.

WB (2008), *Social Protection, Viet Nam Development Report 2008*, Joint Donor Report to the Viet Nam Consultative Group Meeting, Hanoi, 6-7 December 2007.

WEF (2013), *The Global Competitiveness Report 2013-2014*, World Economic Forum.

Wells-Dang, A. (2013), "Promoting Land Rights in Vietnam: A Multi-Sector Advocacy Coalition Approach", paper prepared for presentation at the Annual World Bank Conference on Land and Poverty, World Bank, Washington, DC, 8-11 April.

ORGANISATION FOR ECONOMIC CO-OPERATION AND DEVELOPMENT

The OECD is a unique forum where governments work together to address the economic, social and environmental challenges of globalisation. The OECD is also at the forefront of efforts to understand and to help governments respond to new developments and concerns, such as corporate governance, the information economy and the challenges of an ageing population. The Organisation provides a setting where governments can compare policy experiences, seek answers to common problems, identify good practice and work to co-ordinate domestic and international policies.

The OECD member countries are: Australia, Austria, Belgium, Canada, Chile, the Czech Republic, Denmark, Estonia, Finland, France, Germany, Greece, Hungary, Iceland, Ireland, Israel, Italy, Japan, Korea, Luxembourg, Mexico, the Netherlands, New Zealand, Norway, Poland, Portugal, the Slovak Republic, Slovenia, Spain, Sweden, Switzerland, Turkey, the United Kingdom and the United States. The European Commission takes part in the work of the OECD.

OECD Publishing disseminates widely the results of the Organisation's statistics gathering and research on economic, social and environmental issues, as well as the conventions, guidelines and standards agreed by its members.

OECD PUBLISHING, 2, rue André-Pascal, 75775 PARIS CEDEX 16
(51 2015 09 1 P) ISBN 978-92-64-23514-4 – 2015-17

www.ingramcontent.com/pod-product-compliance
Lightning Source LLC
LaVergne TN
LVHW081402110826
845149LV00010B/1640
* 9 7 8 9 2 6 4 2 3 5 1 4 4 *